T0299160

FUNDAMENTALS OF RADIO ASTRONOMY
Observational Methods

Series in Astronomy and Astrophysics

The *Series in Astronomy and Astrophysics* includes books on all aspects of theoretical and experimental astronomy and astrophysics. Books in the series range in level from textbooks and handbooks to more advanced expositions of current research.

Series Editors:
M Birkinshaw, University of Bristol, UK
J Silk, University of Oxford, UK
G Fuller, University of Manchester, UK

Recent books in the series

Fundamentals of Radio Astronomy: Observational Methods
Jonathan M Marr, Ronald L Snell and Stanley E Kurtz

Stellar Explosions: Hydrodynamics and Nucleosynthesis
Jordi José

Astrobiology: An Introduction
Alan Longstaff

An Introduction to the Physics of Interstellar Dust
Endrik Krugel

Numerical Methods in Astrophysics: An Introduction
P Bodenheimer, G P Laughlin, M Rózyczka, H W Yorke

Very High Energy Gamma-Ray Astronomy
T C Weekes

The Physics of Interstellar Dust
E Krügel

Dust in the Galactic Environment, 2nd Edition
D C B Whittet

Dark Sky, Dark Matter
J M Overduin and P S Wesson

An Introduction to the Science of Cosmology
D J Raine and E G Thomas

The Origin and Evolution of the Solar System
M M Woolfson

The Physics of the Interstellar Medium, 2nd Edition
J E Dyson and D A Williams

Optical Astronomical Spectroscopy
C R Kitchin

Series in Astronomy and Astrophysics

FUNDAMENTALS OF RADIO ASTRONOMY
Observational Methods

Jonathan M. Marr
*Union College, Schenectady,
New York, USA*

Ronald L. Snell
University of Massachusetts, Amherst, Massachusetts, USA

Stanley E. Kurtz
*National Autonomous University of Mexico,
Morelia, Michoacan, Mexico*

CRC Press
Taylor & Francis Group
Boca Raton London New York

CRC Press is an imprint of the
Taylor & Francis Group, an **informa** business

Contents

Preface

A STRONOMY AMUSES AND INTRIGUES many curious minds and is, arguably, the most popular science. Many astronomy enthusiasts, however, have an incomplete appreciation of all that astronomy offers. A concept held by many is that astronomical discoveries are made with observations of light visible to the human eye. In reality, much of astronomy and astrophysics involves observations in other wavelength regimes. Radio astronomy was the first nonvisible-wavelength branch of astronomy developed, and hence is one of the most extensively explored. The contributions of radio astronomy in its 80-year history have contributed greatly to our understanding of the universe. Five Nobel Prizes in physics, for example, have been awarded for the results of research conducted at radio wavelengths.

One can make a strong case, therefore, for course offerings for undergraduate astronomy and astrophysics students to include radio astronomy. The graduate curriculum for astronomy and astrophysics often includes a course in radio astronomy, and textbooks exist for this level. Unfortunately, there is no suitable radio astronomy textbook for undergraduates. All three authors of this book have taught radio astronomy courses for undergraduates, and we have found the need to develop our own teaching material at a level more suitable for students of radio astronomy than what is offered by existing texts.

Our primary goal in converting our teaching material into a textbook is to serve undergraduate courses. However, we also hope that these pages will serve as a reference for graduate students getting involved in radio astronomy. As such, the amount of material presented is more than what would reasonably fit into a one-semester course. Instructors are advised to use their own discretion in selecting the chapters and sections to suit their course.

We have divided the material into two volumes. The first volume, *Fundamentals of Radio Astronomy: Observational Methods*, discusses radio astronomy instruments and the techniques that a radio observer should know in order to make successful observations. The emphasis is on the use and functioning of radio telescopes. It is not a guide to building amateur radio telescopes. In the second volume, *Fundamentals of Radio Astronomy: Astrophysics*, we discuss the more common physical processes that give rise to radio emissions and present examples of astronomical objects that emit radiation by these mechanisms, along with illustrations of how the relevant physical parameters are obtained from radio observations.

A common approach in teaching radio astronomy courses is to teach about the instrumentation first and then discuss the astrophysics learned from radio astronomical studies.

This reflects the philosophy that understanding one's instruments is crucial to the proper analysis of the observations and success in the research. However, we consider it equally feasible, especially at the undergraduate level, to teach the astrophysics first and to finish the course with coverage of the instrumentation. Moreover, a radio astronomy course might cover only instrumentation and observing methods *or* only radio astrophysics, according to other courses offered by the department. To accommodate a wealth of needs, we have structured the two volumes so that they can be presented in either order and are largely independent from one another.

To accompany this text, we have developed a set of laboratory exercises that are available at http://www.crcpress.com/product/isbn/9781420076769. These labs make use of the *small radio telescope* (SRT) and the *very small radio telescope* (VSRT) that were developed by Massachusetts Institute of Technology's Haystack Observatory for educational use.* The SRT is a 2-m-class, single-dish telescope, operating at 21 cm, while the VSRT is a table-top interferometer operating at 2.5 cm. Information about these two telescopes can be found at http://www.haystack.mit.edu/edu/undergrad. Instructions for the VSRT labs are given in Appendix VII. The data analysis software in our zip files are Java packages that can be installed on any operating system (at no cost). Also available at http://www.crcpress.com/product/isbn/9781420076769 is a Java package that demonstrates the principles of Fourier transforms.† These transforms are needed for the analysis of interferometric data; this software application will provide invaluable insights when covering Chapters 5 and 6 of the first volume.

We assume that the reader has completed an introductory physics sequence. More specifically, we assume the reader has experience with electricity and magnetism and an exposure to wave theory. We include some material that will enhance the understanding for readers with more extensive physics and mathematics backgrounds; these sections are indicated with an *asterisk* (*).

A thorough discussion of interferometry and aperture synthesis, in particular, is quite involved and is beyond the scope of a one-semester undergraduate course. With the goal of conveying a conceptual understanding of the fundamental principles, we first discuss a simplified situation in Chapter 5, which avoids many of the inherent complexities. Readers who make actual interferometer observations will need to know more practical details; therefore, we provide in Chapter 6 a more advanced and thorough discussion of real-world interferometry observations, while maintaining the undergraduate level of discussion. Furthermore, the VSRT labs provide a hands-on experience for students to gain an intuitive sense of how aperture synthesis works. These labs are complementary to the presentation in Chapter 5 and can be done without going through the mathematics of Chapter 6. For classes that do not have access to a VSRT, the data analysis package VSRTI_Plotter.jar can be used, which simulate the labs.

---

* The development of the VSRT was funded by the NSF through a CCLI grant to MIT Haystack Observatory.
† The development of the VSRTI_Plotter and TIFT java packages was funded by the National Science Foundation, IIS CPATH Award #0722203.

We expect that this textbook will also be of interest to physics majors with no prior background in astronomy. Thus, the early chapters include explanations of relevant topics in astronomy, with particular emphasis on the physics and measurements of radiation.

There are some mathematical concepts that we do not assume all readers of this book will know, but which can enhance the understanding of the material. We review the relevant aspects of these mathematical topics in the appendices, so they do not distract from the presentation of the radio astronomy but are readily available to both students and instructors.

It is our experience that students greatly desire, and benefit from, examples that demonstrate how the material is used to solve problems. Hence, we have included numerous examples dispersed throughout the chapters. Each chapter also concludes with a list of questions and problems.

Historically, astronomers have predominantly used cgs units. In more recent years, SI units have become commonplace. For this reason, and because we hope these volumes will be of interest to physics majors as well, we have written all equations and examples using SI units. To facilitate comparison with other texts, we also provide the cgs formulations of the fundamental equations and constants. In the second volume, there is a need for astronomy-specific units, such as parsecs, solar luminosities, and solar masses. These units are explained in the beginning chapters of the second volume. The first page of the appendices in both volumes lists all constants and unit conversions.

Although the initial development of radio astronomy occurred in the first half of the twentieth century, with subsequent technological advancements, many technical aspects of radio astronomy continue to change. Some parts of this book, therefore, will become obsolete in time. We encourage the instructor to use updated, supplemental material where appropriate.

We have created all the material here purely as part of our efforts in teaching radio astronomy, which we have enjoyed immensely through the years. We hope that our efforts in making this material available to others enhances both the teaching and the learning of radio astronomy and that you will find the experience as rewarding as we have.

Additional material is available from the CRC Press website: http://www.crcpress.com/product/isbn/9781420076769.

Jonathan M. Marr
Union College, Schenectady, New York

Ronald L. Snell
University of Massachusetts, Amherst, Massachusetts

Stanley E. Kurtz
National Autonomous University of Mexico, Morelia, Mexico

Acknowledgments

WE HAVE RECEIVED GENEROUS and wonderful help from a number of people whose selfless contribution of time deserves to be acknowledged. We are very grateful for the advice, feedback, and other assistance provided by Esteban Araya, Walter Brisken, Barry Clark, Mark Claussen, Chris DePree, Vivek Dhawan, Neal Erickson, Paul Goldsmith, Greg Hallenbeck, James Lowenthal, Carl Heiles, Peter Hofner, Jim Moran, Amy Lovell, Robert Marr, Rick Perley, Thomas Perry, Preethi Pratap, Luis Rodriquez, Alan Rogers, Ken Schick, Greg Taylor, Clara Thomann, Leise van Zee, Francis Wilkin, Min Yun, and all the students who used our lecture notes as their reading material through the years.

CHAPTER 1

Introductory Material

THE DEVELOPMENT OF RADIO astronomy in the mid-twentieth century opened up a new window on the universe. Prior to this development, astronomical observations were limited to the narrow range of visible wavelengths and consequently a limited range of astronomical phenomena. The sky at radio wavelengths is much different from the sky at visible wavelengths. Objects bright at visible wavelengths, such as stars, are not what dominate the emission in the radio sky. At radio wavelengths, we can detect the thermal continuum and spectral-line emission from objects too cold to produce visible light, permitting studies of the cold interstellar medium of our Galaxy and others, as well as the cosmic microwave background, the relic radiation from the early universe. A form of nonthermal radiation called *synchrotron emission* produces prominent radio emission and is seen from a host of interesting astronomical sources such as supernova remnants and quasars. Two of the brightest sources in the radio sky, Cassiopeia A and Cygnus A, are synchrotron-emitting sources and are relatively faint at visible wavelengths; thus, observations at radio wavelengths complement those at optical observations.

Since the development of radio astronomy, many additional wavelength regimes, such as gamma rays, X-rays, ultraviolet light, and infrared light, are now observable and permit astronomers to extend greatly the wavelength coverage of their studies. However, the Earth's atmosphere is only transparent to visible and radio light; therefore, astronomy at other wavelengths must be largely done from space. Since the Earth's atmosphere is transparent to radio waves over four decades of wavelength, nearly all radio telescopes are located on the ground.

This volume is devoted to providing an overview of the techniques and equipment used in observing the sky at radio wavelengths. We discuss the optics of radio telescopes, how receiver systems process and detect the radio emission, and how calibrated measurements of astronomical sources are made. In addition to observations with single-dish telescopes, we extensively discuss the functioning and methods of observing using interferometers and the technique of aperture synthesis.

1.1 BRIEF HISTORY OF RADIO ASTRONOMY

If you are going to read a book on the fundamentals of radio astronomy, it only seems natural that you should come away with some knowledge of its historical development. However, since this information is not crucial to understanding and conducting radio astronomy research, we give only a brief discussion. More thorough histories of radio astronomy are given in *The Radio Universe* by J. S. Hey[*] and *The Early Years of Radio Astronomy; Reflections Fifty Years After Jansky's Discovery*, by W. T. Sullivan,[†] and a history of aperture synthesis is given in *Interferometry and Synthesis in Radio Astronomy* by A.R. Thompson, J.M. Moran, and G.W. Swenson, Jr.[‡]

The study of radio astronomy started in the early 1900s. Before that time, astronomers studied the universe at visible wavelengths only. The study of electricity and magnetism in the 1800s helped to set up the discovery of radio astronomy in the 1900s. First and foremost, the development of Maxwell's equations revealed that any wavelength of light is possible, and that those in the visible window actually comprise only a tiny fraction of the full range. In 1887, Heinrich Hertz successfully produced radio waves in the lab, igniting the inspiration for others to try to detect radio waves from space. In 1890, Thomas Edison and Arthur Kennelly proposed to search for correlations between the number of sunspots and the radio signal detected by a piece of iron ore. In 1894, Sir Oliver Lodge tried this experiment. The attempt failed, which Lodge attributed to interference from local terrestrial sources. Astronomers such as Johannes Wilsing, Julius Scheiner, and Charles Nordmann also failed in their attempts. The real reason for these failures, we now know, is that the radio equipment during that time was not sensitive enough.

The development of radio astronomy required the amplification of the radio signals entering the antenna. This came in the December of 1932, when Karl Jansky of Bell Labs made the first successful detection of astronomical radio emission. Jansky was studying the direction and level of static that interfered with long distance short-wave radio communications. He used a steerable antenna, amplified and converted the radio frequency waves to audible sound waves, and listened to the noise with earphones at a frequency of 20.5 MHz. He heard a steady hiss on top of crackling, the latter due to thunderstorms. He discovered that the strength of the hiss was directionally dependent. With further investigation, he discovered that the direction of the hiss was fixed relative to the *stars*, not relative to Earth or the Sun. Finally, he determined that the hiss was coming from the plane of our galaxy, with the most intense signal from the galactic center (which is located in the constellation of Sagittarius). Few astronomers took note of this discovery and it might have never been considered significant were it not for an individual named Grote Reber.

Reber, a professional engineer greatly interested in astronomy, was intrigued by Jansky's report, and in 1937 built his own 30-foot antenna with his own money in his backyard in Wheaton, Illinois. After failing to detect any signal at 900 and 3,300 MHz, he succeeded in mapping the radio emission from the Galaxy at 160 MHz with a greater resolution than that

---

[*] 2nd Edition, Pergamon Press, Oxford-New York, 1975.
[†] Cambridge University Press, Cambridge, 1984.
[‡] John Wiley & Sons, Inc., New York, 2001.

in Jansky's maps. He identified the locations in the sky of several secondary peaks in signal strength in addition to the strong emission from the center of the Milky Way: one in the constellation Cassiopeia (which we now refer to as *Cas A* and which is known to be a supernova remnant) and another in Cygnus (*Cyg A*, a radio galaxy). His results were published in 1940 and 1944 and were the first radio wavelength observations published in an astronomical journal.

For military reasons, during World War II, there were significant technological developments in radar. J.S. Hey, a civilian scientist for the British Army Operational Research Group, analyzed all reports of jamming of army radar sets. In the February of 1942, there was an especially large number of jamming reports filed. Hey concluded that the cause was related to a large sunspot. This was not unanimously accepted, but later in the same year, Southworth of Bell Labs, observing at gigahertz frequencies, made the first successful detection of radio emission from the quiescent Sun. The combination of these two publications, occurring in two different countries, led to the interest in the development of solar radio astronomy.

After the war, Hey, in collaboration with S. J. Parson, J. W. Phillips, and G. S. Stewart, continued with radar and radio studies and made a number of discoveries. In 1945, Hey and colleagues discovered that meteors leave trails of ionization that reflect radio waves. In 1946, they mapped the radio sky in greater detail than that made by Reber and found that the radio emission observed toward Cygnus varied with time. They recognized Cyg A as a discrete source, distinct from the extended radio emission associated with the Milky Way. In 1948, they determined that the solar radio bursts were associated with sunspots and solar flares.

The realization of a major avenue for study of large pieces of the universe using radio observations was initiated in 1944 when the famous theoretical Dutch astrophysicist Jan Oort suggested to Hendrik van de Hulst that he calculate the wavelength of the emission line of hydrogen due to the spin flip of the electron. The calculations by van de Hulst predicted that this transition should emit radiation at 21 cm wavelength. The first detection of the *21-cm line* was made in 1951 by Harold Ewen and Edward Purcell at Harvard University and confirmed several weeks later by groups in the Netherlands and Australia. In the 1950s, maps of hydrogen gas throughout the Milky Way were made using the 21-cm line in the Netherlands and Australia. The 21-cm line is still used extensively to map the distribution of hydrogen atoms throughout space. In fact, some of the labs associated with this book (and which you may get to do yourself if your school has the proper equipment) involve detection and analysis of the 21-cm line.

In 1946, as the first step in the long development of a high-resolution radio astronomy observing technique, Sir Martin Ryle and D. D. Vonberg made the first astronomical observation using a pair of radio antennas as an interferometer. Later interferometric observations by a number of investigators yielded precise measurements of the positions of the bright radio sources, sufficient for optical identification (of Cyg A, Cass A, and the Crab Nebula). In the late 1950s, with more advanced interferometers, a catalog of the brightest 471 sources at 159 MHz was created and published in 1959. This catalog was known as the *Third Cambridge Catalog of Radio Sources (3C)* (the first two attempts had problems with confusion of sources). A revised version, the *3CR catalog*, including observations at 178 MHz, was produced a few years later. In the 1960s, the 4C catalog, with a smaller minimum flux limit, was created.

Also in the 1950s, an understanding of the exotic nature of the radiation detected from the Milky Way and from many of the brightest discrete radio sources was fostered. Hannes Alfvén and Nicolai Herlofsen, in Sweden, and Iosif Shklovsky, in Russia, theorized that the continuum radio emission from the most intense radio sources results from an exotic process known as *synchrotron radiation*, in which relativistic electrons emit photons as they spiral around magnetic field lines.

During the 1960s, radio astronomy yielded discoveries of quasars, pulsars, and the cosmic microwave background, with the last two discoveries yielding Nobel Prizes (part of the 1978 prize to Arno Penzias and Robert Wilson and part of the 1974 prize to Anthony Hewish). Subsequent studies of pulsars and the cosmic microwave background led to two more Nobel Prizes (1993 to Joseph Taylor and Russell Hulse and 2006 to John Mather and George Smoot).

Ryle and colleagues, meanwhile, continued advancing the use of radio interferometers, including creation of arrays of a number of antennas and making use of the Earth's rotation. The other part of the 1974 Nobel Prize was awarded to Sir Martin Ryle for the development of aperture synthesis. In 1967, the first very long baseline interferometry (an interferometric method in which the individual antennas are independently operated and may be separated by continental distances) observation was made. The modern era of imaging methods in aperture synthesis started in 1974 with the development of clever image enhancement techniques.

Up until about 40 years ago, radio astronomers were primarily engineers. They were the builders of the telescope in addition to being the observers. Today, radio astronomy has reached the same level of operation as visible-wavelength astronomy, so that most observers are astronomers by training who apply for time on already-existing radio telescopes without the need to build their own equipment.

1.2 SOME FUNDAMENTALS OF RADIO WAVES

1.2.1 Electromagnetic Radiation

We now provide a basic discussion of radio waves, assuming a freshman-physics level of familiarity with waves in general. In this discussion, we will also define terms and notation.

In short, radio waves are a form of *electromagnetic (EM) radiation*, just like visible light but at frequencies that are far too low for our eyes to detect. The laws of physics require that EM radiation contains waves of both electric and magnetic fields and that in a vacuum these fields must be perpendicular to each other and to the direction of propagation of the waves. An example of waves of EM radiation is depicted in Figure 1.1. Both electric and magnetic fields oscillate in their own planes.

Additionally, the strengths of the electric and magnetic fields have a definite relation (the equation of which we omit here) and the speed of travel of the waves in vacuum must equal a definite value, known as the *speed of light*, and generally depicted with the variable c ($c = 3.00 \times 10^8$ m s^{-1} or 3.00×10^{10} cm s^{-1}).

The *frequency* of the waves is generally represented by either f or ν. In this text, we will use the more conventional symbol ν (the Greek letter *nu*). The unit of frequency is called *hertz*, abbreviated as *Hz*, and 1 Hz is equal to 1 cycle per s. The *wavelength* is represented by λ, with units of length, often in meter or in centimeter. You should recall from your

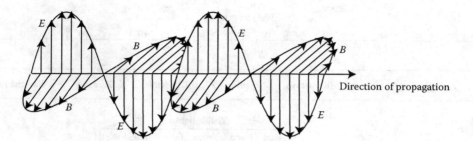

FIGURE 1.1 Schematic of electromagnetic plane waves. The electric field oscillates in the vertical plane; the magnetic field oscillates in the horizontal plane; and the direction of propagation to the right is perpendicular.

freshman physics class that for simple waves with a single frequency (called *monochromatic waves*), the wavelength, frequency, and speed are all related by

$$\lambda v = c \tag{1.1}$$

EM waves (radio or otherwise) are often created by a charged particle (usually an electron) when it undergoes an acceleration. The reason that we usually consider the electron to do the radiating, and not the proton, is that the electron will undergo a much greater acceleration than the proton with the same magnitude of force because of its much smaller mass. Light behaves both as a wave and as a particle. The oscillating electric and magnetic fields, as described earlier, is the wave description of light. We can also describe light by a particle we call the *photon*. The photon has no mass, and its *energy*, E, is given by

$$E = hv = \frac{hc}{\lambda} \tag{1.2}$$

where:
  h is the Planck's constant $= 6.626 \times 10^{-34}$ J s (or 6.626×10^{-27} erg s)

Depending on the circumstance, one may need to describe light as a wave or as a particle.

The different bands of the *EM spectrum* (e.g., radio waves vs. visible light) are defined by the energy, frequency, or wavelength of the waves. The full electromagnetic spectrum, from the lowest energy waves to the highest, is displayed in Figure 1.2.

Example 1.1:

Electromagnetic radiation is emitted from a distant object with a frequency of 1.40 GHz (or 1.4×10^9 Hz).

  1. What is the energy of each photon?
  2. What is the wavelength?
  3. To observe this object, in which band of the EM spectrum must the telescope be sensitive?

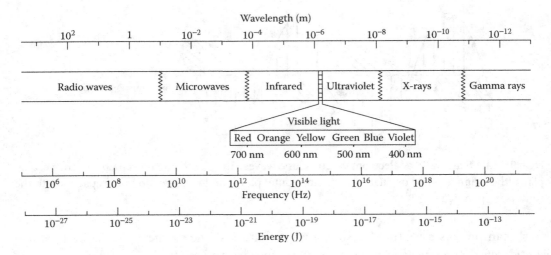

FIGURE 1.2   Display of all the bands of the entire electromagnetic spectrum, shown in order of energy of the waves, with the lowest energy radiation on the left.

Answers:

1. To get the energy, we use Equation 1.2, so

$$E = (6.626 \times 10^{-34} \text{ J s})(1.40 \times 10^9 \text{ Hz})$$

$$= 9.28 \times 10^{-25} \text{ J (Recall that 1 Hz} = 1/\text{s.)}$$

2. Using Equation 1.1, we have

$$\lambda = \frac{3.00 \times 10^8 \text{ m/s}}{1.40 \times 10^9 \text{ Hz}} = 0.214 \text{ m}$$

3. This is in the radio band. (Of course! This is, after all, a book on radio astronomy.)

The radio band contains a large range of frequencies that can travel through the Earth's atmosphere unimpeded and so provides an excellent realm for the detection and study of EM radiation from space. The *radio window*, as it can be called, ranges in frequency from 10 MHz (1 × 10⁷ Hz) up to 300 GHz (3 × 10¹¹ Hz), or in wavelengths from 30 m down to 1 mm. The energies of radio photons range from about 10⁻¹⁹ to 10⁻¹⁵ ergs—that is, *very* tiny!!! The boundaries of the radio window are due to atmospheric and ionospheric processes. At low frequencies (below about 10 MHz), free electrons present in the ionosphere easily absorb and/or reflect radiation. Frequencies below this can be generated and transmitted on the Earth's surface, but they cannot escape through the ionosphere. Likewise, waves with frequencies less than 10 MHz arriving from space cannot penetrate the ionosphere to reach ground-based radio telescopes. Frequencies below about 10 MHz must be observed from space-based radio telescopes.

At the high-frequency boundary, absorption by H_2O and O_2 in our atmosphere blocks the radiation from space. An especially unfortunate aspect of this is that these and many other molecules are abundant in space, and so it is of scientific interest to observe at these higher frequencies to probe the emission and absorption by these molecules in space (such as in star formation regions). Radio observatories that operate at millimeter wavelengths, therefore, are desirable and must be located at high altitude, dry locations (e.g., on mountain tops).

1.2.2 Spectroscopy

A powerful tool for analyzing the detected radiation is to examine its spectrum (a plot or display of the amount of radiation vs. frequency or wavelength), as the details of a source's spectrum contain much valuable information about the physics of the source. For now, we will just give some rudimentary basics.

With a classical spectrograph, diagrammed in Figure 1.3, the different wavelengths of visible light are separated by utilizing the wavelength dependence of refraction or diffraction, the former using a prism and the latter using a grating. To obtain images at different wavelengths that are easily distinguished, the light from the object must first pass through a slit, with its narrow dimension corresponding to the direction of refraction or diffraction. At some distance past the prism or the grating, the radiation is detected with a charge-coupled device in modern spectrographs. The spectrum is then an elongated image in which the wavelengths are spread along the long dimension. Absorption and emission lines (which will be discussed shortly) will show up as dark and bright lines perpendicular to the long dimension.

Spectrographs used at radio wavelengths are quite different. It is cumbersome to refract or diffract radio waves by optical means, so usually the separation of the different frequencies is accomplished electronically. We will discuss this in Chapter 3. Additionally, the radio-frequency spectrum appears to the user as a graph of intensity or flux density (i.e., the amount of radiation) versus frequency or wavelength, as is often the case now for optical spectra taken with modern digital detectors. Therefore, the concept of lines seen in

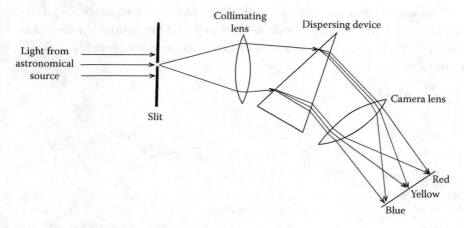

FIGURE 1.3 Schematic of an optical spectrograph.

earlier photographic images of optical spectra is not what we see today; however, we still call these regions of enhanced or reduced intensity as *spectral lines*. As you will see in our discussion below, the term *lines* has become a common part of the spectroscopy lingo.

There are three general types of spectra, as described by *Kirchhoff's rules*. These are named after Gustav Kirchhoff, who, along with Robert Bunsen, initiated the study of spectroscopy in the 1860s. It is helpful to keep in mind that the terminology used to describe spectra comes from the initial experiments with spectra that used visible-wavelength light. The three basic types are as follows:

1. *Continuous spectra*: When a radiation source emits at all frequencies over a range without breaks, the spectrum is called a *continuous spectrum* and the emitting object is called a *continuum source*. A classic example of a continuum source is an incandescent lamp. A plot of a continuous spectrum, for example, will show intensities over a broad range of frequencies, although the intensity of the emission can vary significantly. A qualitative example of a continuous spectrum is shown in Figure 1.4.

2. *Bright line or emission line spectra*: When a radiating object emits radiation only at some very specific frequencies, or wavelengths, the spectrum contains a set of discrete bright lines. These lines of light are called *emission lines*. The reason that a light source would emit only at some very specific frequencies is due to the quantum physics of atomic and molecular structures. An atom or molecule can be in an *excited energy state* and then spontaneously drop to a lower energy state. To conserve energy, it gives off a photon that carries the exact amount of energy that the atom or molecule loses. However, due to the laws of quantum mechanics, the internal energy levels of atoms and molecules are restricted to a set of discrete values. Thus, changes in energy (and hence the emitted photon energy or frequency) can only have certain specific values.

   In a graphical representation of the spectrum, emission lines show up as peaks in the intensity versus frequency graph. A qualitative sketch of an emission line spectrum is shown in Figure 1.5.

   Different atoms and molecules emit different sets of frequencies, and so the composition of a radiation source can be identified by determining the frequencies of the emission lines. This is one example of how spectroscopy is such a powerful tool in astronomy.

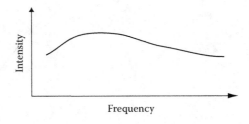

FIGURE 1.4   Graphical representation of a continuous spectrum.

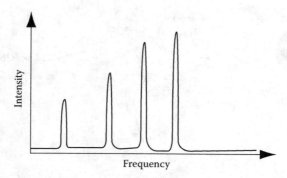

FIGURE 1.5   Graphical representation of an emission-line spectrum.

3. *Dark line or absorption line spectra*: A more complicated and interesting case arises when the radiation from an intense continuum source passes through a cool, transparent gas of atoms or molecules. Some of the atoms or molecules in the gas can absorb photons from the continuum to raise them into a higher allowed energy level. The photons that they absorb must have exactly the same amount of energy that the atom or molecule gains, and so only the photons of certain, specific frequencies will be removed from the continuum. To an observer stationed beyond the cool gas, the spectrum will show the continuous spectrum of the background source with the radiation at these specific frequencies appearing darkened, thus the dark line nomenclature. In a plot of intensity versus frequency, they will appear as dips or decreased intensity, as shown in Figure 1.6. These dips are called *absorption lines*.

    The atomic or molecular absorption frequencies are the same as the corresponding emission frequencies, and so again, the chemical composition of a gas cloud can be identified by the frequencies of the absorption lines that it produces.

Example 1.2:

The light emitted by two glowing objects, a hot solid and a warm gas cloud, is passed through spectrographs by three observers. Observer 1 obtains the spectrum of the first object and finds it to be a continuum. Observer 2 finds that the spectrum of the second object is an emission-line spectrum. Observer 3, though, obtains the spectrum of the second object but from a position where the first object is behind the second. The viewing angle of each observer is depicted in Figure 1.7. What kind of

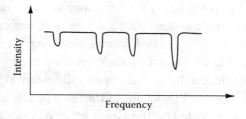

FIGURE 1.6   Graphical representation of an absorption-line spectrum.

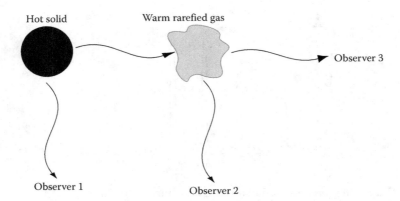

FIGURE 1.7   Three observers each obtain a spectrum of the light emanating from a glowing object. Observer 1 gets the spectrum of one object: a hot solid; Observer 2 gets the spectrum from the second object: a warm gas cloud; while Observer 3 gets the spectrum emanating from the gas cloud with the hot solid in the background.

spectrum does Observer 3 see and why is it different from that obtained by Observer 2 even though they receive the light emanating from the same object?

Answer:

Observer 3 sees an absorption-line spectrum. Observer 3's spectrum is different from that obtained by Observer 2 because the light received by Observer 3 includes the continuum emitted by the hot solid. The light from the heated solid passes through the rarefied gas and then on to Observer 3, and there is an interaction between that continuum radiation and the gas in the second object. If the solid is hotter than the rarefied gas, the gas in the second object causes absorption lines to occur. The spectrum obtained by Observer 2, meanwhile, does not include the radiation from the solid object, and so this spectrum contains only the emission from the gas in the second object, which is an emission-line spectrum.

1.3 FINDING OUR WAY IN THE SKY

1.3.1 Sky Coordinate System: Right Ascension and Declination

To find and observe objects in the sky, we need an established and consistent system of defining their location, that is, a *coordinate system*. To us, on Earth, the surrounding sky appears like a spherical shell—we can see the directions to objects but we do not know their distances. Since Earth, also, is (roughly) a sphere, and we have an established coordinate system for it, the sky coordinate system that is set up is of basically the same construct. Therefore, picture a globe of the sky, called the *celestial sphere*, with a globe of Earth at its center. Figure 1.8 should help with this visualization. Extensions of the lines of longitude on Earth produce similar lines on the sky, which we call lines of *right ascension (RA)*, often represented by the Greek letter α. Extensions of the lines of latitude make lines on the sky called *declination (Dec)*, represented by the Greek letter δ. Since the apparent rotation of the sky is actually due to Earth's rotation, the poles of the celestial sphere are the points

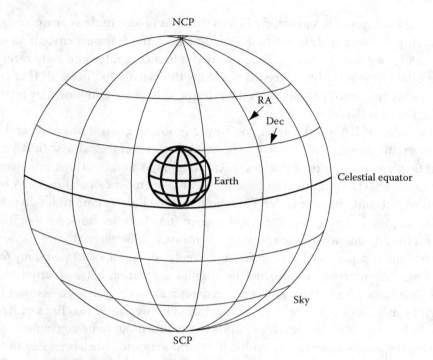

FIGURE 1.8   Celestial sphere, containing lines of RA and Dec, with Earth at the center.

directly overhead the poles on Earth. The sky, then, appears to rotate like a sphere about an axis running through it from the *north celestial pole* to the *south celestial pole*.

The details of the sky coordinate system, though, are a bit more complicated because of Earth's rotation. This rotation produces a constant relative motion between the lines of longitude and RA. Since this does not happen with the lines of latitude and Dec, Dec is easier to handle, and so we will discuss it first.

It is natural to choose the 0°-line of Dec to be the exact extension of the 0°-line of latitude. Just as we call the 0°-line of latitude on Earth as *the equator*, we denote the 0°-line of Dec on the sky as the *celestial equator*. Then, the extension of the 10°-line of latitude to the sky points to the 10°-line of Dec; likewise for the 20° line, the −10° line, and so forth. If you are standing at a latitude of +45°, for example, and you point at the spot straight overhead, you will be pointing at a spot on the sky of Dec +45°.

Now let us consider the lines of RA; remember that they are complicated by Earth's rotation. If at some moment in time we define the 0-line of RA to be the extension of the 0°-line of longitude, a minute later, these lines will be shifted and, as Earth rotates, they will continue to shift relative to each other. Of course, we want the lines of RA to be useful for defining the position of an object on the sky. Therefore, we need the lines of RA to be fixed relative to the stars, not Earth. What astronomers have chosen to do is to pick noon on the first day of spring (approximately March 21) as the moment in time when the 0°-line of RA aligns with the 0°-line of longitude (the latter passing through Greenwich, England). And then, through our knowledge of how Earth rotates during the day, and how it moves around the Sun in its orbit throughout the year, we can figure out by how much the lines of RA are shifted from the lines of longitude at any given moment of time during the year. This is not that difficult a task, but it is burdensome to do every time. Therefore, we usually have it programmed into a computer and only rarely do we make the calculation ourselves. We will not show the calculation here; all that you need to know is what the coordinate lines on the celestial sphere are and how they relate to the coordinate lines on Earth.

Now, the units of RA are not what you would probably guess. Longitude and latitude are both given in degrees, and so it would be natural to expect the same with RA and Dec; Dec is, in fact, given in units of degrees. RA, though, has units of time; this needs some explanation. In short, using units of time for RA makes most calculations involving RA simpler, as will become apparent later. First, note that the Earth's longitudes do not behave like normal angles. In particular, the angular *separation* between adjacent longitude lines depends on the latitude where the separation is measured. Between 0° and 10° longitude, there is an angular separation, as seen from the center of Earth, of 10°, *but only for points on the equator.* Away from the equator, the angular separation between any two lines of longitude decreases with latitude and even becomes zero at the poles. The same, of course, occurs with the lines of RA. The angular separation between any two lines of RA is proportional to the cosine of the Dec. Now, units of time turn out to be convenient to use for RA, because the displacement between RA lines and longitude lines is related to the rotation of Earth. The units of RA are set up as follows. There are 24 h of RA around the sky, and each hour is divided into 60 min of RA, and each minute contains 60 s of RA. Keep in mind, though, that *one second of RA does not equal one second of arc.* In fact, since there are 24 h of RA around the sky and 360 degrees around a full circle, at the equator, 1 h of RA = 15° of arc, and so 1 minute of RA = 15 min of arc and 1 s of RA = 15 s of arc *at the equator.* Away from the equator, the general relation between the number seconds of RA and seconds of arc is given by

$$\text{Seconds of arc} = 15 \cos(\delta) \times \text{seconds of RA} \qquad (1.3)$$

where:
   δ is the Dec

Another apparent oddity is that if you look at a map of the sky with north at the top, then east is to the *left* side of the map. This is not due to some strange choice made by astronomers. This, in fact, is the same east as on maps of the ground. The reason for the apparent difference is that you, the observer, are *between* the ground and the sky and look in opposite directions to see each one. Try the following stretching exercise. Bend over and face the ground with the top of your head pointed to north. Stick your right arm out to point to the east. Now, without moving your arm or your body, twist your head around 180° to look at the sky. (OK, you can't physically do that…but you can imagine it.) If your right arm is still pointed to the east and the top of your head is to the north, with your head turned around 180° to look at the sky, what direction is east? Answer: to the left. As an additional oddity, when you look at a map of the sky, RA increases to the left, which, at first, seems the reverse of what you would expect. We will explain the reason for this in a little bit.

1.3.2 Observer-Centered Definitions

We also need a system that describes the way a particular observer (or telescope) sees the sky at any particular moment in time. Therefore, we need definitions to describe an observer-centered coordinate system. The terms that you need to know for this book are as follows:

1. *Horizon*: This defines the limit of what parts of the sky you can see at any particular moment. It is due to the ground and structures on Earth blocking your view of the sky. If Earth were transparent, you could also view the sky below your feet by looking through Earth. Since Earth is not transparent, you cannot see this part of the sky, because it is below your horizon.

2. *Zenith*: This is the point in the sky directly overhead. Since the sky rotates continuously, this point on the celestial sphere continually changes, unless you are standing at either the North or the South Pole.

3. *Altitude or elevation*: These are synonyms for the angular height of an object above your horizon at any given moment. When a star is on your horizon (so that it is either just rising or just setting), its altitude, or elevation, is 0°, and when it is directly overhead—at the zenith—its altitude, or elevation, is 90°. Keep in mind that this is an *angle*, not a physical distance. Figure 1.9 displays the altitude angle of a star.

4. *Azimuth*: This is the angular position perpendicular to the altitude, and is defined as the angular position of an object along the horizon relative to due north. If the object is located to the north of the zenith, its azimuth is 0°. If it is located to the south of the zenith, it has an azimuth of 180°. An azimuth of +90° is due east and 270° is due west. The angles of azimuth and their relation to the cardinal directions are shown in Figure 1.10. Note that both azimuth and altitude make a pair of angles that completely define an object's position in the sky *relative to the observer*. This last detail is important: all observers will see the same RA and Dec for an astronomical object, but the altitude and azimuth for the object will be different for observers at different locations, and indeed, even for the same observer, at different times of the day.

5. *Meridian*: This is the line of RA that runs through your zenith. Equivalently, it is the line in the sky that runs through both your zenith and the celestial poles. Consider the following. As Earth rotates, you will see the stars move across the sky from east to west. Consider, for a moment, just the stars that are at the same Dec as your latitude;

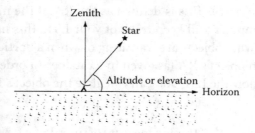

FIGURE 1.9 Schematic showing the relations between zenith, horizon, and altitude or elevation.

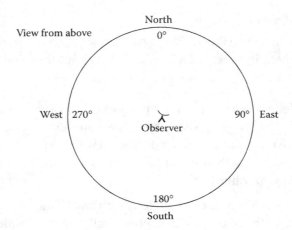

FIGURE 1.10 View from above depicting azimuthal angles along with the cardinal directions around the horizon.

these are the stars that will pass directly overhead (i.e., through the zenith) on their trek westward. The moment that one of these stars is at the zenith is when it is highest in the sky (i.e., when it reaches its maximum elevation). Stars that are at different Decs will never pass directly overhead because the point directly overhead must have the same Dec as your latitude (by definition of Dec). All stars, regardless of their Dec, will be highest in the sky during the moment when they cross your meridian. This is important because it is the best time to observe a particular object, and it also defines the midpoint of the time that an object is visible above the horizon (i.e., the moment in time halfway between rising and setting).

6. *Transit*: (verb) means to pass through the meridian. So, if your friend asks, "When does Mars transit?" he/she means at what time of day is Mars highest in the sky. This is a typical question whose real meaning is, "When is the best time to observe Mars?" Check your understanding: What is the azimuth of an object when it transits? Assuming that the object does not pass through the zenith, then its azimuth will be either 0° or 180° at transit.

7. *Hour angle (HA)*: An object's HA is the amount of time (in hours) since the object transited. For example, if the object is currently at the meridian, its HA at that moment is 0. If the object transited 1 h ago, it has an HA of +1 h, and if it will transit in two and a half hours, then its HA is −2.5 h.

8. *Local sidereal time (LST)*: This is defined as the RA of the meridian. While you are observing, the computer will keep track of your LST. This number is useful to help you keep track of what objects are transiting or when a particular object will transit. You must know an object's RA (as given in a catalog) in order to observe it. In addition, if you also know the LST, you can calculate the object's HA from

$$HA = LST - RA \tag{1.4}$$

The name LST suggests that this is a measure of time. In fact, it is, but it runs faster than the time we normally use. Consider the definition of the units of time that your watch or clock uses. One hour on your watch is defined as 1/24th of a day, which is defined as the average time period between transits of the Sun (i.e., from noon to noon). This method of measuring time is called *solar time*, because the Sun is used as the reference. *Sidereal time* uses the celestial sphere (or stars, other than the Sun) in the same way. One sidereal hour is 1/24th of the time period between transits of the same star. Time on the local sidereal clock is given by the RA of the meridian. For example, when the 12-h line of RA is directly overhead, the LST is 12 h. You do not want to use sidereal time as a real measure of time, though, because 1 h of sidereal time is *not* equal to 1 h of solar time. Each star rises 4 min earlier each day, meaning that a sidereal day is 4 min shorter than a solar day. This is due to the Earth's orbit—because of the Earth's motion about the Sun, in the time it takes to rotate once it has moved 1/365th of the way around its orbit, and so the direction of the Sun has moved slightly. The Earth must rotate an additional 4 min for the Sun to reappear directly over the same place on Earth.

An important point to remember is that a celestial object will transit when your LST is equal to the RA of the object. The Galactic Center, for example, has an RA of about 18 h, so it is highest in the sky (and best observed) at an LST near 18 h. Note, however, that because LST is *slower* than solar time, the time of day when the Galactic Center transits will change throughout the year. In the northern hemisphere, 18 h LST is in the daytime during the winter months, while it occurs at nighttime during the summer months, gradually shifting by 4 min every day.

9. *Universal time (UT)*: This is the solar time in Greenwich, England (i.e., the time on the clock where the longitude is 0°). This is useful for marking the exact time of an observation or an event. One does not need to know, then, from what time zone the observation was made. It defines a time that is the same for everyone on Earth. Universal time is particularly useful for situations in which the different time zones can cause confusion. Airlines, for example, actually schedule their flights using Universal time, but translate to local time for the customers.

Now we can understand why hours, minutes, and seconds make a convenient choice of units for RA and why RA increases to the east. Suppose you wish to observe an object whose RA is 12 h and 10 min. You look up at the sky and note that the object straight overhead has an RA of 4 h and 10 min. Now you can ask, "How long will it be before my desired object transits?" The calculation you need to do is basically to treat the RA of the stars like the tick marks on a clock. Currently, the LST is 4:10 and the question is really asking, "When will the LST be 12:10?" Therefore, the answer is simply 12:10−4:10 = 8 h. (This does not account for the minor difference between the movement of your clock, which is based on the Sun, and the movement in LST, which is based on the stars.) Sources further to the east will transit at a later LST, hence their RA is larger; that is, RA increases to the east.

This section has presented a long list of definitions, which is certainly not easy to absorb all at once. You may find Figure 1.11 helpful in collating and gathering all these definitions into a complete picture.

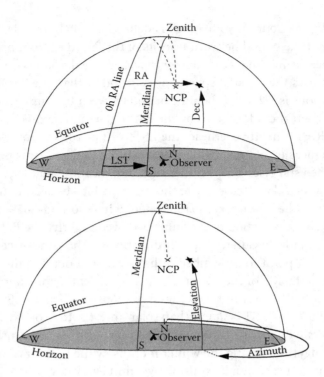

FIGURE 1.11 Depiction of sky coordinates as viewed by the observer (top) and observer-centered coordinates (bottom).

Example 1.3:

A pair of astronomers learns of an interesting radio source at a location in the sky given by the coordinates:

$$RA = 12 \text{ h } 30 \text{ min } 00 \text{ s and Dec} = 0.00°$$

The astronomers decide to check it out and so set out to observe it with their own telescope, which is located at longitude = 75.0° W and latitude = 35° N. They first check their computer and see that the current LST at the telescope is 03h 00m 00s.

    a. What are the RA and Dec of the astronomers' zenith?
    b. What is the current HA of the radio source?
    c. Is the radio source above or below the horizon at this time?
    d. What will the azimuth and altitude of the desired sky location be when it rises?
    e. What will the azimuth and altitude of the desired sky location be as it transits?

Answers:

    a. The RA of the zenith is given by the LST, so RA = 03 h 00 min 00 s, and since the lines of Dec are extensions of the lines of latitude, the Dec of an observer's zenith is the same as the observer's latitude, so the astronomers' zenith is at Dec = +35°.

b. The HA is easily calculated by subtracting the RA from the LST of the target source, so

$$HA = 03\,h\ 00\ min\ 00\,s - 12\,h\ 30\ min\ 00\,s = -9\,h\ 30\ min\ 00\,s$$

This means that the best time to observe this source is in another 9 h or so.

c. Since the Dec of the radio source is 0°, it is on the celestial equator. Hence, it is above the horizon for exactly half the time, that is, 12 h per day. So, it rises 6 h before it transits, or when its HA = −6 h. The HA of the object is more negative than −6 h and so the object is further from the meridian than the horizon is, and hence the object is below the horizon.

d. When an object in the sky rises, it is on the horizon and so its altitude at that time is 0°. To figure out its azimuth, we note that the object is on the celestial equator, and so it rises due east. This happens because the celestial equator passes through the points directly east and west for observers at *all* latitudes, and so its azimuth at that time is +90°.

e. When a body transits, it is on the meridian and so it is either at the zenith, or due south, or due north of the zenith. We can figure out which of the three cases it is by comparing the Dec of the radio source to the Dec of the zenith (which equals the latitude of the observer). In this case, the radio source's Dec is smaller than the Dec of the zenith and so it will be due south of the zenith when it transits. Therefore, its azimuth will be +180°.

Determining the altitude is a little more complicated. We use the Decs of the sky position and of the zenith to calculate the angle between the two, which is sometimes called the *zenith angle*. We then recognize that this angle is the complement of the altitude angle, as shown in Figure 1.12. So, then,

$$Altitude = 90° - \left(Dec\ of\ zenith\ -\ Dec\ of\ sky\ position \right)$$

$$= 90° - (35° - 0°)$$

$$= 55°$$

1.3.3 Apparent Sizes

In addition to an object's position, how we see it also involves its apparent size. Our perception of size for distant objects is determined by the object's angular size, which is related to the object's linear size and distance. However, we do not generally know an astronomical object's distance, so we often cannot know its linear size; but we can perceive its angular size.

For this purpose, the best and most natural unit to measure angle is a radian. Degrees are an arbitrary unit and are awkward to use in calculations. In fact, most calculations require radians as the unit of angle. Therefore, we will first review the mathematical definition of a radian.

For any given circle, an arc length has an angular size as seen from the center of the circle, and the angle, θ, subtended by that arc can be described by the ratio of the arc length, s, to the radius, r, of the circle. This ratio gives the angle in units of *radians*

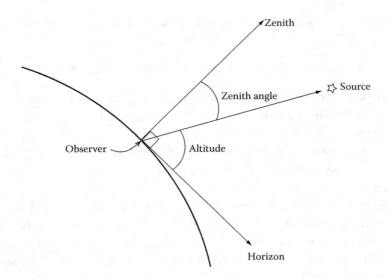

FIGURE 1.12 Pictorial depiction of the relation between the altitude and the zenith angle, that is, the angle between the zenith and the source.

$$\theta \text{ (radians)} = \frac{s}{r}$$

Note that the definition of radian involves a length in the numerator and a length in the denominator. A radian, then, is a ratio of two distances and so is actually dimensionless. One way that you might find radians confusing is that they can appear and disappear in equations without altering the units.

Now, to consider how this relates to the angular size of a distant object, note that if the angle θ is small, then the arc length, s, is approximately equal to the straight line distance, l, between the two end points of the arc (i.e., equal to the length of the chord), as shown in Figure 1.13. For small angles, then

$$\theta \text{ (radians)} \approx \frac{l}{r} \tag{1.5}$$

Now notice that the chord of length l as viewed from the center of the circle is identical to an object of length l at a distance r from you. Therefore, the angular size of an object, in radians, is approximately l/r.

In reality, objects in the sky appear to us as two dimensional, while angles, as we just discussed, are only one dimensional. We need, therefore, a two-dimensional description of angular size. This is called a *solid angle*, which we denote as Ω. To help gain an understanding of solid angle, consider a celestial object with a circular cross-sectional area small enough in angular diameter that Equation 1.5 applies. The calculation of the solid angle of the object, then, is essentially the same as calculating the geometric area of a circle. [Remember that the area of a circle $= \pi r^2 = \pi \text{ (diameter/2)}^2$.] Since the angular diameter

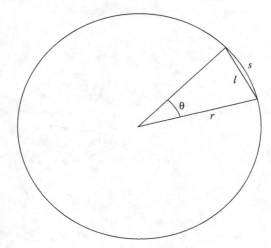

FIGURE 1.13   The measure of angle θ is the arc length, s, divided by the radius of the circle, r, and the units of this angle are radians. For small angles this ratio is approximately the same as the chord length divided by the radius.

of the circular object is θ (in radians), using our analogy between area and diameter of an object, its solid angle, Ω, is

$$\Omega = \pi \left(\frac{\theta}{2}\right)^2 = \left(\frac{\pi}{4}\right)\theta^2 \qquad (1.6)$$

To obtain an equation for Ω of any shape object, we can substitute in the relation between θ (radians) and A, the area of the object, to get

$$\Omega = \frac{A}{d^2} \qquad (1.7)$$

where:
  d is the distance of the object
  A is its cross-sectional area

Keep in mind that Equation 1.6 applies only for small angular sizes, because we made the approximation that the chord length was approximately equal to the arc length. To obtain a general equation for solid angle, then, we can use Equation 1.7 but define A as the area occupied on the surface of the sphere of radius d, centered on the observer, as shown in Figure 1.14.

Unfortunately, Equation 1.7 requires knowledge of both the distance of a body and its area, and so it often is not a very useful equation. Generally, you will need to calculate the angular area in similar ways as calculating geometric areas. Example 1.4 demonstrates how to calculate the solid angle of an elliptical source. A word of caution: do not forget to convert your angles to radians (if you have measured them in degrees, arc minutes, and arc seconds). This is a common mistake.

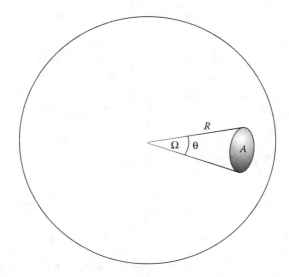

FIGURE 1.14   Solid angle, Ω, of a circle with diameter θ on the surface of a sphere.

The units of solid angle are *steradians* (sr). The number of steradians over the whole sky is 4π, as can be shown by Equation 1.7 and using the fact that the surface area of a whole sphere is $4\pi R^2$. The solid angle of the whole sphere as seen from the center is, then,

$$\Omega = \frac{4\pi R^2}{R^2}$$

$$= 4\pi \text{ sr}$$

Example 1.4:

Consider a radio source that is found to be elliptical in shape with a major axis of $0.3°$ and a minor axis of $0.1°$. What is the solid angle of this radio source?

Answer:

In this case, we use an equation similar to Equation 1.6. Since this source is elliptical instead of circular, we use the equation for the area of an ellipse, that is,

$$\Omega = \pi \left(\frac{\theta_{major}}{2} \right) \cdot \left(\frac{\theta_{minor}}{2} \right)$$

But, we must first convert our angles to radians. $\theta_{major} = 0.3° \, (\pi/180°) = 0.00524$ radians and $\theta_{minor} = 0.00175$ radians. So then,

$$\Omega = \pi \left(\frac{0.00524}{2} \right) \cdot \left(\frac{0.00175}{2} \right)$$

$$= 7.18 \times 10^{-6} \text{ sr}$$

1.4 BASIC STRUCTURE OF A TRADITIONAL RADIO TELESCOPE

Observing with a radio telescope can be very different from using a visible-wavelength telescope. The Sun does not light up the whole sky at radio wavelengths, as it does at visible wavelengths. The blue daytime sky you see is sunlight that is scattered by the atmosphere, which scatters blue light to a greater extent than the longer visible wavelengths. At radio wavelengths, there is no scattering by the atmosphere at all. Therefore, the daytime sky is dark at radio wavelengths, and so radio observations can be made during day or night. At long radio wavelengths, observations can occur even with a cloud-covered sky since clouds are transparent at these frequencies. At shorter radio wavelengths, though, the water in clouds is a significant source of light loss, and so shorter-wavelength observations require clear weather. Radio telescopes operating at these short-radio wavelengths are often placed in high and dry sites to minimize this loss. We will discuss other differences in detail in Chapters 3 and 4. For now, as part of this introductory chapter, we just describe the basic structure of a radio telescope.

A traditional radio telescope is composed of five basic parts, which are depicted in Figure 1.15.

1.4.1 Parabolic Reflector

Most radio telescopes use a *parabolic reflector*, also known as the *dish, to collect and focus the radio light* (although other forms of antennas are often used at low frequencies). Since radio waves are easily reflected by metallic surfaces but not easily refracted, no radio telescopes are refractors (in contrast to visible-wavelength telescopes where lenses are sometimes used). The radiation can then either be detected at the focal point of the dish or be reflected back through the middle of the dish and focused and detected behind the dish. The former case is known as a *prime focus telescope* and the latter is known as a *Cassegrain telescope*. The Haystack Small Radio Telescope (SRT), shown in Figure 3.3 (which you might use for some labs if your school has one), is an example of a prime focus telescope. A color photograph of a Cassegrain telescope is shown in Figure 3.4.

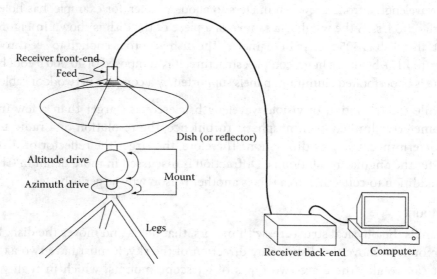

FIGURE 1.15   A prime-focus radio telescope and its parts.

There are some important aspects of the radio telescope reflector (or dish) that are worth discussing at this point.

1. As with a visible wavelength reflector, the *sensitivity* (or the light-gathering power) of the telescope depends on the collecting area and hence on the square of the diameter of the reflector. Larger dishes collect more light, and hence receive greater power, from astronomical sources.

2. Unlike visible light reflectors, radio dishes do not need highly polished surfaces. Radio waves, being much longer than visible waves, can be reflected by a much less precise surface. The general requirement for successful reflection of EM radiation is that irregularities in the reflecting surface be much smaller than the wavelength of the radiation. We explain in Chapter 3 that a successful reflection is obtained when

$$\delta z < \frac{\lambda}{20}$$

where:
δz is the deviations of the surface from that of a perfect reflector
λ is the wavelength of light that we wish to reflect

At longer radio wavelengths, in fact, the reflecting surface can even be a mesh, provided the holes in the mesh are sufficiently smaller than the wavelength of light to be reflected. The dish of the Haystack SRT, for example, is used for observations at 21-cm wavelength and is composed of a see-through mesh. A radio telescope dish, then, does not have nearly as tough constraints as a visible wavelength mirror. However, because they need to be much larger than visible light telescopes, they can be equally challenging to construct. The radio reflector at Arecibo, Puerto Rico, which is used for observations at wavelengths from a few centimeters to about a meter, for example, has holes in the surface to lessen the weight—a sample of a piece of the dish is shown in Figure 1.16— but, this dish is 305 meters in diameter! The dish is porous enough to see through (see Figure 1.17), but it is a huge, complex structure. It is composed of 40,000 3-foot by 6-foot panels of perforated aluminum panels supported by a complex network of cables!

3. While the resolution of visible-wavelength telescopes (larger than a few inches in diameter) is limited by atmospheric turbulence, the resolution of a radio telescope is determined solely by diffraction. Therefore, the larger the reflector or dish is, the better the angular resolution is. Diffraction is discussed in more detail in Chapter 3. In addition to collecting area this is another reason why larger is better.

1.4.2 Mount

The mount is the physical structure, with motors, that holds and moves the dish. In order for the mount to move the dish to any direction of the sky, it must have two axes about which it can rotate. There are two types of telescope mounts, which in reality are the same but with different tilts. All modern radio telescopes have altitude-azimuth or *Alt-Az*

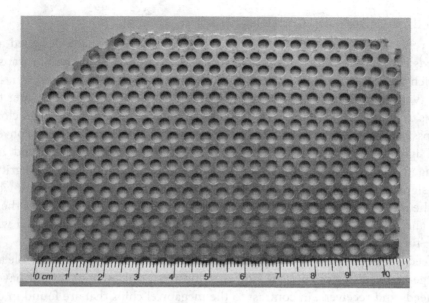

FIGURE 1.16 Small sample of the surface of the Arecibo reflector surface alongside a metric rule. Holes 2 to 3 mm in diameter pervade the surface, which does not hinder the reflection of radiation of centimeter wavelengths and longer.

FIGURE 1.17 **(See color insert.)** View through the Arecibo dish from below. The clouds and sky are clearly visible, as is the feed. (Courtesy of Clara Thomann.)

(also known as azimuth-elevation or *Az-El*) mounts, meaning that the movement of the dish around one axis moves its pointing direction in altitude and movement around the other axis moves the pointing in azimuth.

The other type of mount, which is common for older and smaller visible-wavelength telescopes, is called an *equatorial mount*. With an equatorial mount, the axes correspond to the sky coordinates: movement around one axis moves the telescope in RA and around the other axis moves it in Dec. An equatorial mount is really just an alt-az mount that is tilted, so that the azimuth axis of the telescope points toward the Celestial Pole rather than toward the zenith.

1.4.3 Feeds, Receivers, and Computer

The dish reflects radio light from the sky to its focus where specially designed antennas called *feeds* convert the EM waves in free space into confined EM waves in transmission lines, which carry the signal to *receivers*. Each feed is connected to one receiver, of which there are two parts. The first part is attached at the telescope, immediately after the feed; this part is called the *front-end*. It is important that the receiver front-end be as close to the feed as possible. The receiver front-end provides amplification and frequency conversion of the input signal and restricts the frequency range that will eventually be detected. After this processing, the signal can then travel a reasonable distance without loss of integrity; therefore, the signal is then sent by co-axial cable to the observatory control room, where it is fed into the receiver *back-end*. The back-end contains a *detector* that measures the amount of power and converts it to a digital signal, which is then passed to the computer where it is stored for future analysis. We discuss all these processes in detail in Chapter 3.

Each feed–receiver assembly comprises a single detector, playing a similar role as a pixel in a charge-coupled device camera. However, most radio telescopes do not have a large array of feeds and receivers, in contrast to the megapixel chips that are found in the cameras on visible-wavelength telescopes. Just as there is a maximum size of holes in the radio telescope dish to ensure that the radiation is reflected, there must be a minimum size for the opening of the feeds, to ensure that the radiation is *not* reflected, but rather *passes through* the opening. This minimum size is of order the wavelength of the radiation, so for longer radio wavelengths, the feeds must be fairly large. For many radio observations, the feed horns have openings of order tens of centimeters to meters across. Especially for the longer radio wavelengths, there usually is not enough room to mount a large array of feeds. Additionally, each radio receiver is complex and expensive to build and so constructing a large number of them is often not practical and beyond normal budgets. For instance, some of the larger telescopes operating at a wavelength of 20 cm, which include the Arecibo Telescope, the Green Bank Telescope, the Parkes Observatory, and the Effelsberg Radio Telescope, have managed to fit 7 to 13 feed–receiver assemblies in the focal plane. (The Haystack SRT has only one.) At millimeter wavelengths, since the feed horns can be smaller, a number of telescopes have several dozen receivers; however, this is a far cry from the millions of pixels in a charge-coupled device camera. At the very shortest radio wavelengths (millimeter and submillimeter wavelength), *bolometer detectors* are often used. These are smaller and less expensive; therefore, telescopes with arrays containing thousands of these detectors do exist.

The data from a single observation, therefore, often are quite different with a radio telescope from that with a visible-wavelength telescope. While a simple visible-wavelength observation can easily produce detailed images with brightness values at millions of points, a single observation at centimeter wavelengths yields values at a handful of positions, and sometimes (for the smaller and/or older telescopes operating at longer wavelengths) *only a single value*. However, the radio astronomy technique of *aperture synthesis*, which involves combining the detected signals by multiple telescopes, typically yields megapixel images. We give an extensive explanation of this observing method in Chapters 5 and 6.

1.5 RADIO MAPS

Showing the structure of a source at radio wavelengths (or at other wavelengths), *requires* the presenter to make a choice about how best to display the two-dimensional data. There are three common approaches.

The simplest way is with a *contour map*. An example of a contour map is shown in Figure 1.18. The contours are lines of constant intensity. A familiar example of a contour map is a topographical map, in which the contours indicate the lines of constant elevation. Many closely spaced contours indicate a steep slope in a topographical map. In a similar way, a large number of closely spaced contours in an astronomical map indicate a region where the intensity is changing rapidly with the position.

A slightly more intuitive radio map is a *gray scale map*, in which no color is indicated, and the brightness variations in the source are directly indicated by the shades of gray, where black is the most intense and white is the least. A gray scale map of the same source as in Figure 1.18 is shown in Figure 1.19. However, details in the structure are not revealed

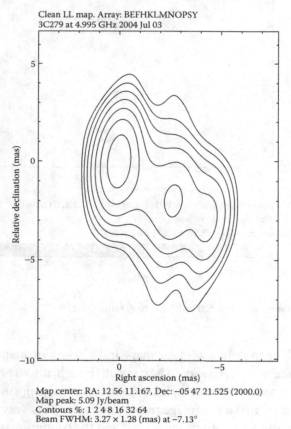

Clean LL map. Array: BEFHKLMNOPSY
3C279 at 4.995 GHz 2004 Jul 03

Map center: RA: 12 56 11.167, Dec: −05 47 21.525 (2000.0)
Map peak: 5.09 Jy/beam
Contours %: 1 2 4 8 16 32 64
Beam FWHM: 3.27 × 1.28 (mas) at −7.13°

FIGURE 1.18 Contour map of the radio emission from the core region of the well-known radio source 3C279 at a wavelength of 6.0 cm. The contours displayed represent lines of intensities equal to 1%, 2%, 4%, 8%, 16%, 32%, and 64% of the maximum intensity in the map. (The map was made with data obtained from an observation with the very long baseline array in 2004.)

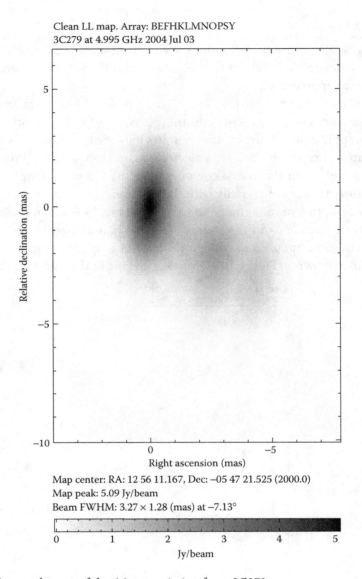

FIGURE 1.19   Gray scale map of the 6.0 cm emission from 3C279.

as well in a gray scale map as in a contour map, and so contour maps are often preferred for good analysis of the map. (One should be careful, though, not to present too many contours, especially for a source with complex structure. Overdoing contours with complex structure can sometimes make the image more confusing.) It is very common, and quite pleasing, to present maps with both the gray scale and the contour maps overlaid, as shown in Figure 1.20.

Many radio astronomers are quite fond of presenting *false color maps*, as shown in Figure 1.21, in which the color does not indicate different wavelengths but different levels

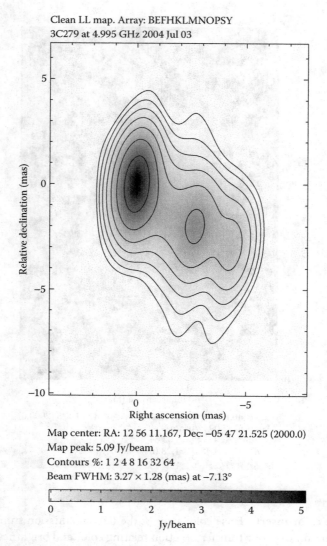

Clean LL map. Array: BEFHKLMNOPSY
3C279 at 4.995 GHz 2004 Jul 03

Map center: RA: 12 56 11.167, Dec: −05 47 21.525 (2000.0)
Map peak: 5.09 Jy/beam
Contours %: 1 2 4 8 16 32 64
Beam FWHM: 3.27 × 1.28 (mas) at −7.13°

FIGURE 1.20   Contour map of Figure 1.18 overlaid on the gray scale map of Figure 1.19.

of brightness. These images are usually accompanied with a color-wedge that indicates the relation between color and brightness. These maps look nice and also can significantly facilitate the analysis.

Finally, we note that sometimes radio astronomers form multicolor images in a similar way to visible light astronomers. If several different radio wavelengths have been observed, their images can be overlaid with one another, with the longest wavelength image assigned the color red, intermediate wavelengths the color green, and the shortest wavelength the color blue.

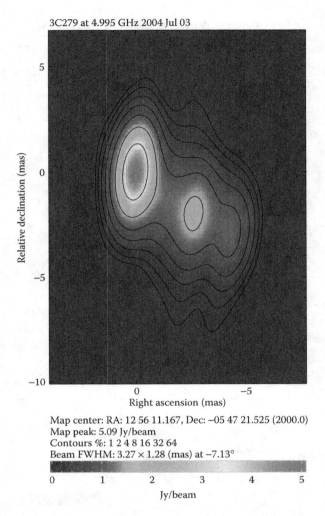

3C279 at 4.995 GHz 2004 Jul 03

Map center: RA: 12 56 11.167, Dec: −05 47 21.525 (2000.0)
Map peak: 5.09 Jy/beam
Contours %: 1 2 4 8 16 32 64
Beam FWHM: 3.27 × 1.28 (mas) at −7.13°

FIGURE 1.21 **(See color insert.)** False color map of the 6.0 cm emission from 3C279. The color wedge at the bottom indicates the transfer function relating color and brightness.

QUESTIONS AND PROBLEMS

1. What was the main reason that the first attempts at detecting radio signals from space failed?

2. a. What is the wavelength of radiation with frequency $\nu = 22.2$ GHz?

   b. What is the energy of a single photon of this frequency?

3. How many radio photons of frequency 2.00 GHz are needed to make up the same energy as a single ultraviolet photon of wavelength 150 nm?

4. a. Name the three basic types of spectra.

    b. Name at least one physical characteristic of the source that can be learned from the analysis of a spectrum that contains lines.

5. Two sources have sky coordinates given by

$$\text{Dec}_1 = 0, \text{RA}_1 = 4\text{h } 0\text{min } 0\text{s; and Dec}_2 = 0, \text{RA}_2 = 5\text{h } 30\text{min } 0\text{s.}$$

    a. Which source is further east?

    b. If source 1 is transiting, what is the LST?

    c. If source 1 is transiting, what is the HA of source 2?

    d. What is the angular separation, $\Delta\theta$, between these two sources?

6. In Section 1.3.1, we took care to stress that 1 s of RA = 15 s of arc *at the equator* and that 1 s of RA = 0 s of arc *at the north and south celestial poles.* In Equation 1.3, without derivation, we gave the general relation between seconds of arc and seconds of RA as involving a cosine of the Dec.

    a. Make a strong justification for why Equation 1.3 must be correct.

    b. Consider a pair of sources located at Dec = 60° that are separated by 50 arcsec. What is the difference in their RAs?

7. To an observer at the North Pole,

    a. What is the altitude of the celestial equator?

    b. What is the altitude of a star at Dec = +30°?

8. To an observer at latitude = +42°,

    a. What is the altitude of the North Celestial Pole?

    b. What is the altitude of the celestial equator?

    c. What is the altitude of a star at Dec = +30° when it is transiting?

    d. What is the azimuth of the North Celestial Pole?

    e. What is the azimuth of a star on the equator at the moment that it is setting?

9. A telescope's pointing position is given by altitude = 40° and azimuth = 120°. Describe the direction that the telescope is pointing.

10. Show that Equation 1.6 is not a generally true equation by considering the solid angle of an entire sphere. Use the fact that there are 2π radians in a full circle and Equation 1.6 to obtain a value for the solid angle of the entire sphere. You should find that this is not equal to 4π sr (which is the correct answer). Why is Equation 1.6 wrong in this situation? What is the correct equation to use to solve for Ω of the whole sphere?

11. Solve for the solid angle of the Sun using,

    a. Equation 1.6 (the angular diameter of the Sun is 0.533°).

    b. Equation 1.7 (the linear diameter of the Sun is 1.392×10^6 km and its distance from Earth is 1.496×10^8 km).

12. Imagine a distant object that appears as a rectangle in projection. If its apparent lengths are 0.0100° and 0.0500°, what is the solid angle of the object in steradians?

13. Why is a radio telescope cheaper to build than a visible-wavelength telescope of the same size?

14. Why can you see through the dish of some radio telescopes but never through the mirror of visible-wavelength telescopes?

15. What is the difference between an alt-az mount and an equatorial mount?

CHAPTER 2

Introduction to Radiation Physics

A RADIO ASTRONOMER MUST ANALYZE and interpret the radiation detected in order to learn about the physics of the object being studied. Therefore, a solid understanding of the basic physics of radiation is critically important. In this chapter, we discuss some introductory radiation physics essential to the rest of the book.

2.1 MEASURES OF THE AMOUNT OF RADIATION

One quantitative measure that astronomers can make when they observe a light source is the amount of radiation that was received. How this amount is to be quantified is not as obvious as you might think. In addition to the radiation that is received, we also want to quantify the amount of radiation an astronomical source emits. As you will see, the amount of radiation we receive and the amount of radiation emitted by the source are quite different, but they are related. With these ideas in mind, we will start this chapter by discussing the various ways in which the amount of radiation can be described.

First, at radio wavelengths (as well as with almost all bands of the EM spectrum except perhaps for X-rays and γ-rays), we generally measure the amount of radiation by its energy, rather than by the number of photons. Our first quantity to consider, then, is the total light energy emitted by a source.

2.1.1 Total Energy Emitted

One can describe a source's light output in terms of the total amount of energy emitted over a source's lifetime, at all frequencies, and in all directions. This, however, is clearly not the kind of measurement that we can readily make. Since one can make measurements only over a finite time period, one can only describe the amount of energy detected in that time period. We wish to make a measurement that is independent of the time interval observed, and so a more useful quantity is luminosity or power, which is energy normalized by the time period.

2.1.2 Luminosity

Luminosity (L) or power is the rate at which energy is emitted. Its SI units are J s^{-1} or watts (1 W = 1 J s^{-1}) and erg s^{-1} in cgs. One calculates a source's luminosity or power by dividing the amount of energy emitted by the length of time over which the energy was emitted. This yields the *rate* of emission.

Luminosity, though, is not directly measurable because we do not detect *all* the radiation that was emitted by the source in any given second. The vast majority of the radiation is emitted in directions other than toward our telescope. This leads to our next quantity.

2.1.3 Flux

The radiation power that we detect depends on the size of our telescope—the larger its cross-sectional area, the more radiation will be detected—so we again wish to normalize our measurement, this time, by dividing by the area of the telescope. This gives the measure of *flux* (F), that is, the amount of light energy per unit time per unit area. The units of flux are J s^{-1} m^{-2} or W m^{-2} (SI) and ergs s^{-1} cm^{-2} (cgs).

We can determine the relation between the detected flux and the luminosity of the source by calculating the fraction of the total emitted radiation that enters our telescope. If the radiation from the astronomical source is emitted isotropically (i.e., equally in all directions) and the source is at a distance d, then the luminosity is spread out over a spherical shell of radius d (see Figure 2.1). The fraction that we detect is given by the ratio of the effective area of our telescope (A_{eff}) to the area of the entire spherical shell. The surface area of the spherical shell is $4\pi d^2$, so the power detected, P, is

$$P = L \frac{A_{\text{eff}}}{4\pi d^2}$$

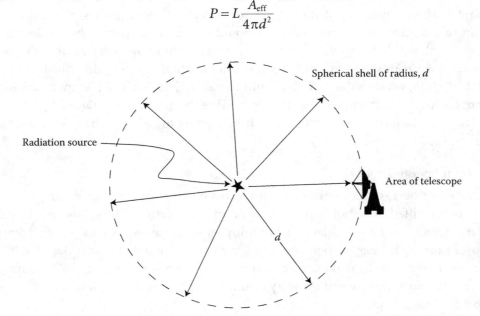

FIGURE 2.1 The power collected by the telescope is the fraction of the total luminosity emitted by the source that enters the small area of the telescope relative to the entire spherical shell of radius d.

The radiation flux, that is, the detected power divided by the area of the telescope (for an isotropic source), then, is related to the luminosity of the source by

$$F = \frac{L}{4\pi d^2}$$ (2.1)

Another way of thinking of flux is that it is a measure of the rate that energy crosses a unit cross-sectional area at a given distance from the source.

Flux, also, is not a truly measurable quantity because we cannot measure the radiation emitted at all frequencies over the entire EM spectrum. There is no detector in existence that can do that. We can only detect the radiation emitted over the tiny fraction of the electromagnetic spectrum to which our detector is sensitive.

2.1.4 Flux Density

Flux density (F_ν or F_λ) is the flux per unit frequency in the observed spectral range, and it equals the detected flux divided by the width in frequency of the observation. Therefore,

$$F_\nu = \frac{F}{\Delta \nu}$$

where:

$\Delta \nu$ is the *bandwidth*, or the range in frequency of the detected EM waves

The symbol S_ν (rather than F_ν) is often used by radio astronomers to represent flux density per unit bandwidth.

When working at visible wavelengths, astronomers tend to measure flux density in terms of wavelength rather than frequency, so flux density is also often defined as flux per unit wavelength.

$$F_\lambda = \frac{F}{\Delta \lambda}$$

Although these two quantities are the same conceptually, they are not the same quantitatively. One distinguishes between the two by using the appropriate subscript, as indicated in the equations above. We will explain the quantitative difference between F_ν and F_λ in Section 2.2. But for the rest of this book, we will only use F_ν, the flux per unit frequency, since that is the standard approach in radio astronomy.

Since the range of frequencies used in an observation can vary, one must again normalize the telescope's measurement by dividing the detected flux by the frequency range over which the observation was sensitive. We call this flux *density* because it describes the density of flux in spectrum space. The mathematical relation between flux and flux density is

$$F = \int F_\nu d\nu$$

Flux density, at last, is a measurement that we can obtain directly and which all observers should agree upon, regardless of the telescope they use. The telescope measures the amount of power it receives, and this depends on the collecting area and the bandwidth. The flux density is the characteristic of the source that we want to infer from these data. The amount of power a telescope collects from a source of given flux density is given by

$$P = F_v A_{eff} \Delta v \tag{2.2}$$

Since flux density is a measurable quantity, we need to define convenient units. For astronomical sources at radio frequencies, the amount of energy we receive per unit time, per unit area, and per unit bandwidth is incredibly small, so both the SI and the cgs systems yield awkwardly small numbers for typical flux densities. To avoid carrying around excessively negative powers of 10 in our calculations, radio astronomers have defined a unit of flux density, named after the father of radio astronomy, Karl Jansky. In units of the SI and cgs systems, a jansky is defined as

$$1 \text{ jansky (Jy)} = 10^{-26} \text{ W m}^{-2} \text{ Hz}^{-1} = 10^{-23} \text{ ergs s}^{-1} \text{ cm}^{-2} \text{ Hz}^{-1} \tag{2.3}$$

This is a very important unit in radio astronomy; it is a good number to commit to memory—along with its units.

If you examine the units of flux density on the right-hand side of Equation 2.3 you might be tempted to cancel the s^{-1} with the Hz^{-1} (since $1 \text{ Hz} = 1 \text{ s}^{-1}$). Although dimensionally correct, this is not a good idea. Remember that the definition of flux density is energy *per second* per unit area *per unit frequency*. To be clear that we are discussing units of flux density, we leave the units with the s^{-1} and Hz^{-1} clearly displayed. In terms of F_λ, depending on the choice of unit for wavelength, one might express flux density with units of W m^{-2} nm^{-1}.

Example 2.1:

You use a radio telescope that has a collecting area of 1.20 m^2 to observe a radio source at a frequency of 1420.4 MHz with a bandwidth of 2.00 MHz. You measure a detected power $= 1.20 \times 10^{-19}$ W. What is the flux density, F_v, of the source, in janskys, at the observed frequency?

Answer:

This is a straightforward example of a simple measurement. Using Equation 2.2, we have that total detected power equals the flux density times the collecting area of the telescope and the bandwidth of the observation, so

$$F_v = \frac{P}{A_{eff} \Delta v}$$

$$= \frac{1.20 \times 10^{-19} \text{ W}}{1.20 \text{ m}^2 \times 2.00 \times 10^6 \text{ Hz}}$$

$$= 5.00 \times 10^{-26} \text{ W m}^{-2} \text{ Hz}^{-1} = 5.00 \text{ Jy}$$

Note that we did not need to use the observing frequency in this calculation. This frequency is important, though, because this is the frequency where the calculated flux density is measured.

2.1.5 Intensity

The last important quantity regarding the amount of radiation is also the most fundamental. *Intensity* (I_v or I_λ) *(or specific intensity)*, which is often referred to as *surface brightness* and sometimes just *brightness*, is the flux density per unit solid angle. (Solid angles were introduced in Chapter 1; remember that the units of solid angle are steradians and are abbreviated as sr.) Intensity has units of W Hz^{-1} m^{-2} sr^{-1} (for I_v) or W nm^{-1} Hz^{-1} m^{-2} sr^{-1} (for I_λ). If you know the solid angular size of a source, you can calculate the average intensity of the source by dividing the measured flux density (F_v) by the solid angle of the source (Ω). A general expression of intensity, then, is given by

$$I_v = \frac{F_v}{\Omega} \tag{2.4}$$

where:

$$F_v \sim \frac{L}{4\pi d^2 \Delta v}$$

This might appear to be a somewhat arcane quantity, but it is actually one with which you are quite familiar. Intensity is, in fact, the correct description of the quantity that your eye measures and your brain interprets. When a light bulb appears especially bright (note the word that we use here: *bright*), your eye receives an especially large intensity of light. Consider how the appearance of the light is affected when you surround the bulb with a much larger soft-light globe. The total amount of light emitted is the same, but it comes from a larger surface, or larger solid angle, so the light seems less intense.

There are several aspects of intensity, which are important to know:

1. Flux density, F_v, does not distinguish between the directions that the photons come from or travel to, whereas I_v does. The directionality of the photons is important when studying the transfer of radiation through material, such as clouds of interstellar gas.

2. Intensity is independent of distance! This is hard to believe at first, but is easy to show. Intensity has two separate dependences on distance and they cancel with one another. Intensity depends on flux density and solid angle, as in Equation 2.4

   But Ω also depends on distance. Recall from Chapter 1 that

$$\Omega = \frac{\text{Cross-sectional area of source}}{d^2}$$

So, if A is the cross-sectional area of the source, then for a small source

$$I_v \sim \frac{F_v}{\Omega} \sim \frac{Ld^2}{4\pi d^2 \, \Delta v A}$$

and the d^2s cancel, leaving

$$I_v \sim \frac{L}{4\pi \Delta v A}$$

To help appreciate this fact, imagine looking at the Sun from Mars in comparison to looking at the Sun from Earth. In both cases, the lens in your eye will focus the Sun's rays, so that an image of the Sun appears on the retina. Since the flux density of light depends on the distance from the Sun, the amount of sunlight reaching the eye on Mars will be smaller. But since the solid angle of the Sun at the distance of Mars is smaller by the same factor as the reduction in flux density, the image of the Sun on the retina will be smaller by the same factor, so the amount of energy received per photoreceptor will be the same. The photoreceptors at the center of the image will be damaged just as easily on Mars as on Earth. The total number of damaged photoreceptors will be less, but that will not be much consolation for the person going blind.

3. Intensity is a direct measure of the object's surface brightness, that is, the amount of energy radiated per second per unit area of the surface per unit solid angle *right at the surface*. Since intensity is independent of distance, the intensity one measures on Earth (or on Mars) will be the same as that measured right at the surface. For example, consider two different objects—let us call them A and B—at the same distance with the same luminosity (L), flux (F), and flux density (F_v), but which have different intensities. Let $I_v(A) > I_v(B)$. This can only occur when one of these sources is larger than the other, so that $\Omega(B) > \Omega(A)$. Now consider what this means about the difference between the two sources. For B, to have the same flux density even though it has a larger surface, it must be less intense. Intensity, we see, is directly related to the microscopic radiation processes of the object. If you measure a source's flux density, and you can also measure the solid angle of the source (i.e., the source's size is large enough that you resolve its two-dimensional angular size), then you can get a direct handle on the microscopic radiation processes that produced the radiation.

Example 2.2:

A radio source at a distance of 20.0 Mpc (1 Mpc $= 10^6$ pc and 1 pc $= 3.09 \times 10^{16}$ m) is observed with a telescope, which has a collecting area of 300 m², at a frequency of 22.2 GHz, and with a bandwidth of 250 kHz. The source is determined to have a flux density, F_ν, at this frequency of 20.0 Jy, and is found to be a uniform circle with an angular diameter of 30.0 arcsec.

1. What is the intensity, I_ν, of the radiation from this source at this wavelength?
2. What is the total power of the radiation detected by the telescope?
3. What is the luminosity of the source over the observed spectral range (assuming that the radiation is isotropic)?

Answers:

1. As given in Equation 2.4, the average intensity is the flux density divided by the solid angle of the source, and since the source is found to be a uniform circle, the intensity equals the average intensity. We first need to find the solid angle, since we are only given the angular diameter. Using Equation 1.6, the source subtends a solid angle of

$$\Omega = \frac{\pi}{4}\left(30'' \frac{1°}{3600''} \frac{\pi \text{ radians}}{180°}\right)^2$$

$$= 1.66 \times 10^{-8} \text{ sr}$$

The intensity, then, is

$$I_\nu = \frac{20 \text{ Jy } (10^{-26} \text{ W m}^{-2} \text{ Hz}^{-1} \text{ Jy}^{-1})}{1.66 \times 10^{-8} \text{ sr}}$$

$$= 1.20 \times 10^{-14} \text{ W m}^{-2} \text{ Hz}^{-1}\text{sr}^{-1}$$

2. The total detected power equals the flux density multiplied by the bandwidth and the collecting area of the telescope

$$P = F_\nu A_{\text{telescope}} \Delta\nu$$

Therefore,

$$P = 20.0 \text{ Jy } (10^{-26} \text{ W m}^{-2} \text{ Hz}^{-1} \text{ Jy}^{-1}) \ (300 \text{ m}^2)(250 \times 10^3 \text{ Hz})$$

$$= 1.50 \times 10^{-17} \text{ W}$$

3. The luminosity of the source, over the observed spectral range, is given by the observed flux density times the bandwidth, and then using Equation 2.1. Therefore,

$$L = 20.0 \text{ Jy}(10^{-26} \text{ W m}^{-2} \text{ Hz}^{-1} \text{ Jy}^{-1}) \, 250 \times 10^3 \text{ Hz}$$

$$\times \left[4\pi(20 \times 10^6 \text{ pc} \times 3.09 \times 10^{16} \text{ m pc}^{-1})^2\right]$$

$$= 2.40 \times 10^{29} \text{ W}$$

Example 2.3:

An alien civilization at a distance of 7.00 pc (1 pc $= 3.09 \times 10^{16}$ m) uses a radio antenna of diameter 100 m, acting as a transmitter, to beam a radio signal toward Earth. The civilization converts 5.00×10^6 J of energy into this signal, which is transmitted at a constant power for 100 s, centered at a frequency of 1.000 GHz with a bandwidth of 1.00 MHz, and in a beam with a solid angle of 9.30×10^{-6} sr.

  1. What is the luminosity of the beamed signal?
  2. What is the intensity of the beamed signal?

The signal is detected with an antenna of diameter 7 m at a frequency of 1.00 GHz with a bandwidth of 500 kHz.

  3. What is the measured flux density of the signal?
  4. What is the measured power of the signal (at Earth)?

Answers:

  1. $L = 5.00 \times 10^6$ J/100 s $= 5.00 \times 10^4$ W $= 50$ kW.
  2. Here, we can use the equivalence of intensity and surface brightness. The intensity is the power emitted per area of the emitting surface per solid angle of the transmitted beam per bandwidth, that is,

$$I_v = \frac{5.00 \times 10^4 \text{ W}}{(A \text{ of transmitting antenna}) \times (\Omega \text{ of beam}) \times \Delta v}$$

We are given that the transmitting area, beam solid angle, and bandwidth are as follows:

$$A = \pi(50.0 \text{ m})^2 = 7.85 \times 10^3 \text{ m}^2, \Omega = 9.30 \times 10^{-6} \text{ sr, and } \Delta v = 1.00 \times 10^6 \text{ Hz}.$$

Therefore,

$$I_v = \frac{5.00 \times 10^4 \text{ W}}{(7.85 \times 10^3 \text{ m}^2) \times (9.30 \times 10^{-6} \text{ sr}) \times 1.00 \times 10^6 \text{ Hz}}$$

$$= 0.685 \text{ W Hz}^{-1} \text{ m}^{-2} \text{ sr}^{-1}$$

  3. This is actually a tricky question and must be approached with care. First, let us ask the question: Can we use the intensity we found in (2), say that it is independent of distance, and therefore we know the intensity of the radiation entering the telescope, and then multiply by the solid angle of the source to get the flux density? The answer is both yes and no. *Yes*, the intensity is independent of distance, even though the signal is not isotropic, but *no*, we cannot use this to get the measured flux density because we do not know the solid angle of the source. We will first explain why the intensity is still distance-independent and then we

will show how to calculate the flux density. Recall that intensity is independent of distance, because the distance dependences of flux and solid angle cancel one another. The solid angle of the source still depends on $1/d^2$. So, how about the flux? The distance dependence of flux comes from the fact that an isotropic signal, at a distance d, is spread out over an area of a spherical shell, that is, $4\pi d^2$, so the energy per unit area is $L/4\pi d^2$. But in this case, the signal is not isotropic, so the radiation is *not* spread out over the whole area $4\pi d^2$, so the normal equation for flux, $F = L/4\pi d^2$, is not true. To figure out the flux of the signal at any distance, we need to determine the area that the beam is distributed over at a given distance, d. By beaming the radiation, the alien civilization directs and confines the radiation to a cone of small angle Ω_{signal}. Now, as the radiation travels outward, the area of the end of the cone is given by $\Omega_{signal} d^2$, as depicted in Figure 2.2. The flux, then, is given as

$$F = \frac{L}{\Omega_{signal}\, d^2}$$

The flux still has the same $1/d^2$ dependence on distance, and therefore intensity is still independent of distance. (Note: when the emission is isotropic, then the solid angle of the beam is 4π steradians, which, when substituted in for Ω, recovers the familiar equation $F = L/4\pi d^2$.) Therefore, the flux still diffuses at the same rate, even when the signal is beamed.

Now we can calculate the measured flux density. The flux density of the beamed radiation at the distance of Earth is

$$F_\nu = \frac{L}{\Omega_{emitted}\, d^2 \Delta\nu_{emitted}}$$

$$= \frac{5.00\times10^4 \text{ W}}{9.30 \times 10^{-6}\,\text{sr}\, (7.00\,\text{ly} \times 3.09 \times 10^{16}\ \text{m pc}^{-1})^2\, 1\times10^6 \text{Hz}}$$

The flux density, then, is

$$F_\nu = 1.15 \times 10^{-31}\ \text{W Hz}^{-1}\ \text{m}^{-2} = 1.15 \times 10^{-5}\ \text{Jy}$$

This is very small and would be challenging to detect.

4. Detected power = detected flux density × area of telescope × bandwidth

$$= (1.15 \times 10^{-31}\ \text{W Hz}^{-1}\ \text{m}^{-2})\pi(3.50\ \text{m})^2 (5 \times 10^5\ \text{Hz})$$

$$= 2.21 \times 10^{-24}\ \text{W}$$

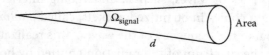

FIGURE 2.2 Even non-isotropic radiation diffuses through space as $1/d^2$.

2.1.6 Relation between Intensity and the Electric Field and Magnetic Field Waves

In Chapter 1, we reviewed the concept that radio signals are waves of electric and magnetic fields, while here we talk about radiation in terms of energy, that is, measures of power, flux, flux density, and intensity. These two concepts, of course, must be related; the energy that we measure is contained in the electric and the magnetic fields. The energy flux of the radiation in terms of those fields is described by a quantity known as the *Poynting vector*, generally represented by $\vec{S}$. In addition, since the strength of the magnetic field in the radiation is directly proportional to the strength of the electric field, the energy flux in the radiation can be expressed in terms of just the electric field strength. Skipping the lengthy derivation from electrodynamics, the magnitude of the flux in the radiation is given by

$$S = \frac{1}{2} c \varepsilon_0 E_0^{\,2} \text{ (SI) and } S = \frac{c}{8\pi} E_0^{\,2} \text{ (cgs)}$$

where:

E_0 is the amplitude of the electric field waves

(Note: A constant, ε_0, the permittivity of free space, appears in the SI expression, but not in the cgs expression. This constant is not dimensionless; the definition of charge and electric field are quite different in the cgs system compared to the SI system, so changing systems is not as simple as just converting units.)

Now, the intensity of radiation, as explained above, is related to the flux. We do not worry about the details, though, for we will rarely need to analytically convert between the electric field in the waves and the intensity of the radiation. The important point to remember is that the intensity of the radiation is proportional to the square of the electric field in the waves, that is,

$$I_v \propto E_0^{\,2}$$

Therefore, most radio telescope detectors produce a signal proportional to the square of the amplitude of the electric field so that the measured output is proportional to the power in the radiation. This detection process is explained further in Chapter 3. The overall calibration of the flux density and the intensity of any incident radiation is discussed in Chapter 4, where we also discuss the quantitative relation between intensity and measured power.

2.2 BLACKBODY RADIATION

One of the major breakthroughs in physics at the start of the twentieth century was the realization that light has characteristics of both waves and particles. It was well known that light exhibits properties of waves, such as undergoing interference and diffraction, but the energy of the light comes in quantized packets, called *photons*, with the energy of each packet proportional to the frequency of the waves. This realization started with Max Planck's discovery that the spectrum of the radiation emitted by hot, opaque objects can be fit by a model that assumes quantized units of energy.

The emission of light is one means by which a hot body can cool by transferring energy to its surroundings. The light we see coming from objects, though, usually involves other light as well. A blue shirt, for example, looks blue because it *reflects* primarily the blue frequencies of all the light that hits it, while its emitted light is primarily in the infrared. A body may also allow some light to pass through from the background. To model only the light *emitted* by the body, we can imagine that the body absorbs all light that hits it, reflecting nothing, and preventing the transmission of all light. Such a body is called *opaque* and is considered black. The radiation modeled by Planck, that is, the radiation emitted by a body that absorbs all light incident on it, is therefore called *blackbody radiation*. The name *blackbody* can be confusing since black implies an *absence* of light, while in discussions of blackbody radiation, we are interested in the radiation that is *emitted* by the body. Remember, though, that with the vast majority of objects that we see, the color of the object is determined by the frequencies of light that the object *reflects* (except for the light bulbs, computer screens, and any object that glows).

One way of understanding blackbody radiation is as a description of the maximum amount of radiation that an ordinary body of a particular temperature will emit solely as an attempt to cool. The total radiation emitted by a cooling body is not correctly described merely as the sum of all the individual emission processes, as the particles can also absorb radiation. In the end, the statistics of a large number of emission *and* absorption events determine the resultant radiation.

To start with, consider that the amount of radiation emitted by a hot body must depend on the number of particles in the hot body—the more particles the body contains, the more photons that can be emitted in each second. But having a larger number of particles also increases the likelihood that some of the emitted photons will be absorbed by other particles in the body. Therefore, as more and more photons are created, there starts to be an exchange of energy from the photons back to the particles. This is the key issue—that the photons and particles have numerous interactions involving exchanges of energy. If the condition of numerous interactions is not met, then the resulting radiation will be quite different. We can think of the large number of photons, which are created in the body, as a second body of particles, so that there are two bodies present: the radiating matter (i.e., the atoms, molecules, or electrons) and a cloud of photons, both of which occupy the same volume. The particles in each body interact with those in the other, exchanging energy back and forth. The cloud of photons, though, is created by the matter particles and if more photons are created, then more energy is contained within the photon cloud. Now recall the fundamental laws of thermodynamics. The body of matter particles wants to cool by giving energy to the photon cloud (which will, in turn, carry the energy away into space). But, a hot body cannot heat another body to a temperature higher than itself. Therefore, the hottest the photon cloud can get is when it has the same temperature as the radiating body. In other words, if there are enough particles in the radiating body to produce a very large number of photons and photon-matter interactions, then the radiating body and the cloud of photons will achieve thermal equilibrium. The resultant radiation emitted from the body, then, is a cloud of photons at the same temperature as the body itself.

You may reasonably wonder what is meant by describing the radiation as a cloud of photons at a particular temperature. Well, photons are real particles and they fall into a particular category of particles called *bosons*. (For those who have taken statistical mechanics, you may recall a discussion of Bose–Einstein statistics, which describe the statistical behavior of photons, along with Fermi–Dirac statistics, which describe the behavior of fermions, like electrons and protons, and Maxwell–Boltzmann statistics, which describe the non-quantum behavior of larger bodies.) Therefore, when the photons have been thermalized, statistics can describe the distribution of energies of the photons, that is, the relative number of photons carrying each small range of energy. In other words, with blackbody radiation, the photons have achieved a *thermal distribution* of energies. The gas molecules of the air in your room, for example, have an assortment of energies; because of collisions, their energies are randomized. Over a large number of collisions, the fraction of particles with energy falling in each small energy range can be described by a statistical function. For gas molecules, one uses the Maxwell–Boltzmann function. Likewise, for thermalized photons, the Bose–Einstein function describes the number of photons in each small-energy range. The main point here is that the numerous interactions that the photons experience with the particles in the opaque medium accomplishes this. The photons and the radiating particles achieve *thermal equilibrium*, meaning that they both have thermal energy distributions, which we can write as $N(E)$ (representing the number of particles as a function of their energy), that are described by the same temperature.

Now, the energy distribution function of the photons, $N_{ph}(E)$, is directly related to its spectral distribution function, that is, intensity versus frequency. The frequency is proportional to the energy (since for photons $E = h\nu$) and the intensity at a given frequency is proportional to the flux of photons at that frequency. Therefore, starting with the energy distribution function for an ensemble of thermalized photons, it is straightforward to convert this into intensity as a function of frequency. We will not go through that derivation here, as we only wish to make clear that the energy distribution function of the photons is directly related to the spectrum. The result of this conversion is the Planck function.

The Planck function provides a mathematical description of the spectrum of the light emitted by blackbodies. In terms of the emitted flux per unit frequency interval per unit steradian, the Planck function is given by

$$B_\nu(T) = \frac{2h\nu^3}{c^2} \frac{1}{\exp(h\nu/kT)-1} \tag{2.5}$$

where:
  $h = 6.626 \times 10^{-34}$ J s $= 6.626 \times 10^{-27}$ erg s is Planck's constant
  $k = 1.38 \times 10^{-23}$ J K^{-1} $= 1.38 \times 10^{-16}$ ergs K^{-1} is Boltzmann's constant
  c is the speed of light
  ν is the frequency of the observation
  T is the temperature of the radiating body in Kelvins

In the Planck function given in Equation 2.5, written as $B_\nu(T)$, the subscript ν indicates that the spectral measure is per unit frequency and B represents the intensity, or brightness, of the blackbody radiation and has units of intensity, W m^{-2} Hz^{-1} sr^{-1}. Since this is an intensity, the Planck function can also be expressed as flux per unit *wavelength* per steradian, which is denoted as $B_\lambda(T)$. Radio astronomers generally express the Planck function using $B_\nu(T)$. The equation for $B_\lambda(T)$ is

$$B_\lambda(T) = \frac{2hc^2}{\lambda^5} \frac{1}{\exp(hc/\lambda kT) - 1} \qquad (2.6)$$

It is very important to understand and appreciate that, even though B_ν and B_λ represent the same concept, they are *not* the same numerical quantity or even the same function. We discuss in more detail about the difference between these functions later in this section. We first discuss some important features of the Planck function.

Figures 2.3 and 2.4 display the log-log plots of these functions. Blackbody emission is a continuous spectrum that reaches zero at $\nu = 0$ (owing to the ν^3 term), increases to some peak value as ν increases, and then decreases and reaches zero again at $\nu = \infty$ (owing to the exponential term). Note that the Planck function depends only the body's temperature and the frequency of the radiation. No other characteristic of the body is relevant. In other words, the intensity of radiation that a blackbody emits at any given frequency depends only on its temperature.

The total flux of radiation (W m^{-2}) emitted by the body can be obtained by integration of the Planck function over frequency and solid angle. The result shows that the total flux is proportional to the fourth power of the body's temperature:

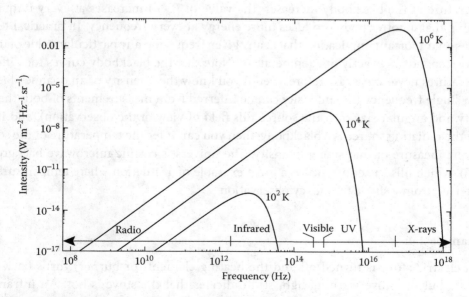

FIGURE 2.3 Log-log plot of $B_\nu(T)$ versus ν for three different temperatures.

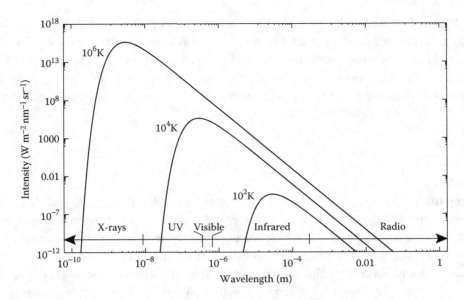

FIGURE 2.4   Planck function, plotted as B_λ, for the same three temperatures as in Figure 2.3.

$$F = \sigma T^4 \qquad (2.7)$$

This is known as the *Stefan–Boltzmann law* and the constant is known as the *Stefan–Boltzmann constant* ($\sigma = 5.67 \times 10^{-8}$ W m^{-2} K^{-4}).

A key feature of the blackbody radiation, apparent in Figures 2.3 and 2.4, is that as the temperature of the blackbody increases, the value of $B_\nu(T)$ increases at every frequency, meaning that a hotter body produces more energy at every frequency. In practical terms, this has an important implication that at any given frequency, any particular value of intensity corresponds to exactly one temperature. Note that the blackbody curves for different temperatures never cross. Therefore, even if you know the intensity of an opaque object at only a single frequency (the intensity can be inferred from measurements of both the flux density and angular size or if the source fills field of view of the observation), and if you know the radiating source is a blackbody, then you can infer the temperature of the source from the measure of that single intensity. The universe's cosmic microwave background (CMB), which fills the entire sky, is a great example of a situation where temperature can be inferred from a single frequency observation.

Example 2.4:

An electric stove is turned off and the heating element (or burner) turns back to black, but the stove warning light still indicates that the stove is hot. An infrared sensor is aimed at the burner and the radiation at a frequency of 3.33×10^{14} Hz emanating from the burner is determined to have an intensity of 1.46×10^{-28} W Hz^{-1} m^{-2} sr^{-1}.

1. Estimate the temperature of the stove burner.
2. Estimate the total flux of electromagnetic radiation emitted by the stove burner.

Answers:

1. The stove burner is solid and opaque, so the Planck function is a good approximation to the intensity emitted. Therefore, we can use Equation 2.5 to solve for the temperature.

$$1.46 \times 10^{-28} \text{ W Hz}^{-1} \text{ m}^{-2} \text{ sr}^{-1} = \frac{2h\nu^3}{c^2} \frac{1}{\exp(h\nu/kT) - 1}$$

By substituting in the values for h, c, k, and ν, rearranging we get

$$\exp\left(\frac{1.60 \times 10^4 K}{T}\right) - 1 = \frac{5.44 \times 10^{-7}}{1.46 \times 10^{-28}} = 3.73 \times 10^{21}$$

This gives

$$T = \frac{1.60 \times 10^4 \text{ K}}{\ln(3.73 \times 10^{21})}$$

$$= 322 \text{ K (or } 120°\text{F)}$$

2. Since the burner is opaque at almost all wavelengths, we can approximate the total flux emitted by the Stefan–Boltzmann law. Using Equation 2.7, we have

$$F = \sigma T^4 = 5.67 \times 10^{-8} \text{ W m}^{-2} \text{ K}^{-4} (322 \text{ K})^4$$

$$= 609 \text{ W m}^{-2}$$

The average photon energy can be calculated from the ratio of the total energy flux to the total photon flux, as given by

$$\langle E_{\text{ph}} \rangle = \frac{\int B_\nu(T) d\nu}{\int \left(B_\nu(T)/h\nu \right) d\nu}$$

The result of this calculation shows that the average photon energy is given by

$$\langle E_{\text{ph}} \rangle = 2.70 \, kT = (3.73 \times 10^{-23} \text{ J K}^{-1}) \, T \tag{2.8}$$

Since this is a thermal distribution of photons, it should not be surprising that the average energy only depends on the temperature.

Figures 2.3 and 2.4 also show that the location of the peak of the Planck spectrum depends on the body's temperature. Just as the average photon energy increases linearly with temperature (as shown in Equation 2.8), the peak of the blackbody spectrum from a hotter body occurs at a higher frequency. The peak of the curve, of course, is obtained by determining where the derivative of the Planck function equals zero, and the equation relating to the location of the peak to the body's temperature is called the *Wien displacement law*. For $B_\nu(T)$, the frequency of the peak of the curve is given by

$$\nu_{peak} = (5.879 \times 10^{10} \text{ Hz K}^{-1}) \, T \qquad (2.9)$$

while for $B_\lambda(T)$, the wavelength of the peak of the curve is

$$\lambda_{peak} = (2.898 \times 10^{-3} \text{m K})/T \qquad (2.10)$$

It is also important to appreciate that Equations 2.9 and 2.10 do not give the same information. This relates to the issue that intensity defined as flux per unit frequency per steradian, I_ν, and intensity defined as flux per unit wavelength per steradian, I_λ, are not the same function. Equating I_ν and I_λ is a common misconception that often leads to mistakes. If you try substituting $\nu = c/\lambda$ in the equation for $B_\nu(T)$, you will find that you do not get the correct equation for $B_\lambda(T)$. The functions $B_\nu(T)$ and $B_\lambda(T)$ cannot be the same quantity, for they even have different units. To really drive the point home, try using Equations 2.9 and 2.10 to calculate the peak frequency of $B_\nu(T)$, for a given T, and the peak wavelength of $B_\lambda(T)$ for the same T, and then see if ν_{peak} of $B_\nu(T)$ and λ_{peak} of $B_\lambda(T)$ correspond to the same point in the spectrum, that is, they are related by $c = \lambda_{peak}\nu_{peak}$. You will find that they are not. The peaks of these curves for the same temperature might not even occur in the same spectral band.

Let us apply the Wien displacement law to the Sun as an example. Where does the spectrum of a blackbody with the temperature of the Sun's surface ($T = 5800$ K) occur? Figure 2.5 displays the Planck curves for B_λ and B_ν for $T = 5800$ K. According to Equation 2.10, the peak of B_λ for the Sun's temperature occurs at $\lambda_{peak} = 5 \times 10^{-7}$ m $= 500$ nm, which corresponds to *green*, in the middle of the visible band. Now, if we use Equation 2.9 to do the same calculation for B_ν, we find that the peak of B_ν occurs at $\nu_{peak} = 3.4 \times 10^{14}$ Hz, which corresponds to a wavelength of $\lambda_{peak} = c/\nu_{peak} = 880$ nm, which is in the *infrared*!! The peaks of B_ν and B_λ are in different regimes of the EM spectrum.

So which one is right? Which function, B_ν or B_λ, should you use? It depends on the means by which the spectrum is measured and/or displayed. What color the Sun *appears to be* is actually a very complex question, which depends on the sensitivity of the eye at different colors and how the brain interprets the signals it receives from the eyes. So, it is really OK to say that the Sun *looks* yellow, because that is how *we see it* after all the complex biochemistry and neurophysics come into play. But, it is not really fair to say that the Sun's blackbody spectrum peaks in the yellow. In fact, the spectrum of the Sun's radiation across

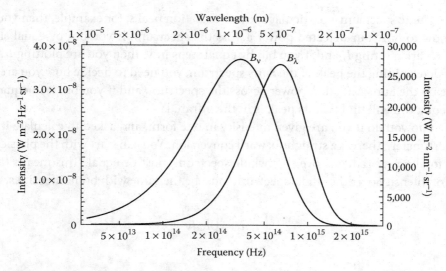

FIGURE 2.5   Curves of the Planck function for a 5800 K blackbody shown as both B_λ and B_ν. Even though both of these curves represent the intensity emitted by the same blackbody, they peak in different parts of the electromagnetic spectrum.

the visible band is actually quite flat.[*] The entire visible band fits quite easily somewhere under the broad peak of the Sun's Planck spectrum; a better statement about the Sun's spectrum is to say that it is white, that is, it contains about an equal amount of all colors of the rainbow, and our brain interprets it as yellow.

An alternative description of the *color* of the Sun's radiation is to use the average energy of the photons, as described by Equation 2.8. By substituting in the expression for photon energy, we find that the wavelength of the average energy photon emitted by the Sun is

$$\left\langle \frac{hc}{\lambda} \right\rangle = 2.70 \; k \; (5800 \text{ K})$$

or

$$\lambda_{\langle E \rangle} = \frac{hc}{2.70 \; kT} = 9.20 \times 10^{-7} \text{ m}$$

or 920 nm, which is in the infrared.

It is common to find the distinction between B_ν and B_λ (or between I_λ and I_ν) confusing. A reasonable concern you might have is that the energy in emitted radiation is a physical quantity that is measurable, and so it should not matter how you define your units, right? Well, the trick is how you actually make a measurement. One cannot actually measure I_λ or I_ν directly. One detects *power*, which requires a finite range of wavelengths or frequencies. Therefore, what we actually measure is $\int I_\lambda d_\lambda$ or $\int I_\nu d_\nu$ (times the area of our telescope), and these are, in fact, equal (as we explain below). But, if you want to know where in the

---

[*] For a full discussion, see Marr and Wilkin, 2012, *Am. J. Phys.*, 80, p. 399.

electromagnetic spectrum a particular source's radiation peaks, for example, then you must choose how to bin your detected power across the spectrum. If you bin by equal steps in λ, then you are plotting I_λ and if you bin by equal steps in v, then you are plotting I_v. So, in terms of determining the peak of the Sun's spectrum, you need to decide how you are going to divide up the Sun's radiation power across the spectrum, and if you bin it in equal steps of frequency you will find that its peak is in the infrared.

So what do you do if you are given intensity in one form, and asked to calculate it in the other? Fortunately, there is a straightforward conversion. You can start with the physical idea that the total power radiated over the whole spectrum must be equal. This means that the integral of intensity over the whole spectrum must be the same with both intensities, that is,

$$\int_0^\infty I_\lambda d\lambda = \int_0^\infty I_v dv$$

We can be even more specific and say that the total power carried by photons over a fixed part of the spectrum must be the same, so that the integrations will be equal when the limits are λ_1 and λ_2 and v_1 and v_2 as long as $\lambda_1 = c/v_1$ and $\lambda_2 = c/v_2$. We can choose the limits to be very close to each other and hence conclude that for each little piece of the spectrum,

$$I_\lambda d\lambda = -I_v dv$$

The negative sign is needed because the changes in frequency and wavelength go in opposite directions. Therefore,

$$I_\lambda = I_v \left(-\frac{dv}{d\lambda} \right)$$

Now to get $dv/d\lambda$, we use

$$v = \frac{c}{\lambda}$$

from which we infer that

$$\frac{dv}{d\lambda} = \frac{d}{d\lambda}\left(\frac{c}{\lambda} \right) = -\frac{c}{\lambda^2}$$

Substituting the right hand side of this equation for $dv/d\lambda$ above, we get

$$I_\lambda = \frac{c}{\lambda^2} I_v \tag{2.11}$$

If you use this conversion in either of the Planck functions, you will find that one does convert to the other.

Example 2.5:

Imagine measuring the spectrum of the light emitted by a fellow human being, at a temperature of 98.6°F. Assume that there is no other light source in the room, so that there is no reflected light, and ignore any additional emission or absorption of light from the molecules at the surface of the skin; the observed spectrum, then, fits the Planck function.

1. At what frequency will the intensity, I_ν, of the radiation emitted be greatest? (In reality, the temperature of the outer layer of the skin will be less than 98.6°F, but we can use this temperature for demonstration purposes.) Convert this to a wavelength.
2. At what wavelength will the intensity, I_λ, be greatest?
3. What is the average energy of the photons emitted?
4. If the body's temperature were to increase, how would we detect this with a radiation detector that operates over a single small frequency range?

Answers:

1. This question is really asking at what frequency is the peak of B_ν at this temperature, hence we use Equation 2.9. First, though, we must convert the temperature to units of Kelvins.

$$T(\text{K}) = 273 + \frac{5}{9}(98.6 - 32) = 310 \text{ K}$$

Now using the Wien displacement law in Equation 2.9, we have

$$\nu_{\text{peak}} = (5.879 \times 10^{10} \text{ Hz K}^{-1})\, 310 \text{ K}$$

$$= 1.82 \times 10^{13} \text{ Hz}$$

This corresponds to a wavelength of

$$\lambda = \frac{c}{\nu_{\text{peak}}} = \frac{3.00 \times 10^8 \text{ m s}^{-1}}{1.82 \quad 10^{13} \text{ Hz}}$$

$$= 1.65 \times 10^{-5} \text{ m} = 16.5 \; \mu\text{m}$$

which is in the infrared band. Although a 310 K object has its peak emission in the infrared, it also is easily detectable at radio wavelengths.

2. Now we want the peak of I_λ, which requires Equation 2.10. We have then

$$\lambda_{\text{peak}} = (2.898 \times 10^{-3} \text{m K})/310 \text{ K}$$

$$= 9.33 \times 10^{-6} \text{ m} = 9.33 \; \mu\text{m}$$

3. The average energy of the photons is given by Equation 2.8, and so

$$\langle E_{ph} \rangle = (3.73 \times 10^{-23} \text{ J K}^{-1})310 \text{ K}$$

$$= 1.16 \times 10^{-20} \text{ J}$$

which corresponds to a wavelength of

$$\lambda = \frac{hc}{E} = \frac{(6.63 \times 10^{-34} \text{ J s})(3.00 \times 10^{8} \text{ m s}^{-1})}{1.16 \times 10^{-20} \text{ J}}$$

$$= 1.72 \times 10^{-5} \text{ m} = 17.2 \text{ μm}$$

4. If the temperature of the radiating surface increases, we will see an increase in intensity at all frequencies.

Example 2.6:

In an article about a visible-wavelength spectral observation of a nearby galaxy, the measured flux density centered at a wavelength of 4250 Å (1 Å = 10^{-10} m), and with a bandpass of 50.0 Å, is reported to be 2.30 × 10^{-8} W m^{-2} Å^{-1}. Meanwhile, a radio observation of the same galaxy, at 22.2 GHz with a bandwidth of 50.0 MHz, yields a flux density of 42.0 Jy.

1. In which band does this galaxy have a larger flux density?
2. Compare the detected fluxes in the two bands.

Answer:

1. We first note that the conversion between F_λ and F_ν is the same as between I_λ and I_ν, which is given in Equation 2.11. Therefore, we first convert the F_λ from the visible-wavelength observation to F_ν and then to Jy. We have, then, that the flux per Hz at 5500 Å is

$$F_\nu = \frac{\lambda^2}{c} F_\lambda = \frac{(4250 \times 10^{-10} \text{ m})^2}{3.00 \times 10^8 \text{ m s}^{-1}} 2.30 \times 10^{-8} \text{ W m}^{-2} \text{ Å}^{-1}(10^{10} \text{ Å m}^{-1})$$

$$= 1.38 \times 10^{-29} \text{ W m}^{-2} \text{ Hz}^{-1}$$

which equals 1.38 × 10^{-3} Jy. This is 4 orders of magnitude smaller than the radio-frequency flux density.

Let us now convert the radio-frequency flux density measurement to F_λ to compare with the measured visible flux density value. We get,

$$F_\lambda = \frac{c}{\lambda^2} F_\nu = \frac{\nu^2}{c} F_\nu = \frac{(2.22 \times 10^{10} \text{ Hz})^2}{3.00 \times 10^8 \text{ m s}^{-1}} 4.20 \times 10^{-25} \text{ W m}^{-2} \text{ Hz}^{-1}(10^{-10} \text{ m Å}^{-1})$$

$$= 6.90 \times 10^{-23} \text{ W m}^{-2} \text{ Å}^{-1}$$

This is 15 orders of magnitude smaller than the visible-wavelength flux density!

How can we find that each measurement is many orders of magnitude smaller than the other one? Does not one have to always be greater than the other? The reason for this apparent inconsistency is the difference between the quantities F_λ and F_ν. The amount of the spectrum contained in 1 Hz is tiny at visible wavelengths, so a large value for F_ν at visible wavelengths would require an incredible amount of total flux. Similarly, each angstrom at radio wavelengths is an absurdly small sliver, so the values of F_λ will naturally be very small. Although it is good to know how to convert between these quantities, Question 1 is ambiguous and not really of much value and Question 2 is more interesting.

2. The detected fluxes are given by multiplying the measured flux densities by the bandwidths. For the radio, this is

$$F = 4.20 \times 10^{-26} \text{ W m}^{-2} \text{ Hz}^{-1} \times 50.0 \times 10^{6} \text{ Hz} = 2.10 \times 10^{-18} \text{ W m}^{-2}$$

while for the visible, we find

$$F = 2.30 \times 10^{-8} \text{ W m}^{-2} \text{ Å}^{-1} \times 50.0 \text{ Å} = 1.15 \times 10^{-6} \text{ W m}^{-2}$$

We see that the total detected flux at visible wavelengths is much larger.

Finally, let us return to the consideration of the body's particles creating photons as an attempt to cool. We explained that if the body has enough particles so that the photons are trapped, then the body will create enough photons such that the temperature of the photons equals the temperature of the body. Let us now ask: What if the body does not have enough particles to produce a distribution of photons equal in temperature to the body? In this case, because the cooling body cannot produce a photon cloud with a higher temperature than itself, the photon cloud *must* have a lower energy density than the radiating body. This situation also requires that some photons escape the body before thermal equilibrium is achieved, so this body would not be completely opaque, and therefore would not qualify as a blackbody. We conclude, then, that an opaque body emits more intense radiation than a transparent body, and that the Planck function describes the *maximum* intensity that a radiating body of given temperature will emit by thermal processes at each frequency. In short, the maximum intensity that can be produced by thermal radiation from an object at any wavelength is $B_\nu(T)$, and $I_\nu \leq B_\nu(T)$.

Example 2.7:

The Orion nebula is an HII region containing hot ionized gas that emits thermal radiation. At a frequency of 400 MHz, the Orion nebula is found to have a flux density of 220 Jy and comes from an area on the sky with solid angle of 2×10^{-5} sr. Use these data to infer a lower limit to the temperature of the gas in this nebula.

Answer:

Although the Orion nebula is known to be opaque at lower frequencies, it is not opaque at 400 MHz, so the detected emission is not blackbody radiation. But, since this radiation is thermal, we know that the maximum intensity it could have is given by the Planck function. Therefore, we set the observed intensity to be less than or equal to the Planck function. Since the observed intensity cannot be larger than the Planck function, and the Planck function is larger at higher temperatures, this will give us a lower limit to the temperature of the gas.

We must first calculate the intensity from the flux density and angular size. Using Equation 2.4.

$$I_v = \frac{200 \text{ Jy } (10^{-26} \text{ W m}^{-2} \text{ Hz}^{-1} \text{ Jy}^{-1})}{2\times10^{-5} \text{ sr}} = 1\times10^{19} \text{ W m}^{-2} \text{ Hz}^{-1} \text{ sr}^{-1}$$

Setting the Planck function as the upper limit to this intensity and substituting in the values for h, v, c, and k, then, we have

$$1\times 10^{-19} \text{ W Hz}^{-1} \text{ m}^{-2} \text{ sr}^{-1} \leq 9.42\times10^{-25} \text{ W Hz}^{-1} \text{ m}^{-2} \text{ sr}^{-1} \frac{1}{\exp(0.0192\text{K}/T)-1}$$

Rearranging, we find

$$\exp(0.0192\text{K}/T)-1 \leq 9.42\times10^{-6}$$

which requires that

$$T \geq 2040 \text{ K}$$

2.3 RAYLEIGH–JEANS APPROXIMATION

At most radio wavelengths, the Planck function can be approximated by a much simpler expression, which makes for much easier math when using it. This approximation will also lead to an extremely important definition (presented in Section 2.4). At most radio wavelengths, the frequency, v, is so small that $hv/kT \ll 1$ for any reasonable temperature (see Problem 9). The exponential in the denominator of the Planck function then can be approximated by a Taylor series expansion (Problem 9 guides you through this expansion), yielding

$$\frac{1}{\exp(hv/kT)-1} \approx \frac{kT}{hv} \text{(for } h/kT \ll 1)$$

and so,

$$B_v(T) \approx \left(\frac{2hv^3}{c^2}\right)\frac{kT}{hv} = \left(\frac{2kv^2}{c^2}\right)T \qquad (2.12)$$

or

$$B_\nu(T) \approx \frac{2kT}{\lambda^2} \tag{2.13}$$

Note how simple this expression is in comparison to the Planck function (Equation 2.5). This expression is very useful, provided you are in the realm where $h\nu/kT \ll 1$. This expression is known as the *Rayleigh–Jeans approximation*, often referred to as the *Rayleigh–Jeans law*.

At centimeter wavelengths, most of the time we can use the Rayleigh–Jeans approximation, avoiding the far more complex Planck function. As demonstrated in Problem 9, parts b and c, only at the highest radio frequencies and with observations of cold objects does the approximation start to differ from the full expression. This is one major convenience of working at radio wavelengths. Equations 2.12 and 2.13 are so important in radio astronomy that you will want to memorize them!

Example 2.8:

A radio observation at a wavelength of 6.00 cm yields the determination that a particular radio source has a solid angle of 7.18×10^{-6} sr, is opaque and thermal, and has a flux density of 350 Jy.

1. What is the temperature of the radio source?
2. What is the intensity of this source at 2.70 cm?
3. An observation is made of another opaque, thermal radio source that is twice as hot as the first source. What is its intensity at 2.70 cm?

Answers:

1. Since the flux density is measured at 6.00 cm, which is still in the realm where the Rayleigh–Jeans approximation works well, we solve for the temperature using Equations 2.13 and 2.4.

$$F_\nu = \frac{2kT}{\lambda^2}\Omega$$

Solving for T gives us

$$T = \frac{3.50 \times 10^{-24}\ \text{W m}^{-2}\ \text{Hz}^{-1}(0.060\ \text{m})^2}{(2 \times 1.38 \times 10^{-23}\ \text{J K}^{-1})7.18 \times 10^{-6}\ \text{sr}}$$

$$= 63.6\ \text{K}$$

2. Again using Equation 2.13, we find that the intensity at 2.70 cm, which is 11.1 GHz, is

$$I_{11.1\text{GHz}} = \frac{2\,(1.38\times10^{-23}\,\text{J}\,\text{K}^{-1})\,63.6\,\text{K}}{(0.0270\,\text{m})^2}$$

$$= 2.41\times10^{-18}\,\text{W}\,\text{m}^{-2}\,\text{Hz}^{-1}\,\text{sr}^{-1}$$

3. As Equation 2.13 shows, in the Rayleigh–Jeans part of the blackbody spectrum, the intensity is directly proportional to the temperature of the radiating body. Therefore, if the second source is twice as hot as the first, then its intensity at the same wavelength will be twice as great. The intensity of the second source, then, is 4.82×10^{-18} W m^{-2} Hz^{-1} sr^{-1}.

2.4 BRIGHTNESS TEMPERATURE

We now introduce an extremely important parameter—*the brightness temperature*.

First, consider the Rayleigh–Jeans approximation. This approximation provides radio astronomers with an extremely convenient, alternative way of describing the intensity of radiation. As discussed in Section 2.3, the radiation intensity from a thermal source is related to, and is a rough measure of, the source temperature. In particular, the Rayleigh–Jeans approximation shows that at radio wavelengths, the intensity of the radiation emitted by a blackbody is directly proportional to the temperature of the body, to wit, $B_\nu(T) \propto T$. As a result, at radio wavelengths, the intensity and the temperature of blackbody sources can be used interchangeably: they are linearly related by the proportionality constant $2k/\lambda^2$. Moreover, intensity is a measure of the radiation emitted, while temperature refers to a physical condition of the source. Temperature is often the more interesting parameter; it adds to our understanding of the physics of the source. In the Rayleigh–Jeans approximation, we define the *brightness temperature*, T_B, as

$$T_B = \left(\frac{\lambda^2}{2k}\right) I_\nu \tag{2.14}$$

We stress that T_B is a property of the *radiation*, not the emitting object. For an opaque thermal source of radiation, T_B is a very useful parameter because conversion from intensity to temperature is not necessary. If we told you that an opaque object has a brightness temperature of 300 K, then you know immediately that the object has a temperature of 300 K. However, we do not, *a priori*, know whether a particular source is opaque or partially transparent, so you must keep in mind that the brightness temperature is a direct measure of the radiation intensity, and only equals the temperature of the radio source when the source is both thermal and opaque.

At higher frequencies and/or for lower temperature objects, where $h\nu$ is not much smaller than kT, the Rayleigh–Jeans approximation is not applicable. In fact, at the high frequency end of the radio window and when observing relatively low temperature thermal sources, the Rayleigh–Jeans approximation may be insufficient (see Problem 9, part b). To maintain the concept that the brightness temperature equals the temperature for which the blackbody radiation equals the observed intensity, the full Planck function must be used. In this case, the brightness temperature would have to be defined as follows:

$$I_\nu = B_\nu(T_B) = \frac{2h\nu^3}{c^2} \frac{1}{\exp(h\nu/kT_B)-1} \tag{2.15}$$

A similar equation exists in terms of I_λ, which we do not show, since I_λ is rarely used at radio frequencies.

In summary, brightness temperature is a measure of intensity, and is equal to the temperature that the source would have if it were opaque and a thermal source. At low frequencies, where the Rayleigh–Jeans approximation applies, the brightness temperature is directly proportional to intensity.

Additionally, we must mention that for non-thermal radiation (such as synchrotron radiation, in which relativistic electrons emit photons while accelerated by magnetic fields), brightness temperature is not related to the temperature of the source. However, it does still describe the intensity of the radiation, which is related to the physical conditions of the source, and therefore is still a useful quantity.

Finally, the use of temperature to describe intensity or radiation power occurs in a number of other circumstances in radio astronomy. In Chapters 3 and 4, you will learn of antenna temperature, noise temperature, receiver temperature, and system temperature. These are all quantities, expressed in Kelvins, that represent power per unit frequency of signals involved in radio telescope observations. When you see these terms introduced, hopefully you will recall our discussion here, showing how intensity, which is proportional to radiation power, can be equated to a temperature.

Example 2.9:

1. What is the brightness temperature of the source discussed in Example 2.8? (At a wavelength of 6.00 cm, the source is opaque with thermal radiation and has a temperature equal to 63.6 K.)
2. A spotlight has an intensity of 2.00×10^{-11} W Hz^{-1} m^{-2} sr^{-1} at a frequency of 5×10^{14} Hz. What is its brightness temperature (regardless of the emission process)?

Answer:

1. Since the source is opaque and thermal and has no reflected light, its intensity equals the Planck curve intensity for the source's temperature. We could solve for the brightness temperature using Equation 2.14 (since we are considering a long enough wavelength, the Rayleigh–Jeans approximation is applicable), or we could realize that this is the same thing as saying that $T_B = T$. We know then that $T_B = 63.6$ K.
2. The only measurement of the spotlight beam we are given is that at the frequency of 5.00×10^{14} Hz, which is far too high a frequency to use the Rayleigh–Jeans

approximation. Instead, we put T_B into the full Planck function and set it equal to the intensity of the spotlight beam. We then have

$$2.00 \times 10^{-11} \, \text{W} \, \text{Hz}^{-1} \, \text{m}^{-2} \, \text{sr}^{-1} = \frac{2h\nu^3}{c^2} \frac{1}{\exp(h\nu / kT_B) - 1}$$

By substituting in the values for h, c, k and ν, we get

$$\exp\left(\frac{24000\,\text{K}}{T_B}\right) - 1 = \frac{1.84 \times 10^{-6}}{2.00 \times 10^{-11}} = 92000$$

or,

$$T_B = \frac{24000\,\text{K}}{\ln(92000)} = 2100\,\text{K}$$

2.5 COHERENT RADIATION

The radio-wavelength radiation emitted by an individual electron is undetectable by any radio telescope; the radiation we detect is the sum of many individual emission events by many individual electrons (or other charged particles). We will describe two very different ways that electromagnetic waves emitted by separate electrons can combine; the way they add together makes a huge difference in how we treat the radiation mathematically and also in its physical properties. This issue is conceptually subtle and relevant to fundamental issues regarding the functioning of radio telescopes and interferometers. We will discuss this issue here to greater depth, with the goal of conveying a conceptual understanding.

To start with, let us imagine a single electron that is continuously oscillating at a fixed frequency, so that it continually emits electromagnetic waves at a single frequency or wavelength. A chain of sine waves, then, propagates away from the electron. Now imagine that after traveling some distance, this single chain of sine waves is joined by another chain of sine waves exactly in phase with it. In addition, a little further along, it is joined by another wave chain, and then another, as depicted in Figure 2.6. Eventually, the initial wave chain has gathered many other identical wave chains, all in phase with each other. Since all these wave chains are identical and with exactly the same phase, they add constructively, thereby amplifying the initial wave chain by a very large factor. This is, basically, what happens in a laser and the result is that at any specific position in space at any given instant in time, one can assign a frequency, wavelength, and phase. The resultant radiation is said to be *coherent*.

In contrast, imagine the light created by adding the radiation emitted by many unrelated electrons. Each electron emits its radiation at some random time, independent of the radiation emitted by the other electrons. This is what happens in an incandescent light bulb, for example, and in most astronomical sources. The composite radiation in this case will involve numerous sine wave chains, but the frequencies, directions of travel, and phases of all the separate wave chains at any given location in space and at any given moment in

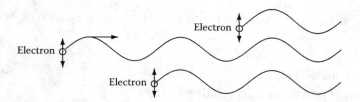

FIGURE 2.6   Schematic of the creation of coherent radiation. An oscillating electron emits a chain of sine waves, which is joined by another identical chain of sine waves with exactly the same phase, and then another wave is added, in phase with the first two.

time will be completely unrelated. The resulting radiation, in this case, cannot be modeled by a single chain of sine waves. The composite radiation in this case is called *incoherent*.

In mathematical terms, *coherence of radiation* is often defined as when radiation at any location in space and time has a specific phase. However, the issue of coherence of radiation is actually more complex; its specific meaning depends on the context in which it is used. There are different forms of coherence and varying degrees of coherence. In Section 2.6, we hope to convey a broader understanding of this phenomenon.

To demonstrate the difference between the two situations described above, we first review how one, in general, adds two electromagnetic wave chains. For simplicity, we consider a beam of electromagnetic radiation composed of just two identical wave chains—with identical frequencies and directions of travel—but with a phase difference. Representing the sinusoidal waves as cosines, the first wave chain can be represented by

$$\vec{E}_1 = E_0 \cos(\omega t)\hat{y}$$

where:

ω is the angular frequency of the oscillation and is related to the wave frequency by $\omega = 2\pi v$ and the radiation propagates in either the x- or z-direction

The second wave chain, with the same amplitude, frequency, and direction, but with different phase, is given by

$$\vec{E}_2 = E_0 \cos(\omega t + \Delta\phi)\hat{y}$$

Their sum is simply

$$\vec{E}_1 + \vec{E}_2 = E_0 \cos(\omega t)\hat{y} + E_0 \cos(\omega t + \Delta\phi)\hat{y}$$

Remember that the intensity of the light is proportional to the square of the electric field, averaged over time. The intensities of the two input electromagnetic wave chains, separately, then, are

$$I_1 = \alpha E_0^{\;2} \left\langle \cos^2(\omega t) \right\rangle \text{ and } I_2 = \alpha E_0^{\;2} \left\langle \cos^2(\omega t + \Delta\phi) \right\rangle$$

where:

⟨ ⟩ indicates a time average

α is a constant of proportionality relating I to $\langle E^2 \rangle$

The time average of the square of a cosine function with any non-zero frequency is ½, so the intensity of each wave chain is

$$I_1 = I_2 = \frac{1}{2}\alpha E_0^{\;2} \qquad (2.16)$$

The intensity of the beam containing both wave chains is

$$I_{\text{total}} = \alpha \left\langle \left(\vec{E}_1 + \vec{E}_2 \right)^2 \right\rangle = \alpha E_0^{\;2} \left\langle \left[\cos(\omega t) + \cos(\omega t + \Delta\phi) \right]^2 \right\rangle$$

$$= \alpha E_0^{\;2} \left\langle \cos^2(\omega t) + 2\cos(\omega t)\cos(\omega t + \Delta\phi) + \cos^2(\omega t + \Delta\phi) \right\rangle$$

We can simplify this equation using the trigonometric identity

$$\cos^2(A) = \frac{1}{2}\left[1 + \cos(2A) \right]$$

The intensity of the total beam, then, is given by

$$I_{\text{total}} = \alpha E_0^{\;2} \left\langle \frac{1}{2} + 2\cos(\omega t)\cos(\omega t + \Delta\phi) + \frac{1}{2} \right\rangle = \alpha E_0^{\;2} \left\langle 1 + 2\cos(\omega t)\cos(\omega t + \Delta\phi) \right\rangle$$

Now, using the additional trigonometric identities,

$$\cos(A \pm B) = \cos(A)\cos(B) \pm \sin(A)\sin(B)$$

we have that

$$\cos(A)\cos(B) = \frac{1}{2}\left[\cos(A+B) + \cos(A-B) \right]$$

and so,

$$\cos(\omega t)\cos(\omega t + \Delta\phi) = \frac{1}{2}\left\{ \cos\left[\omega t + (\omega t + \Delta\phi)\right] + \cos\left[\omega t - (\omega t + \Delta\phi)\right] \right\}$$

$$= \frac{1}{2}\left[\cos\left(2\omega t + \Delta\phi\right) + \cos\left(-\Delta\phi\right) \right]$$

Thus, the intensity of the total beam, using the fact that $\cos(-\Delta\phi) = \cos(\Delta\phi)$, is

$$I_{\text{total}} = \alpha E_0^2 \left\langle \left[1 + \cos(2\omega t + \Delta\phi) + \cos(\Delta\phi) \right] \right\rangle$$

Since the average of a cosine function with any non-zero frequency is zero, and since $\cos(\Delta\phi)$ is constant in time, we have, finally,

$$I_{\text{total}} = \alpha E_0^2 \left\langle \left[1 + \cos(\Delta\phi) \right] \right\rangle$$

Using Equation 2.16, we can rewrite this in terms of the intensity of the individual wave chains, which we represent by I_1 (since I_2 equals I_1). The intensity of the total beam, then, is,

$$I_{\text{total}} = 2I_1 \left\langle \left[1 + \cos(\Delta\phi) \right] \right\rangle \tag{2.17}$$

The term $\cos(\Delta\phi)$ is the *interference term* and dictates whether the interference is constructive or destructive when the waves are added. If the waves are in phase, then $\Delta\phi = 0$ and $\cos(\Delta\phi) = +1$, so the total intensity is twice as large as the sum of the two individual wave chains. However, if the wave chains are out of phase, then $\cos(\Delta\phi)$ must be less than 1, and if they are out of phase by $\pm\pi$, then $\cos(\Delta\phi) = -1$ and the total intensity is zero.

Therefore, for *two* wave chains that add coherently and are in phase, the total intensity will be *four* times larger than the intensity of the individual wave chains. Now consider adding N identical wave chains coherently. You can probably surmise that the resulting intensity will be N^2 times larger than the intensity of each individual wave chain. Now imagine summing N wave chains, in which the phase of each individual wave chain is random relative to all the others, as is the case for incoherent light. There will be just as many pairs of wave chains with zero phase difference, producing an interference term of $+1$, as pairs of wave chains with relative phase of π radians, yielding an interference term of -1, and there will be the whole gamut of other possible phase differences. In summing all these wave chains with random phases together, the interference term averages to zero, at all times and at all locations. The total beam will then have an intensity proportional to N times the intensity of an individual wave chain. For simplicity, we have assigned the same electric field amplitude to all waves; when the waves have different amplitudes, the expression for the sum will be somewhat more complicated, but the conceptual result is the same. In conclusion, with incoherent light, the total intensity equals the sum of the component intensities, whereas with coherent light, the total intensity grows as the square of the sum of the component intensities.

Does this mean that we now have a problem with conservation of energy with coherent radiation? Should not the intensity of N beams simply be N times the intensity of one beam, whether they are combined coherently or incoherently? Or do light beams added coherently magically have more energy than light beams added incoherently? The answer is no; the *intensities* of the coherent and incoherent beams differ, but their total *energies* are the same. Recall that intensity is flux of energy *per unit solid angle per unit bandwidth*. The total flux in a beam is the intensity integrated over solid angle and over the bandwidth, that is,

$$F_\nu = \int I_\nu \, d\Omega \, d\nu$$

The high intensity of a coherent light beam is only present over a small range of wavelengths and the beam is very narrow in solid angle. Therefore, when the two intensities are integrated over frequency and solid angle, they yield the same flux, or the same rate of energy flow. The high intensity, small range of wavelengths, and narrow beam are the distinguishing properties of laser light.

2.6 INTERFERENCE OF LIGHT

Interference of light is an important concept that has a number of applications that are relevant for our later discussion regarding the functioning of radio telescopes. You have likely been introduced to this concept by the famous double-slit experiment of Thomas Young. This experiment helped to establish the wave nature of light in the early part of the nineteenth century. Although a highly coherent light source makes classroom reproduction of Young's experiment easier, as we will explain, highly coherent light is not essential, and was not used by Young himself.

The standard demonstration in the classroom uses an optical laser beam (which is highly coherent) shining on a double slit, and the light from the two slits combines and illuminates a background screen producing the famous interference pattern. The interference pattern is produced because the path lengths from the two slits to most places on the screen differ, causing the waves to have different phases when they combine. Imagine a point on the screen midway between the two slits; then the path lengths from either slit to the screen are the same, so the phases will be the same, and from our discussion in Section 2.5, we found that the light will add constructively, producing a bright spot. But if you look at a position on the screen offset from the midpoint along the direction of the slit separation, the light paths from each of the slits to the screen are now of different lengths and therefore the phases will differ. The light in the offset position will not add constructively, so the intensity will be diminished. If the offset is selected so that the path difference is exactly half the wavelength of the light, then the phase difference will be π radians, and, as we showed earlier, the waves will cancel and the intensity will be zero, causing a dark region called a *null*. However, at a point further offset, where the path difference is one wavelength, the phase difference will be 2π radians, and the waves will again add constructively, producing another peak. Thus, along the screen, there will be a series of bright and dark regions (peaks and nulls) occurring wherever the waves add constructively and destructively. A schematic of the double-slit and the resulting interference patterns is shown in Figure 2.7. Young recognized that such a behavior required light to have wave properties.

Seeing a demonstration of the double-slit experiment with a laser light source might lead you to believe that you only get the interference pattern when using highly coherent light sources. Incoherent light is made up of many light beams with random phases, so why would you get an interference pattern? In fact, Young performed this experiment with sunlight, which is not coherent, so what are we missing?

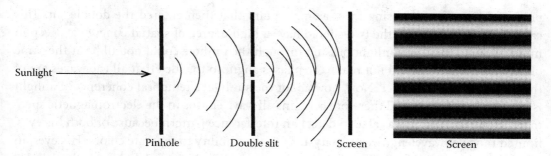

Sunlight ⟶

Pinhole   Double slit   Screen   Screen

FIGURE 2.7 Schematic representation of Young's double-slit experiment is shown on the left and the display of the bright and dark lines that appear on the screen due to interference is shown on the right. Note that the curved lines represent only a small fraction of the wave crests and the distance between them does not represent the wavelength that results in the interference pattern shown.

First, we need to understand and address three separate issues about the combining of the waves: the relative phases, the direction of travel, and the wavelengths of the different wave chains. With regard to the phases, the interference pattern does not require that all the wave chains be in phase. Each individual wave chain that approaches the double slit will be broken up into two pieces—one piece passing through each slit—and the two pieces will interfere with each other, and hence each individual wave chain will produce an interference pattern on the screen. Other wave chains, with different phases, will also produce interference patterns and since the interference depends solely on the slit geometry and the wavelength of the light, these wave chains produce the same interference pattern as the first, regardless of the starting phase. The total intensity pattern due to these wave chains is simply the sum of the same pattern many times over.

If the light contains either a range of frequencies and/or a range of travel directions, then the interference pattern will be degraded. We will call these two aspects of coherence as *temporal coherence* and *spatial coherence*. A range of frequencies reduces the temporal coherence and a range of travel directions degrades the spatial coherence. Now, in reality, no light source is perfectly coherent or incoherent, but has differing degrees of partial coherence. Laser beams have an extremely high degree of both spatial and temporal coherence, but no laser is 100% coherent. So, imagine a beam of light with a broad range of wavelengths entering the double slit. Since the phase difference between the two wave fronts arriving at the screen depends on the wavelength, the locations of the nulls and peaks will differ for differing wavelengths, and the total interference pattern will be smeared out. Additionally, a lack of spatial coherence will mean that wave fronts will approach the double slits from a range of angles, which will also affect the locations of the nulls and peaks. To produce well-defined bright and dark lines in the double-slit experiment, it does require some degree of coherence.

So, how, then, did Young succeed with this experiment without using a laser beam? For starters, we should clarify that Young first passed the sunlight through a pinhole, and this produced wave fronts arising from a single point that then entered the double slit. This caused the light entering the two slits to have a high degree of spatial coherence. Keep in mind, though, that the sunlight passing through the pinhole could not all be at the same phase and the pinhole did not make the phases align, so the idea that all the waves are of the same phase is incorrect. Now, how about the sunlight's temporal coherence? Sunlight is blackbody radiation and therefore contains all wavelengths in the electromagnetic spectrum. However, Young was able to detect an interference pattern because of both his eyes' limited range of wavelength sensitivity and his eyes' ability to separate colors. His eyes, in effect, created a small level of temporal coherence by filtering the light by wavelength. He could, then, see different colored fringes.

The reason that double-slit demonstrations almost always use a coherent light source instead of reproducing Young's experiment with sunlight is, primarily, to produce light with a narrow range of wavelengths. The resultant interference pattern, then, has well-defined nulls and peaks and does not require the viewer to mentally separate the colors.

The concept that light does not have to be 100% coherent, but only partially coherent, to produce an interference pattern is very important to appreciate when learning about radio telescopes and doing radio astronomy observations. Radio astronomy observations must involve a range of wavelengths, causing a finite degree of temporal coherence, and most astronomical radio sources subtend small angles on the sky, yielding a finite degree of spatial coherence. The phases of the radiation, though, are usually random, and understanding that this does not affect the interference of the light is crucial for appreciating the limiting factors in the resolution of a radio telescope (Chapters 3 and 4) and for understanding the technique called *interferometry* or *aperture synthesis* (discussed in Chapters 5 and 6).

2.7 POLARIZATION OF RADIATION

One last characteristic of the radiation that we need to discuss in this chapter involves the direction of the electric field in the propagating electromagnetic wave. To describe this aspect, one talks about the *polarization* of the radiation. As an electromagnetic wave travels through vacuum, the directions of the wave propagation, the electric field, and the magnetic field must all be mutually perpendicular. Consider electromagnetic waves traveling along the $+z$-axis. The oscillating electric field vector is perpendicular to this direction; we will define these to be the $\pm y$-directions, and the magnetic field vector must be perpendicular to both of these, or the $\pm x$-directions, as shown in Figure 2.8. Note that we must be sure to define the x-, y-, and z-axes following the right-hand rule, so that the cross product of $\vec{E} \times \vec{B}$ points in the positive z-direction, hence the x-axis in Figure 2.8 points into the page. There are, however, infinite numbers of directions that are perpendicular to the direction of travel, and there is no law that determines in which of these directions the electric field must point—only that it be perpendicular to the magnetic field vector and to the direction of travel. So, for our $+z$-traveling waves, the electric field vector can also oscillate in the $\pm x$-direction and the magnetic field vector then would oscillate in the $\pm y$-direction. In fact, the electric field can oscillate in *any* other direction in the xy-plane, while the magnetic field oscillates along the perpendicular line in the xy-plane. Moreover,

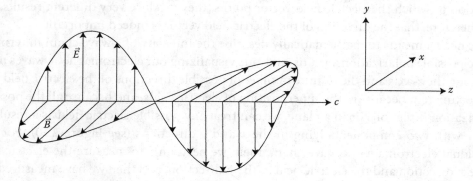

FIGURE 2.8 Propagation of plane electromagnetic waves along the z-axis with the electric field oscillating along the y-axis.

the electric field is not required to stay pointing in the same direction. For example, the electric field could rotate around the z-axis as the wave propagates; the magnetic field would then also rotate so as to stay perpendicular to the electric field.

To see why the direction of the electric field matters, consider the effect that the different electromagnetic waves discussed above have on an electron along the path of the radiation. The oscillating electric field will accelerate the electron, causing it to oscillate in the direction of the electric field. (The magnetic field only accelerates the already moving electrons and the resultant acceleration of electrons with typical velocities is insignificant in comparison to that produced by the electric field.) Now imagine that the electron is constrained to move only in the vertical direction (perhaps the electron is part of a very thin vertical wire antenna). We depict this situation in Figure 2.9. Then, a chain of electromagnetic waves whose electric field points in the horizontal direction *cannot* accelerate this electron at all, so these waves will pass by the electron with no effect, whereas electromagnetic waves whose electric field oscillates in the vertical direction *will* cause the electron to oscillate. Even though these two sets of waves seem very similar in every way except for the

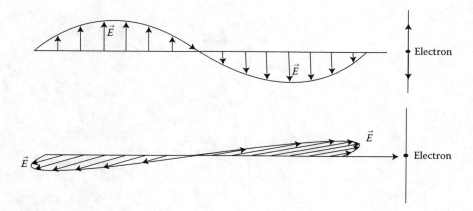

FIGURE 2.9 An electron that can move only in the vertical direction is accelerated by the electromagnetic waves with vertical polarization but is unaffected by waves with horizontal polarization.

direction in which the electric field vector points, they produce very different results. We see, therefore, that the direction of the electric field vector is indeed important.

We need a means to mathematically describe the intensity of a wave chain in terms of all the possible polarizations. Let us continue visualizing our electromagnetic waves moving along the z-axis and the infinite number of possible directions of the electric field vector. Fortunately, because of the superposition of electric fields, and because all the possible polarization directions lie in a plane, we can treat any possible electric field as a resultant vector with two components lying in the x and y directions (see Figure 2.10). For any individual electromagnetic wave chain, then, we need only to measure the electric field in the x-direction and the electric field in the y-direction, and then we have measured the vector's components, which is sufficient for a complete description of the whole vector. These two base electric field measures we can call horizontal polarization and vertical polarization.

Now, the components of the electric field along the x- and y-axes are independent of each other, so they are not required to oscillate in phase (provided that a similar phase difference occurs with the oscillating magnetic field components). The consequence of differing phases for the x- and y-components is that the total electric field vector rotates as it travels. In Figure 2.11, we depict snapshots of the electric field vector of an electromagnetic wave propagating to the left in which the electric field components oscillate out of phase. The wave starts at the right end, where the y-component of the electric field is at its maximum, while the x-component is momentarily zero. As the wave moves to the left, the x-component grows while the y-component decreases. When the x-component reaches its maximum, the y component is zero. In addition, the y-component grows in the negative y-direction, while the x-component decreases down to zero again. Since E_y reaches zero when E_x reaches a maximum, and vice versa, the phase difference between the components in Figure 2.11 is one quarter of a cycle, that is, $\pm\pi/2$ radians.

Now focus on the total electric field vector in Figure 2.11 by considering the vector sum of the components. If the magnitudes of oscillation of the electric field in the x- and

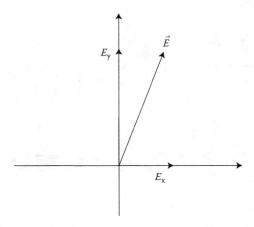

FIGURE 2.10 Total electric field vector of an electromagnetic wave can be fully described by its components along the x- and y-axes.

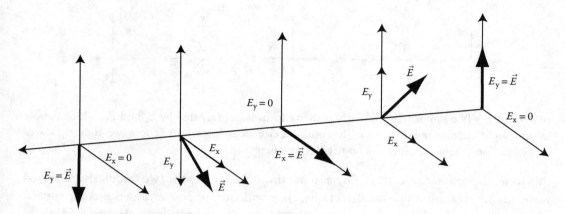

FIGURE 2.11   In an electromagnetic wave propagating to the left, electric field components in the x- and y-directions oscillate out of phase, causing the total electric field vector to rotate around the direction of propagation.

y-directions are equal and the phase difference between them is exactly $\pm\pi/2$ radians, then the total electric field vector will be constant in magnitude but will rotate around the z-axis, tracing out a circle (as will the magnetic field vector). This case is called *circular polarization*. We will not show in detail here, but you can imagine that by allowing the magnitudes of the component electric fields to vary, as well as the phase difference between them, the total electric field vector can also trace out an ellipse. This situation is called *elliptical polarization*. When the phase difference is zero or 180°, the total electric field vector will oscillate in a fixed direction in the xy-plane, a case that is called *linear polarization*. The orientation of the linear polarization will depend on the relative magnitudes of the component fields—if they are equal, the total electric field will oscillate along a line at a 45° angle relative to the x- and y-axes. As you hopefully remember from your high school geometry class, a circle and a line are just the opposite extremes in the range of ellipses. Therefore, you can consider elliptical polarization as a general form of polarization, with circular and linear polarization as special cases.

An alternative, and equally correct, approach is to view two different circular polarizations—one rotating clockwise and the other rotating counterclockwise—as the base components for describing any polarization. Just as a circularly polarized wave chain can be described as the sum of horizontally and vertically polarized wave chains, as we showed above, a linearly polarized wave chain can also be described by the sum of two oppositely rotating circularly polarized wave chains. In addition, the phase difference between the circular polarizations will determine the orientation of the resultant linearly polarized wave chain. Let us follow through with a particular example to see how this works. Consider Figure 2.12 as we discuss this example. Imagine two oppositely rotating circularly polarized wave chains in phase such that both their electric field vectors point upward (in the +y-direction) at the same time. Then, at this time, the sum, or the total electric field, will point upward and will equal the linear sum of the two component electric fields. At

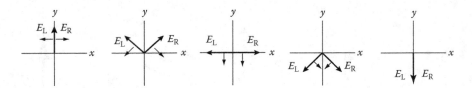

FIGURE 2.12   Two oppositely circularly polarized waves, represented by E_L and E_R, which denote left and right circular polarizations. The resultant electric field vector of these two rotating waves oscillates along the y-axis, and has no net x-component.

this time, the resultant electric field has a maximum value. As the two circularly polarized wave chains rotate in opposite directions, they will gain opposite valued x-components, which will cancel, while their y-components decrease. The resultant electric field, then, will still point in the $+y$ direction, but will be smaller. As the two waves continue to rotate in opposite directions, the total $\vec{E}$ field vector will continue to decrease and stay in the $+y$-direction. The total $\vec{E}$ field will continue to decrease until both of the rotating waves are horizontal, one pointing in the $+x$-direction and the other in the $-x$-direction. At this moment, the two component $\vec{E}$ fields completely cancel out, so the resultant $\vec{E}$ field vector is zero. As the two waves continue to rotate, they both gain $-y$-components and the x-components still cancel. The sum will reach a maximum again when the two vectors are lined up along the $-y$-axis. We see, then, that the sum of these two opposite circularly polarized wave chains produces an oscillation only in the y-direction, and so describes a linear polarization. In similar ways, a wave chain of *any* polarization can be described either by a sum of two perpendicular linear polarization functions of time and phase or by a sum of two oppositely rotating circular polarization functions of time and phase.

The two possible directions for circular polarizations are generally called *left circular polarization* and *right circular polarization*. However, there is, unfortunately, an inconsistency in these definitions between different disciplines, which often causes a fair bit of confusion. The reason for the inconsistency involves the different images one gets from different perspectives. One who imagines the view of an approaching circularly polarized wave chain will define the rotation of the electric field in the opposite sense from one who imagines the view of a receding wave chain as seen from the source. Now, since the initial development of radio astronomy was accomplished by electrical engineers, especially those who are practiced in designing radio transmitters, the definitions and conventions used by radio astronomers to describe radio waves were established from the perspective of the source, viewing the radio waves traveling away. The radio astronomer's definitions of right and left circular polarization (in the conventions established by the International Astronomy Union), therefore, are the same as that given in the conventions of the Institute of Electrical and Electronics Engineers. In this convention, right circular polarization describes an electromagnetic wave whose electric field rotates clockwise as the wave comes out of the transmitter when viewing the wave from the transmitter and in the direction of propagation. However, this means that *the observer*, who views the wave from the opposite direction, *will see the electric field rotate counterclockwise*. So, in terms of the observation of a circularly polarized radio signal, right circular polarization means that the electric field vector, as it enters the antenna, is observed to turn counterclockwise with time. Another easy way to remember this is that a *right* circularly polarized wave obeys the *right*-hand rule in that when the thumb points in the direction of propagation, the curl of the fingers indicates the direction that the electric field turns.

There is confusion in these definitions because in most other fields of physics, right and left circular polarization are defined in terms of the perspective of an approaching wave and so are given the opposite definitions from those used by radio astronomers. Since radio astronomy is now considered a branch of physics, there is a natural tendency to hold onto the definitions found in physics texts. In addition, since radio astronomers naturally think of themselves as observing the radio waves as they enter the antenna, not creating them and sending them off to space, the physicist's definition, conceptually, makes more sense. But, this is not the standard definition used in the field. The lesson for us here is to remember the origin of the field that we are working in, which in this case is electrical engineering, and that the standard definitions in the field are influenced by the field's history.

In an actual observation of radiation from space, not all the wave chains that enter your telescope are going to have exactly the same polarization. What you detect may have no net polarization, in which case it is called *unpolarized* radiation, or it may have a greater intensity in one polarization (horizontal vs. vertical, or left circular vs. right circular), in which case a *percent polarization* of the radiation is a useful measure.

Observations in which there is a non-zero net polarization signal contain additional information about the radiation source. In these special cases, the measure of polarization is an important and powerful tool to add to one's repertoire for inferring the physics of the observed source.

Analysis and discussion of polarized radiation is easier with the use of another set of polarization terms, which we will discuss in Section 2.7.1.

2.7.1 Stokes Parameters

In 1852, Sir George Stokes developed a set of four parameters as a means of quantifying polarization in a more useful way. Here, we will give the mathematical definitions of the Stokes parameters, and also explain what they really mean physically and why they are useful.

The first parameter, labeled I, is equal to the total intensity of all the radiation. So, I equals the sum of the intensities of all orthogonal polarizations. With the $+z$-axis assigned to the direction of propagation of the wave, the electric field vector must be in the xy-plane, and is described using only two orthogonal bases. We have introduced two commonly used sets of independent bases (horizontal and vertical linear polarizations or left and right circular polarizations), so we can write I as

$$I = I_x + I_y \tag{2.18}$$

or

$$I = I_L + I_R \tag{2.19}$$

where:

I_x and I_y are the intensities of any two orthogonal linear polarizations

I_L and I_R are the intensities of the left circular and right circular polarizations

The second parameter, labeled Q, is equal to the difference in the intensities of the two linear polarizations, so

$$Q = I_x - I_y \tag{2.20}$$

The third parameter, U, is very similar to the second, but involves a rotation of the x- and y-axes by 45°. If we call these shifted axes a and b, shown in Figure 2.13, so that I_a is the intensity of light that is linearly polarized along an axis halfway between the x- and y-axes and I_b in the intensity of light polarized 90° relative to a, then

$$U = I_a - I_b \tag{2.21}$$

Note: Now that we have defined axes a and b, you can see that Stokes I can also be set equal to the sum of I_a and I_b.

The fourth parameter, V, is equal to the difference in the intensities of the left and right circular polarizations, so

$$V = I_R - I_L \tag{2.22}$$

Stokes V, alone, is a measure of the amount of net circular polarization.

Q and U determine the amount of linear polarization. To be more precise, the intensity of the net linear polarization, which we will call L, is given by summing Q and U in quadrature, that is,

$$L = \sqrt{Q^2 + U^2} \tag{2.23}$$

One convenience of the Stokes parameters should already be apparent; they provide a means of quantifying the linear and circular polarization intensities at the same time, as indicated by Equations 2.22 and 2.23. In addition, if the detected radiation has no net polarization, linear *or* circular, then

$$Q = U = V = 0$$

FIGURE 2.13 Two sets of orthogonal axes for measuring polarization of radiation. The x-y axes are used to calculate the Stokes Q and the a-b axes are used for the Stokes U.

If there is no net circular polarization, but there is non-zero linear polarization, then $V = 0$, and the fractional amount of linear polarization is given by L/I. If there is no net linear polarization but a non-zero circular polarization, then $Q = U = 0$, and the fractional amount of circular polarization is $|V|/I$. If Q and/or U is non-zero *and* V is non-zero, then the net polarization is elliptical. The total fractional polarization can in general be expressed as

$$\frac{\sqrt{Q^2 + U^2 + V^2}}{I}$$

When there *is* a net polarization, then there are other parameters that you will want to determine as well. We will skip the mathematical development and jump to the final equations.

With elliptical polarization, the orientation angle of the major axis of the ellipse, relative to the x- and y-axes (as defined above), is called the *polarization angle*, and is often denoted by θ, where θ is the angle starting from the $+x$-axis and moving toward the $+y$-axis. The polarization angle is related to Q and U by

$$\theta = \frac{1}{2}\tan^{-1}(U/Q) \tag{2.24}$$

If there is a net linear polarization, the direction of the polarization in the xy-plane is the same as this angle (since a line is just a flattened ellipse). Circular polarization, by its nature, has no major axis, in this case, so the polarization angle is nonsensical.

You will also want to quantify the degree of ellipticity, which is commonly described by an angle denoted as β, in which $\tan\beta$ is equal to the ratio of the amplitude of the electric field along the minor axis to that along the major axis of the ellipse, that is, $\tan\beta = E_{minor}/E_{major}$. For pure circular polarization $\beta = \pm \pi/4$ radians and for pure linear polarization $\beta = 0$. The parameter β is determined from Stokes parameters by

$$\beta = \frac{1}{2}\tan^{-1}(V/Q) \tag{2.25}$$

Stokes parameters are advantageous to use in describing polarization largely because they are all measures of intensity. As such, there is no phase of the waves to worry about and they can easily be added, averaged, and manipulated algebraically. In contrast, consider the difficulty in calculating the average position angle of linearly polarized radiation from a number of measurements, keeping in mind that polarization position angles that differ by 180° represent the same radiation. Imagine you had two measurements yielding polarization position angles of 1° and 179°, which in terms of the physical orientation of the electric fields differ by only 2°. When you algebraically average them, you infer a position angle of 90°, which is orthogonal to the correct answer. The proper approach is to keep track of the Stokes parameters, average their values, and then calculate the polarization angle from Equation 2.24.

Example 2.10:

The data from a radio telescope that can detect both left and right circular polarizations at the same time are processed through a program that calculates the Stokes parameters. The results indicate that $I = 0.500$ Jy/beam, $Q = -0.003$ Jy/beam, $U = +0.004$ Jy/beam, and $V = 0.0$ Jy/beam.

1. Does this radiation have a non-zero polarization? If so, what kind of polarization does it have?
2. Characterize the polarization.

Answers:

1. The radiation is unpolarized only if Q, U, and V are all equal to 0. Since Q and U are not zero, the radiation *is polarized*. Q and U are measures of the linear polarization, while V is a measure of circular polarization. Since $V = 0$ and Q and U are not zero, the radiation is *linearly polarized*.
2. The percent polarization of linear polarization is given by L/I, where L is given by Equation 2.23. Therefore, the percent polarization is

$$\frac{\sqrt{(-0.003)^2 + 0.004^2}}{0.5} = \frac{0.005}{0.5}$$

$$= 1\% \text{ polarized}$$

The position angle is given by Equation 2.24, so

$$\theta = \frac{1}{2}\tan^{-1}\left(\frac{0.004}{-0.003}\right) = \frac{1}{2}(180° - 53.1°)$$

$$= 63.4°$$

Since the numerator of the arctan is positive and the denominator is negative, the angle of the arctan must be in the second quadrant.

The definitions of Stokes parameters, as listed above, involve combinations of the intensities of six different polarizations, that is, I_x, I_y, I_a, I_b, I_L, and I_R. Recall, though, that the intensities are proportional to the squares of the electric fields, and the electric fields in any pair of complementary polarizations (E_x and E_y, E_a and E_b, or E_L and E_R) can be used to obtain the electric fields in any other polarization: linear, circular, or elliptical. For example, E_a, the electric field with linear polarization at position angle 45°, relates to the horizontal and vertical polarizations, E_x and E_y by

$$E_a = \frac{E_x + E_y}{\sqrt{2}}$$

How the Stokes parameters relate to all pairs of complementary polarizations is beyond the scope of this book. Instead, we will choose the horizontal and vertical linear polarizations as the base pair to use as an example. For the discussion here, we will treat the conversion from the square of the electric field to intensity as accomplished by the telescope calibration, and we will omit the conversion factors, that is, we will write the relation simply as intensity $= E^2$. In terms of just E_x and E_y, then, we can rewrite the Stokes parameters as

$$I = E_x^2 + E_y^2 \tag{2.26}$$

$$Q = E_x^2 - E_y^2 \tag{2.27}$$

$$U = 2 E_x E_y \tag{2.28}$$

and

$$V = 2 E_x E_y \sin \phi \tag{2.29}$$

where:

$\phi = \phi_x - \phi_y$ is the relative phase between E_x and E_y. (Note: A more accurate description of how the electric fields of the incoming radiation are manipulated in radio astronomy involves the complex exponential form of waves, which we will discuss in Chapter 6. The products of Equations 2.26 through 2.29, for example, are actually time averages of complex numbers, and hence are actually calculated by a mathematical process called *correlations*, which is also discussed in Chapter 6.)

QUESTIONS AND PROBLEMS

1. Assuming that there are about 100 radio telescopes in the world, with an average effective collecting area of 500 m² and that they are used 80% of the time to observe sources of flux density 0.1 Jy, calculate the total energy collected from all sources by all radio telescopes over the history of radio astronomy. Use a typical bandwidth of $\Delta v \sim 100$ MHz and assume that, on average, these 100 telescopes were built around 1970. To put your answer in comprehensible terms, if this energy equals the amount of gravitational energy released in the fall of an object onto the Earth's surface from a height of 1 m, calculate the mass of the object.

2. An observation of 100 s yields a total detected energy of 4.00×10^{-17} J. The diameter of the radio antenna used is 2.00 m and the observation used a bandwidth of 500 kHz.

   a. What is the observed flux density?

   b. The source is at a distance of 10.0 light years $= 9.45 \times 10^{16}$ m. What is the luminosity per Hz emitted by the source?

3. An observation of a uniform, circular object yields an observed flux density of 5.00 Jy. If the object has an angular width of 0.100°, what is the intensity of the emitted radiation?

4. a. The Sun emits a total luminosity of 3.90×10^{26} W and is at a distance of 1.50×10^{11} m from Earth. What is the flux of solar radiation hitting the top of the Earth's atmosphere?

   b. About 37% of that flux is fairly evenly distributed across the visible band, which runs from 400 to 700 nm. What is the approximate visible F_λ (the flux per unit *wavelength* interval) of the solar radiation at the top of the Earth's atmosphere?

   c. What is the approximate visible, F_ν, (the flux per unit *frequency* interval) of the solar radiation at the top of the Earth's atmosphere? Compare to the answer in (b) and explain.

   d. The Sun has an angular diameter of 0.533°. What is the visible I_λ of the solar radiation?

   e. What is the visible I_ν of the solar radiation?

5. Use Equation 2.5 to calculate the visible I_ν of the solar radiation at the Sun's surface. The Sun's surface temperature is about 5800 K. How does this compare to the answer to Question 4, part e?

6. a. The Stefan–Boltzmann law tells us that the radiation flux from a blackbody is proportional to the body's temperature to the fourth power (Equation 2.7). Equation 2.8 indicates that the average photon energy in blackbody radiation is linearly proportional to the body's temperature. Use these two equations to infer the number of photons emitted per second per unit area as a function of the body's temperature.

   b. In general, a hotter body will emit energy at a higher rate than a cooler body. Considering your answer to (a) in combination with Equation 2.8, which contributes more to the higher rate of emission from a hotter blackbody, the energy of the photons or the number of photons emitted per second?

7. Consider two blackbodies of identical size. Body A has a temperature twice that of body B.

   a. For low-frequency radiation, such that $h\nu/kT \ll 1$, what is the ratio of the intensities of these two bodies?

   b. What is the ratio of the total flux radiated by these two bodies?

   c. What is the ratio of the average photon energy emitted by these two bodies?

8. Infrared observations at 12.0 microns of the nearby galaxy M82 (which is known as an especially infrared luminous galaxy) result in a measure of $F_\lambda = 6.00 \times 10^{-7}$ W m⁻³. How many janskys does this equal?

9. Rayleigh–Jeans approximation ($h\nu/kT \ll 1$)

   a. Derive the Rayleigh–Jeans approximation by doing a Taylor series expansion of $e^{h\nu/kT}$, where $h\nu/kT \ll 1$, and substitute into the $1/[\exp(h\nu/kT) - 1]$ part of the Planck function. Algebraically, manipulate and show that you get the Rayleigh–Jeans approximation.

b. Considering that the radio window goes down to a wavelength of 1 mm, determine the minimum temperature that a source can have for the Rayleigh–Jeans law to apply at all radio wavelengths. Estimate the minimum temperature, first by using the assumption implicit in the Rayleigh–Jeans law by requiring $h\nu/kT \lesssim 0.1$, and then check your answer by solving for $B_\nu(T)$ using both the Planck function and the Rayleigh–Jeans law and requiring that the two agree to better than 10%.

c. Considering that the entire universe bathes in the CMB, which is at a temperature of 2.73 K, determine the observing frequency range in which the Rayleigh–Jeans law is always a safe approximation.

10. What is the mathematical definition of brightness temperature? What is its physical significance? In observations of what types of sources is it generally used? Why is its use primarily limited to radio astronomy?

11. The discovery of the CMB radiation resulted from the persistent detection of a signal with a radio antenna in any direction that the telescope was pointed. It is now known that the CMB has a perfect blackbody spectrum at a temperature of 2.73 K. The antenna was operating at 7.35 cm wavelength.

a. Calculate the intensity, I_ν, of the signal that was being detected.

b. What is the frequency of the peak of B_ν of the CMB? What wavelength does this correspond to?

c. More recent observations of the CMB have revealed that the temperature of the CMB varies by 1 part in 10^5, meaning that there are variations in the temperature where, $\Delta T \sim 10^{-5}\ T$. Calculate the accuracy needed in measurements of the intensity at ν_{peak}—answer to (b)—to measure these fluctuations. That is, what is the difference in W m^{-2} Hz^{-1} sr^{-1} between two points that differ in temperature by 1 part in 10^5 for a blackbody of $T = 2.73$ K?

12. Consider radiation that has 5% net linear polarization in the horizontal plane.

a. What are the intensities of the horizontal and vertical linear polarizations, relative to the total intensity?

b. What percent of the total intensity is unpolarized?

c. One way of producing linear polarization is via a scattering screen. When the radiation reflects off a surface, it gets polarized in the plane of the surface. If that is the cause for the polarization of the radiation in this problem, discuss a scenario in which you get 5% of the radiation to be linearly polarized.

13. a. Show that Equation 2.23 does correctly describe the total linear polarization by considering linearly polarized radiation with random position angle, θ, and calculating the value of Q and U, in terms of the total intensity I, and show that Equation 2.23 then leads to the correct total linearly polarized intensity.

b. Show that U is needed to be sure of determining whether radiation is linearly polarized by considering a linearly polarized wave with PA = 45° and calculating Q. That is, show that Q alone would not reveal that such radiation is polarized.

14. Polarization-sensitive observations of a radio source yield the following measures of the Stokes parameters.

$$I = 5.00 \times 10^{-18} \text{ W Hz}^{-1} \text{ m}^{-2} \text{ sr}^{-1}$$

$$Q = 0$$

$$U = 0$$

$$V = 2.50 \times 10^{-20} \text{ W Hz}^{-1} \text{ m}^{-2} \text{ sr}^{-1}$$

a. What is the total intensity?

b. What is the total intensity of the horizontally polarized radiation?

c. What is total intensity of the right-circularly polarized radiation and of the left-circularly polarized radiation?

CHAPTER 3

Radio Telescopes

THIS CHAPTER PROVIDES AN overview of the principal components of a radio telescope and describes what role each of these components play in detecting astronomical radio signals.

A typical radio telescope consists of a primary reflector (or dish), feed, transmission line, and receiver; these components are shown schematically in Figure 1.15. We note that radio telescopes operating at very long wavelengths (typically 1 m or longer) can take a very different form as discussed in Section 3.6. Most radio telescopes are fully steerable, mounted on Alt–Az, (also called Az–El) mounts (described in Section 1.4.2), and can point to any direction in the sky. A computer controls the motion of the telescope and continuously translates the sky coordinates (see Section 1.3.1) of an astronomical object into current altitude and azimuth positions. We will not discuss further the mechanics of how a radio telescope is mounted or how it moves.

Some aspects of a radio telescope are common to telescopes used at ultraviolet, visible, or infrared wavelengths, such as the use of a primary reflector (and often a secondary reflector) to collect and focus the light. We discuss the optics of radio telescopes in Section 3.1. However, there are also aspects that are quite different from telescopes at other wavelengths. One of the biggest differences is the method by which light is detected. At shorter wavelengths, one detects the particle nature of light, meaning the individual photons. The energy of a visible-wavelength photon ($E = h\nu = hc/\lambda$) is of order a few electron volts (an electron volt is 1.6×10^{-19} J), which is sufficient to either excite a valence electron to a conduction band electron in a semiconductor (as occurs in photo-conductive devices) or to produce electron-hole pairs in a semiconductor (in photovoltaic devices). Charge-coupled devices (CCDs), which are at the heart of digital cameras, for example, work by these principles. Radio photons, which have energies of order 10^{-5} eV, are usually insufficient to produce a measurable effect in a semiconductor device. At very short radio wavelengths (1 mm and shorter), with photon energies of order 10^{-3} eV, there are new semiconductor and superconductor devices, such as kinetic inductance detectors, that are capable of detecting individual photons. In general, at radio wavelengths we must understand the detection of the radiation in terms of large ensembles of photons and make use of the wave nature of light.

Another important difference is that at radio wavelengths, we generally use coherent signal processing (which maintains the phase information of the wave). This processing is carried out in devices called *heterodyne receivers*, which are discussed in Section 3.2. In Section 3.4, we briefly mention an incoherent power detector, often used at millimeter and submillimeter wavelengths, called a *bolometer*.

Depending on the astronomical object and the goals of the science project, we might wish to measure the flux of the radio source over a very broad frequency range. This type of observation is similar to what visible and infrared astronomers call *broadband* photometry, in which they might detect the thermal radiation from a star over broad ranges of wavelength. Alternatively, the astronomical radio signal may contain narrow-band features, such as spectral lines, in which case, we use a *spectrometer* to measure the power across many very narrow adjacent frequency bands. We discuss radio telescope spectrometers in Section 3.5; as you will see, this is another device that is quite different from those used at visible and infrared wavelengths.

3.1 RADIO TELESCOPE REFLECTORS, ANTENNAS, AND FEEDS

The component of a radio telescope responsible for collecting the radio signals is often referred to as an *antenna* or a *reflector* and these terms are often used interchangeably. There is a difference between the precise meanings of these two terms, however. An *antenna* is a device that couples electromagnetic (EM) waves in free space to confined waves in a transmission line, while *reflectors*, which are usually parabolic in shape, collect and concentrate the radiation. Most large radio telescopes employ a reflector as the first element, but they still need an antenna to couple the EM waves into a transmission line, which then carries these waves to the receiver. At long radio wavelengths, simple dipole antennas can be used as the first element; they will be discussed in Section 3.6.

Two other words that are commonly used, and which help to avoid the misuse of the word antenna, are the *dish*, which is often used to refer to the reflector, and the *feed*, which is the device that couples the radiation concentrated by the reflector into a transmission line.

3.1.1 Primary Reflectors

The dishes of most radio telescopes are parabolic reflectors, similar to the primary components in telescopes used in the infrared, visible, and ultraviolet regimes. The parabolic shape causes all waves approaching the dish from the direction perpendicular to the entrance plane to come to a single point, known as the *focus* of the telescope. The EM waves emitted by an astronomical object, as they approach our telescope many light years away, are well approximated as plane waves and so enter the telescope in parallel paths. Figure 3.1 depicts the reflection of radio waves off a parabolic reflector and those arriving at the focus point. We show two methods for representing light paths: ray tracing and wave-front tracing. In the ray-tracing approach, we treat the light like bullets traveling in straight lines—this has traditionally been referred to as a *corpuscular radiation* model. Alternatively, we can follow the wave fronts, which are surfaces perpendicular to the direction of travel and in which the EM waves have a constant phase. If the direction of the astronomical source is off the

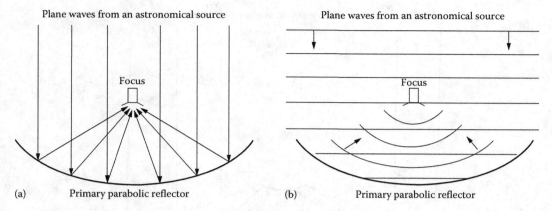

FIGURE 3.1 Two depictions demonstrating the focusing of light to a single point due to reflection of a parabolic surface. The left panel (a) shows rays of light coming from a distant astronomical source and converging to a single focus, while the right panel (b) depicts a wave-front representation, showing how a parabolic reflector converts plane waves from a distant astronomical source to spherical waves that converge at the focus.

central axis of the reflector, the waves will still converge approximately to a point, but offset from that of an on-axis astronomical source. Therefore, in fact, the parabolic reflector has a *focal plane* and not just a single focus. In a visible-wavelength telescope, a CCD camera, which contains a two-dimensional array of tiny detectors, is placed at the focal plane and we get an image of the object. At the focal plane of a radio telescope are the *feed horns* (see Section 3.1.3), which convert the EM waves from free space to *transmission lines* (see Section 3.2.1), through which the signal travels to *receivers* (discussed in Section 3.2).

The telescope depicted in Figure 3.1 is called a *prime focus telescope* because the feeds (and receivers) are placed at the focus of the primary reflector. Figure 3.1 shows just one feed; however, multiple feeds are often used to collect the power from different directions simultaneously. The prime-focus configuration can be inconvenient, because the feed and receiver are in an awkward position, located high above the primary reflector and hence not easily accessible when the telescope is aimed at the sky. For this reason, most radio telescopes (and visible-light telescopes as well) are of *Cassegrain* design. In a Cassegrain telescope, a second reflector (or mirror) is placed before the focal plane of the primary reflector to redirect the waves to another focal point at or behind the vertex of the primary reflector. This arrangement is shown in Figure 3.2. Of course, if the focus is located behind the primary reflector, then the reflector must have an opening at its center to allow the light beam to pass through. Most large radio telescopes are classical Cassegrain telescopes, which use a parabolic primary and hyperbolic secondary. In Figures 3.3 and 3.4, we show photos of a prime-focus radio telescope (the Haystack SRT) and a classical Cassegrain radio telescope (the Green Bank 20-m telescope).

The primary reflector serves two important functions. First, it collects and focuses the radiation from astronomical sources, making faint sources more detectable. The amount of radiation collected depends on the telescope's *effective area* (A_{eff}), which is closely related

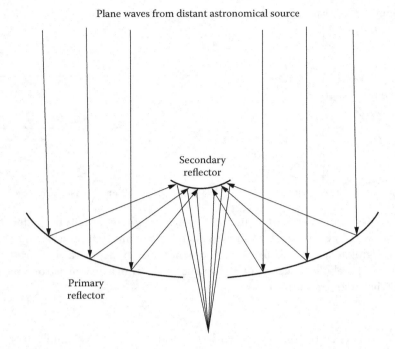

FIGURE 3.2   Optical layout of a Cassegrain radio telescope. Ray tracing illustrates how plane parallel light rays are brought to a common focus, in this case, at a point behind the primary reflector.

FIGURE 3.3   (**See color insert.**) Example of a prime focus radio telescope; Union College's Haystack Small Radio Telescope (SRT). This 2-m diameter radio telescope is made available, at fairly low cost, through MIT's Haystack Observatory radio astronomy education outreach programs. Your school may have a similar one.

FIGURE 3.4 **(See color insert.)** Example of a classical Cassegrain radio telescope. This 20-m diameter radio telescope of the National Radio Astronomy Observatory is located in Green Bank, West Virginia. (Courtesy of B. Saxton, NRAO/AUI/NSF.)

to the physical area of the primary reflector. As given by Equation 2.2, the power, P, of radiation collected from an astronomical source of flux density F_v is given by

$$P = F_v A_{\text{eff}} \Delta v \qquad (3.1)$$

where:
    the bandwidth, Δv, is the range of frequencies being detected

The bandwidth, as we will discuss in Section 3.2, is determined by the receiver. The larger the physical area of the reflector, the more power is collected from an astronomical source and more readily faint sources can be detected. As we will see, a number of factors affect the amount of radiation that enters the receiver and so the effective area of a radio telescope is always smaller than its geometrical area.

The second function of the primary reflector is to provide *directivity*, which is a telescope's ability to differentiate the emission from objects at different angular positions on the sky. When using a single radio telescope to make a map (as discussed in Chapter 4), the directivity determines the *resolution* in the map. The directivity of a telescope depends, largely, on the diameter of the primary reflector and is governed by the principle of diffraction. Diffraction, in fact, limits the directivity of all telescopes. The directivity of a radio telescope is commonly described as the telescope's *beam pattern*, which is the topic of Section 3.1.2.

3.1.2 Beam Pattern

The *beam pattern* is a measure of the sensitivity of the telescope to incoming radio signals as a function of angle on the sky. This is similar to what optical astronomers often call the point-spread function. The term *beam pattern* derives from the idea of a beam of radio waves leaving a *transmitting* antenna. Because the sensitivity pattern is the same, whether the antenna receives or transmits—a principle known as the *reciprocity theorem*—we are free to describe the pattern either way. Ideally, we would like each feed in our telescope to collect radio signals from only one direction in the sky, so that when we point the telescope in a specific direction, the power detected through each feed corresponds to the radiation coming from only that spot in the sky. Unfortunately, this is not possible due to the *diffraction* of light. Diffraction is a phenomenon of all types of waves, whether light waves, sound waves, or water waves, that occurs when the waves pass through an aperture (or opening in which only a fraction of the wave front is not blocked).

A helpful visualization of diffraction is given by the *Huygens–Fresnel principle*, which is depicted in Figure 3.5. In this model, the propagation of a wave is viewed by imagining that each point along the wave front is the source of a secondary spherical wave, and the resulting total wave front is the sum of all these secondary wave fronts, taking into account wave amplitude and phase. You may have seen in a physics class the pattern of light produced when it passes through a single slit. In this case, phase is important, as the secondary waves produced at the slit interfere and add constructively or destructively depending on position. The behavior of light as it reflects off the primary reflector of a telescope is closely related to the behavior of light passing through a single slit. As rays of light from an astronomical source approach the telescope, some rays reflect off the mirror and meet at the focus, while others miss the telescope and hit the ground. The reflector, then, acts like a circular aperture, and the interaction of the light waves meeting at the focus is identical to that when light passes through a circular aperture and meet at a spot on a wall some distance away. In the following paragraphs, we provide a qualitative understanding of how diffraction determines the beam pattern of a reflector. For those who desire a more mathematical proof, Appendix II contains a derivation for the simplified case of a uniformly illuminated rectangular aperture.

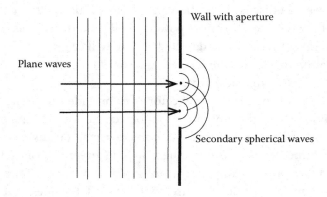

FIGURE 3.5 Huygens–Fresnel principle, in which diffraction of waves is modeled by imagining spherical secondary waves emanating from each point along the front of plane waves.

Consider a prime focus telescope as depicted in Figure 3.6. Light from a distant astronomical source appears as a continuous chain of plane waves and the parabolic reflector brings these light rays to a single focus. Figure 3.6 shows the light rays from an astronomical source located along the optical axis of the telescope, that is, in the direction where the telescope is pointed. To simplify the calculations, we consider first only the contributions from two points on the reflector. The plane waves reflecting off the primary reflector at points b and d, which are separated by distance L, arrive at a common focus at point f. Note that for an on-axis source, the wave fronts are perpendicular to the optical axis and therefore the path length abf is equal to the path length cdf. Therefore, the phases of the waves from these two points are identical (see Figure 3.1, which shows the behavior of the wave fronts) so the waves add constructively at the focus.

What happens to slightly off-axis plane waves, such as those coming from a source that is not along the optical axis? The ray tracing for off-axis waves is shown in Figure 3.7. The plane waves are now not perpendicular to the optical axis, but are tilted by an angle θ. In this case, the path length abf is slightly longer than the path length cdf. Since lines ac and ab are perpendicular, $\sin\theta = \Delta s/L$, and so the path length difference, Δs, is given by the distance from point a to the horizontal dashed line (see Figure 3.7),

$$\Delta s = L \sin \theta$$

For a source located at a small angle, θ, from the telescope axis, we can use the small angle approximation for angles expressed in radians, that is,

$$\sin \theta \approx \theta$$

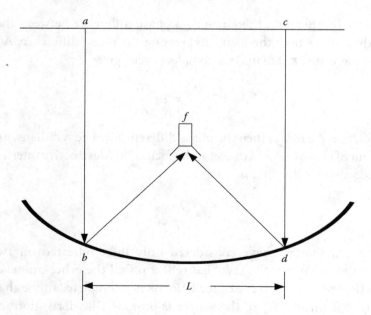

FIGURE 3.6 Path of two light rays, which reflect off two points on the primary reflector (b and d) and meet at the prime focus.

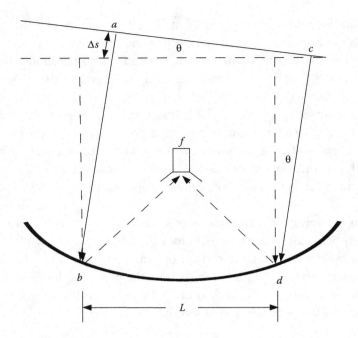

FIGURE 3.7   Ray tracing paths for light coming from a direction offset by angle θ from the optical axis of the telescope.

Therefore, the path difference is approximately

$$\Delta s \approx L\theta$$

Because of this path difference, there is now a phase difference between these two rays of light when they arrive near the focus. Expressing the phase difference, $\Delta\phi$, in radians (remember that there are 2π radians in a complete cycle) gives

$$\Delta\phi = 2\pi\frac{\Delta s}{\lambda} = 2\pi\frac{L\theta}{\lambda}$$

If the path difference, Δs, is $\lambda/2$, then the phase difference will be π radians, and the waves will be exactly out of phase and will cancel at the focus. This destructive interference occurs when the off-axis angular distance is

$$\theta = \frac{\lambda}{2L}$$

Remember, though, that we have considered only light reflected off two opposing points on the reflector. When the rays that reflect off all the other points are included (see Appendix II), one finds that the cancellation is not complete, although the intensity is greatly reduced from that when the source is on-axis. The derivation in the appendix shows that with the case of a uniformly illuminated rectangular aperture, total destructive interference occurs for the sum of *all* the rays when the source is located at λ/D radians from the central axis, where D is the distance across the aperture. For a uniformly illuminated circular aperture, like that of a typical optical telescope, the total collected power is zero when the source is $1.22(\lambda/D)$ radians from the central axis. A sample plot of the sensitivity of a parabolic reflector as a function of angle is shown in Figure 3.8.

In Figure 3.8, there are also some subtle, but important, features well off the central axis. To understand these, return to the situation shown in Figure 3.7, and note that when $\theta = \lambda/L$, the phase difference is 2π radians (one complete cycle), so the waves from these two points are in phase again and add constructively. Considering the entire aperture, though, one finds that there is only partial constructive interference. The sensitivity of the telescope, therefore, has a small increase after it first reaches zero, as shown in Figure 3.8. A telescope, then, can detect power from a source a good ways off-axis, although with much less sensitivity. As the off-axis angle increases further, the response of the telescope goes through a series of peaks and valleys in which there is partial constructive and destructive interference. These off-axis responses are called *sidelobes* and are undesirable as they can add confusion to observations. For instance, we could detect as much power from a bright source located in one of these sidelobes as we detect from a faint source on-axis. The central peak of the sensitivity pattern, in contrast, is often called the *main beam*.

The telescope's sensitivity pattern, like that shown in Figure 3.8, is called the *Airy pattern*. The width of the central peak of the Airy pattern is used to define the angular resolution of a single-dish telescope. By custom, we measure the angular width of this peak between the two points where the received power falls to one-half of the on-axis value. We call this angle the full width at half maximum (FWHM) of the *main beam* of the telescope. Recall that the beam is two dimensional, and so the circular solid angle with a diameter equal to the FWHM of the main beam is considered to be the range over which the telescope can detect radio emission; that is, we can detect off-axis objects anywhere within this solid angle, albeit with reduced sensitivity, if they are located close to or inside the half-power points. Stronger sources may be detectable even

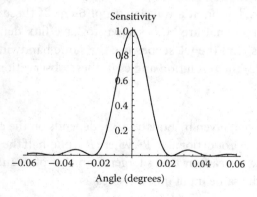

FIGURE 3.8 Typical profile of the relative sensitivity, or response pattern, of a radio telescope as a function of angle, θ, relative to the telescope's central axis. The beam shown is for a uniformly illuminated reflector with a diameter of 40 m observing at 1.4-cm wavelength.

further from the on-axis direction, but customarily we use the half-power points to define the angular dimensions of the beam. This same angle defines the angular resolution of the telescope, because we cannot discern detailed structure of objects at smaller angular sizes. For example, two sources with an angular separation less than the main beam FWHM cannot be distinguished from one another; they appear as a single (and brighter) source.

Our definition of the resolution angle, the FWHM of the main lobe of the antenna pattern, which is fairly standard in the radio astronomy community, is *not* the same quantity used in visible-wavelength astronomy. The canonical resolution angle of a visible-wavelength telescope is $\theta_{res} = 1.22 \, \lambda/D$, which is the expression we provided previously as the angle where the sensitivity pattern (of a circular, uniformly illuminated aperture) first is equal to zero. This is also defined as the radius of the Airy disk. (Note that the plot shown in Figure 3.8 is a one-dimensional profile, while there are two angles in the sky, and so the central peak can be thought of as a disk and the angle to the first null is its radius.) The radius of the Airy disk, in fact, is larger than the FWHM of its central peak, which is equal to $1.02 \, \lambda/D$. (The visible-wavelength astronomy convention follows the *Rayleigh criterion*, which defines the minimum angle of separation between two sources to be resolved as when the center of the Airy disk of one source is in the first null of the other.) Usually a small angular resolution is desirable, as it means that astronomical sources close together in angle on the sky can be distinguished or that fine angular detail can be discerned within a source. Since the FWHM of the main lobe is inversely proportional to the diameter of the reflector, we have that *large diameter telescopes not only collect more power from an astronomical source, but also provide better angular resolution.*

Example 3.1:

Two radio telescopes, *A* and *B*, with perfect reflectors are identical in every regard except for their diameters and observing wavelengths. Telescope *A* has a diameter of 5 m and detects radiation at a wavelength of 2 cm, while Telescope *B* has a diameter of 10 m and detects radiation at a wavelength of 6 cm. If these telescopes are used to observe radio sources that are both known to have flux densities of 1 Jy at the observed wavelengths, and the telescopes use the same bandwidth, compare (1) the power collected and (2) the resolution angles in these observations.

Answers:

1. The power collected, given by Equation 3.1, depends on the collecting area of the telescope, which is proportional to R^2, where R is one-half the diameter. Therefore, telescope B has an area four times larger than that of A and the ratio of power collected (with the same input flux densities) is

$$\frac{P_A}{P_B} = \frac{(5 \text{ m})^2}{(10 \text{ m})^2} = \frac{1}{4}$$

2. The resolution angle is proportional to λ/D. The ratio of the resolution angles, then, is

$$\frac{\theta_A}{\theta_B} = \frac{\lambda_A / D_A}{\lambda_B / D_B} = \frac{\lambda_A}{\lambda_B}\frac{D_B}{D_A} = \frac{2\ cm}{6\ cm}\frac{10\ m}{5\ m} = \frac{2}{3}$$

We see, then, that the larger telescope yields a larger power measurement, but its resolution angle is not as good in this case, and this is because the twofold increase in diameter is more than compensated by the threefold increase in wavelength.

Note that since we only needed to calculate ratios, we did not need to use the full equations relating the calculated values to the parameters. We only needed to set the ratios in terms of their dependences on the variables; the constants in the equations would cancel anyway.

Finally, you may be aware that the Earth's atmosphere has a significant impact on the angular resolution of visible light telescopes. The angular resolution expected from the optics of these telescopes cannot be achieved because of refraction of the light within the atmosphere. In a large portion of the radio window (excluding the lowest and highest frequencies), this refraction effect is small, so radio telescopes generally have *diffraction-limited resolution*, which depends only on the optics of the telescope and the wavelength being observed. At shorter wavelengths, the atmosphere can affect even radio waves, while at longer wavelengths, the ionosphere has a significant impact.

The sensitivity pattern shown in Figure 3.8, and in our discussion so far, is that of a circular aperture that is *uniformly illuminated*, meaning that the same amount of power is collected from equal areas everywhere on the reflector. Although this is generally true for visible-wavelength telescopes, radio telescopes do *not* typically have uniform illumination, as we discuss in Section 3.1.3. We will then revisit the resolution and give modified equations.

3.1.3 Feeds and Primary Reflector Illumination

At the focus of a radio telescope, we need antennas to couple the EM waves in free space into waves confined to transmission lines, so that we can send the waves to the receivers. Each feed is connected to one receiver, each of which produces a single measure of detected power. Typically, these feeds are horn antennas, which can have either rectangular or circular cross sections. They are often flared, with the radiation entering the larger end and tapering down to the proper size (comparable to the wavelength being observed) for a type of transmission line called a *waveguide*. An example of a circular horn antenna is illustrated in Figure 3.9. The flared end of the horn has a size at least as large as the wavelength of the light, though the size will depend on the optical design of the telescope. The minimum size of a feed horn opening at long wavelengths, then, can be quite large, thus limiting the number of feeds that can fit in the focal plane of a radio telescope. For smaller telescopes operating at longer wavelengths, such as the Haystack Small Radio Telescope (SRT), only one feed will fit it the focal plane, and hence the power can be measured at only one position for each pointing of the telescope. In larger telescopes, operating at millimeter and submillimeter wavelengths, an array of feeds can often be used, permitting many positions to be observed simultaneously in a single pointing.

FIGURE 3.9 Example of a circular feed horn used to couple the radio light to a transmission line. In this case, a waveguide, connected to the rear of the horn, serves as the transmission line.

Each feed and receiver work well only for a certain range of frequencies. If observations involving a number of different frequency ranges are needed, then different feed horns and receivers must be employed for each frequency band.

The radiation reflected off the dish enters the feed horn through a finite-sized opening, approaching from many different angles, and is then combined inside the feed. Diffraction, therefore, again determines the amount of power the feed collects and passes onto the receiver. Using the reciprocity theorem, we can model the feed horn's sensitivity to the incident radiation in terms of its own beam pattern. Considering our discussion in Section 3.1.2 of the telescope's beam pattern, in which we found that radiation entering from different directions has different amounts of constructive or destructive interference, we see that the feed horn's sensitivity to radiation reflecting off different parts of the dish is *necessarily nonuniform*. The reciprocity theorem also enables us to visualize how the feed horn and reflector work together by imagining a signal generated in the receiver and considering how it would be transmitted to the sky. This angular pattern of transmitted power is exactly equivalent to how the system responds to radiation power entering the system from the sky as a function of angle. Diffraction determines the pattern of power transmitted by the feed horn, and since the feed horn is relatively small, the power is transmitted over a large angle, illuminating the primary reflector. The beam pattern of the feed, then, determines what is known as the *illumination pattern* of the primary reflector.

Ideally, we would like a feed-horn beam that had as close to uniform sensitivity out to the edge of the dish as possible, as this would yield a maximum response to the radiation from the source. However, we do *not* want the feed horn to detect contaminating background radiation coming from beyond the edge of the dish, that is, that coming from the ground. This, though, would require a feed-horn beam that is approximately flat for a large range of angle and then suddenly (and discontinuously) falls to zero at the edge of the dish, and there is no physical beam pattern that can do this. Therefore, we must either *under-* or *over-illuminate* the dish. In the former case, the sensitivity falls so rapidly that we do not make full use of the area of the dish. In the latter case, we have good sensitivity over the full dish area, but allow radiation from beyond the dish to enter the feed horn.

A quantity that describes how the feed horn's beam is distributed on the primary reflector is called the *edge taper*, which is defined as the ratio of the sensitivity at the center of the reflector to that at the edge. This is determined by the angular width of the feed horn's beam pattern, which, like a telescope's beam pattern, is proportional to λ/D, where D, now, is the size of the large flared end of the horn. Therefore, we obtain a large edge taper, meaning that the telescope is less sensitive to the power per unit area reflecting off the edge of the reflector than that at the center, by using a *large* feed horn. This is because a large feed horn produces a small angular pattern, so the power falls off more quickly with increasing distance from the center of the reflector. The concept of edge taper is illustrated in Figure 3.10, where Figure 3.10a shows the result of having a large edge taper (under-illumination) and Figure 3.10b shows the result of a small edge taper (over-illumination).

The shape of the illumination pattern on the primary reflector affects (1) the angular resolution of the telescope, (2) the sensitivity level in the sidelobes, and (3) the effective collecting area of the telescope. Unfortunately, there is no illumination pattern that optimizes all of these attributes simultaneously.

Of these three issues, the most important, usually, is maximizing the effective collecting area of the telescope, since this determines the telescope's ability to detect faint sources. On one hand, with a large edge taper much of the power coming from the outer regions of the

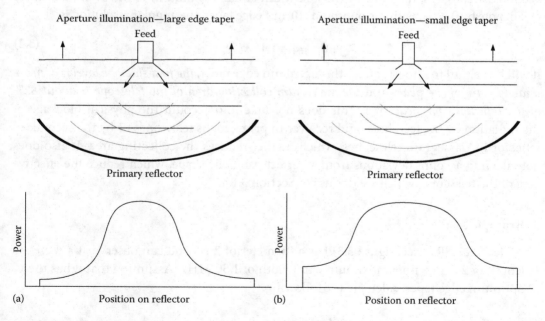

FIGURE 3.10 Schematic showing plots of the illumination pattern (the pattern of power transmitted by the feed horn) resulting from (a) a large feed opening and (b) a small feed opening. The illumination pattern in (a) is said to have a large edge taper because it has a large ratio of power transmitted to the center relative to the edge.

reflector is not detected, and so the effective collecting area of the telescope is reduced. The ratio of the effective collecting area considering this effect to the physical area is called the *illumination efficiency*. On the other hand, with a small edge taper, since the illumination is still fairly large at the edge of the reflector, some of the feed's beam pattern misses the reflector, and so the signal entering the feed is diluted by its sensitivity to area beyond the physical reflector. The illumination of the feed beyond the reflector is called *spillover*. Thus, a large edge taper optimizes the spillover efficiency, while a small edge taper optimizes the illumination efficiency. The maximum collecting area results from a compromise between spillover and illumination efficiency. The edge taper that maximizes the collecting area of the telescope is one in which the power per unit area transmitted to the center of the reflector is 10 times larger than that at the edge; this is called a *10-dB edge taper.*

With regard to the other two issues—resolution and sidelobe levels of the telescope beam pattern—there are, again, two competing factors. The more uniform illumination produced by a smaller edge taper yields a higher angular resolution, as this is equivalent to having effectively a larger primary reflector. However, a small edge taper means that the sensitivity of the feed to power reflected off the primary reflector stays large right up the edge of the reflector and so there must be a sharper truncation of the illumination pattern at the edge, and this produces larger sidelobe levels. A larger edge taper, on the other hand, minimizes the sidelobe level, but produces poorer angular resolution. The edge taper that maximizes the effective collecting area, the 10-dB taper, fortunately, is also a good compromise between good resolution and low sidelobe level. Using the definition of resolution of a radio telescope, given in Section 3.1.2, that is, the FWHM (for full width at half maximum) of the beam, which we will denote by θ_{FWHM}, with the optimum (10-dB) edge taper, the angular resolution is

$$\theta_{\mathrm{FWHM}} = 1.15\lambda/D \qquad (3.2)$$

It will be helpful to know that, for this optimum edge taper, *the first sidelobe level is approximately 0.4% of the peak,* and *the maximum collecting area of the telescope is about 82% of the reflector's physical area.* This does not take into account the physical blockage of the radiation by the feed horn and receiver in prime focus telescopes or by the secondary reflector in Cassegrain telescopes, which further reduces the collecting area. It also does not take into account deviations from a perfect parabolic shape, which reduce the effective area of the telescope, which we discuss in Section 3.1.4.

Example 3.2:

A Haystack SRT (see Figure 3.3) has a diameter of 2 m, and can observe at a wavelength of 21 cm with a maximum bandwidth of 1.50 MHz. Assuming that it has the optimum edge taper, calculate the following:

1. The angular resolution, in degrees
2. The maximum collecting area
3. The maximum detected power from a 1-Jy source located at the peak of a sidelobe

Answers:

1. The resolution with optimum edge taper is given by Equation 3.2; therefore, we have

$$\theta_{FWHM} = 1.15(0.21\text{ m})/(2.00\text{ m}) = 0.121\text{ radians} = 6.92°$$

2. We are given the edge taper but not any information about the physical blockage by the feed horn and receiver, nor about the imperfections in the reflector, so the maximum collecting area is the geometric area, πR^2, multiplied by 0.82 due to the edge taper. We have then

$$\text{Max } A_{eff} = 0.82\pi(1.00\text{ m})^2 = 2.58\text{ m}^2$$

3. At the center of the beam, the detected power is given by Equation 3.1. At the peak of the sidelobe with the optimum edge taper, the sensitivity is 0.4% of that at the peak. So, the maximum detected power due to the 1-Jy (we must remember to convert Jy to SI units) source in the sidelobe is

$$P = 0.004 \times 1\text{ Jy } \left(10^{-26}\text{ W m}^{-2}\text{ Hz}^{-1}\text{ Jy}^{-1}\right) \times 2.58\text{ m}^2 \left(1.5 \times 10^6\text{ Hz}\right)$$

$$= 1.55 \times 10^{-22}\text{ W}$$

This is the maximum detected power, because the true effective area is smaller than our calculated value in 2.

3.1.4 Surface Errors

The primary reflector of a radio telescope is never a perfect parabola. There are always manufacturing imperfections that limit its surface accuracy. We can characterize an imperfect reflector by the root mean square (rms) deviations, δz, of the real surface from that of an ideal parabola measured parallel to the optical axis. Such deviations will cause the path length to the focus to be slightly different for various parts of the reflector. This effect is sketched in Figure 3.11. As we saw in Section 3.1.2, path differences cause phase differences that produce less than full constructive interference; therefore, these deviations reduce the power collected by the telescope. Because the light is reflected off the surface, the total path difference is twice the deviation, $2\delta z$; therefore, these deviations produce *rms* phase errors of $4\pi\, \delta z/\lambda$.

The presence of surface errors therefore reduces the on-axis sensitivity of the telescope, and this can be viewed as a loss in the collecting area. The effect of surface errors on the collecting area is described by the *Ruze equation*, which is given by

$$A_\delta = A_0 e^{-(4\pi\delta z/\lambda)^2}$$

where:

A_δ and A_0 are the collecting areas of a telescope with and without surface errors, respectively

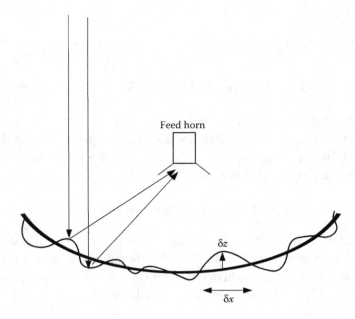

FIGURE 3.11   Surface irregularities on the reflector cause deviations from a perfect parabola with *rms* values δz and δx in the directions parallel and perpendicular, respectively, to the optical axis.

The Ruze equation enables us to compute the fractional loss in collecting area in terms of the surface deviations in units of the observing wavelength. When $\delta z/\lambda = 0.1$, the surface is said to have *1/10th wave accuracy*. Table 3.1 illustrates the relationship between $\delta z/\lambda$ and the decrease in effective collecting area.

It is clear by examining Table 3.1 that a reflector should have *rms* surface errors less than 1/20th of the wavelength of the light being detected to have a reasonable performance. Although the surface accuracy needed for a radio telescope to perform well is substantially more relaxed than that for a visible light telescope, radio telescopes need to be much larger to achieve a useful resolution, and therefore meeting this constraint turns out to be of similar difficulty.

We can gain a better understanding of how the surface irregularities cause a loss of sensitivity by, again, considering the radio telescope as a transmitter, and asking where the lost power goes when there are surface errors. These deviations can also be described by their *rms* length, δx,

TABLE 3.1   Ratio of the Effective Collecting Area Taking into Account Losses due to Surface Irregularities of Size δz to the Geometrical Area for an Assortment of Values of $\delta z/\lambda$, According to the Ruze Equation

$\delta z/\lambda$	A_δ/A_0	$\delta z/\lambda$	A_δ/A_0
1/4	0.00005	1/15	0.50
1/6	0.012	1/20	0.67
1/8	0.085	1/30	0.84
1/10	0.21	1/50	0.85

perpendicular to the optical axis, which are also depicted in Figure 3.11. The length, δx, for manufacturing errors is typically much smaller than the diameter of the reflector and so reflection of radiation off these deviations will cause some of the power to be transmitted into a very large beam, with angular size of order $\lambda/\delta x$. When the telescope is receiving radiation from an on-axis source, then the response of the telescope is diluted by its sensitivity to much larger angles. The large beam due to the surface irregularities is called the *error pattern*. Therefore, the sensitivity patterns of radio telescopes involve not only unwanted sidelobes, but also these error patterns. Because the error pattern typically has a much larger angular scale than the main beam, the sensitivity in the error pattern to any one direction in the sky is very small, but since it has a large angular extent, a significant amount of total power can be lost. In addition to manufacturing errors, there may be larger scale deformations of the reflector due to gravity. These deformations introduce errors with much larger length scales than manufacturing errors and thus produce error patterns on smaller angular scales.

In summary, there are a number of reasons as to why the effective collecting area of a telescope is smaller than the physical area of the reflector. These include the loss due to limited surface accuracy, the loss due to the illumination pattern, and the loss due to physical blockage by the feed horn and receiver or by the secondary reflector. One should not be surprised that large radio telescopes have effective collecting areas that are only about one-half of the physical area of their primary reflector.

Example 3.3:

Assume that the SRT discussed in Example 3.2, which is a prime-focus telescope of 2-m diameter, has an *rms* of the reflector irregularities of approximately 0.7 cm, and that the total loss due to blockage due to the feed, receiver, and feed legs is 12% of the collecting area. Assuming optimal feed taper again and observations at 21 cm with a bandwidth of 1.5 MHz, calculate the following:

1. The effective collecting area
2. The detected power when observing a 1-Jy source at the center of the beam

Answers:

1. In Example 3.2, we calculated the maximum effective collecting area of the SRT taking into account only the illumination efficiency to be 2.58 m². Now we can determine and include the reduction in effective collecting area including the other two factors. The blockage introduces a factor of 0.88. We calculate the reduction due to the surface irregularities using the Ruze equation,

$$\frac{A_\delta}{A_0} = e^{-[4\pi(0.7 \text{ cm}/21 \text{ cm})]^2} = 0.84$$

Our final estimate of the effective area of the SRT, then, is

$$0.88(0.84)(2.58 \text{ m}^2) = 1.91 \text{ m}^2$$

2. We can now use this area in Equation 3.1 to calculate the detected power of a source at the center of the beam. We get

$$P = (1 \text{ Jy})(10^{-26} \text{ W m}^{-2} \text{ Hz}^{-1} \text{ Jy}^{-1})(1.91 \text{ m}^2)(1.5 \times 10^6 \text{ Hz}) = 2.86 \times 10^{-20} \text{ W}$$

3.1.5 Beam Pattern Revisited*

Our discussion of diffraction in Section 3.1.2 provides a qualitative understanding of why a radio telescope has a beam pattern, of its general shape, and how the width of the beam is related to the size of the telescope. We have not, though, provided a prescription for determining the beam pattern. A more precise calculation of the beam pattern of a radio telescope can be made using the Fourier transform. Modeling the telescope as a transmitter, we can imagine a signal generated in the receiver and transmitted out the feed horn and reflected off the dish and into the sky. That signal, of course, involves an electric field, and so one can describe the transmitted signal in terms of the electric field as a function of angle far from the telescope, often referred to as the *far-field*. Furthermore, the beam of the feed horn produces an electric field as a function of position in the front plane of the reflector, which is the same as the *aperture* in the diffraction considerations we discussed in Section 3.1.2. We state, but do not prove, that the electric field as a function of angle in the far field, which we can write as $E_{ff}(\theta)$, is the Fourier transform of the electric field as a function of position in the aperture divided by the wavelength, or $E_{ap}(x/\lambda)$, where x is the position relative to the center of the reflector. This Fourier transform relation is a result of the Huygens–Fresnel principle and the far-field solution is often called Fraunhofer diffraction. Note that the independent variables of the Fourier transform functions are angle in the sky, θ, and linear distance from the center of the aperture divided by wavelength, x/λ, which we will denote as x_λ. However, both the sky and the aperture are two dimensional, and so the electric fields defined in the *aperture plane* and the *sky plane* are related by a two-dimensional Fourier transform.

Since the power in an EM wave is proportional to the square of the electric field, the square of the electric field in the aperture plane is just the power distribution that we referred to earlier as the illumination pattern. For the 10-dB edge taper that we discussed in the previous section, the power at the edge of the reflector is one-tenth at the center; therefore the electric field at the edge of the reflector would be about one-third at the center. Likewise, the distribution of transmitted power in the sky is just the square of the electric field in the far field. By the reciprocity theorem, the distribution of transmitted power in the sky is the same as the angular sensitivity of the radio telescope to receiving the power from an astronomical source. In general, the Fourier transform of the electric field in the aperture plane can be very complicated, and only in a few simple cases is there an analytical solution.

To demonstrate the power of this Fourier transform relation, we consider a simple example—a uniformly illuminated one-dimensional aperture of width a, a plot of which is illustrated in Figure 3.12.

Writing the electric field in the aperture plane as $E_{ap}(x_\lambda)$ and the angular distribution of the electric field in the far-field as $E_{ff}(\theta)$, the Fourier transform relationship is represented by

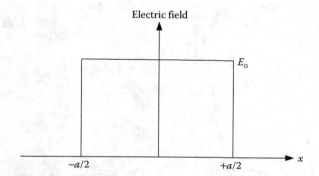

FIGURE 3.12　Electric field distribution in a uniformly illuminated one-dimensional aperture.

$$E_{\text{ff}}(\theta) = \int\limits_{-\infty}^{\infty} E_{\text{ap}}(x_\lambda)e^{-i2\pi x_\lambda \theta}dx_\lambda$$

The electric field for an aperture of width a with uniform illumination is

$$E_{\text{ff}}(x_\lambda) = E_0 \text{ (a constant), for } -a_\lambda/2 < x_\lambda < +a_\lambda/2$$
$$= 0 \text{ for } x_\lambda < -a_\lambda/2, \text{ and } x_\lambda > +a_\lambda/2$$

where:

$$a_\lambda = a/\lambda$$

Our equation becomes, then,

$$E_{\text{ff}}(\theta) = \int\limits_{-a_\lambda/2}^{a_\lambda/2} E_0 e^{-i2\pi x_\lambda \theta}dx_\lambda$$

Integrating and applying the appropriate limits of integration, we find

$$E_{\text{ff}}(\theta) = \frac{i}{2\pi\theta}E_0\left(e^{-i\pi a_\lambda \theta} - e^{+i\pi a_\lambda \theta}\right)$$

Using Euler's equation, we can rewrite this as

$$E_{\text{ff}}(\theta) = \frac{i}{2\pi\theta}E_0\left(\cos(\pi a_\lambda \theta) - i\sin(\pi a_\lambda \theta) - \cos(\pi a_\lambda \theta) - i\sin(\pi a_\lambda \theta)\right)$$

or

$$E_{\text{ff}}(\theta) = a_\lambda E_0 \frac{\sin(\pi a_\lambda \theta)}{\pi a_\lambda \theta}$$

which is identical to Equation II.8 in the appendix ($E_{\text{max}} = a_\lambda E_0$ and occurs at $\theta = 0$). The function $\sin(x)/x$ is called the sinc function and has a value of 1 for $x = 0$. Therefore, the far-field power pattern or beam pattern is given by $P(\theta) \propto [\sin(\alpha)/\alpha]^2$. A plot of the relative

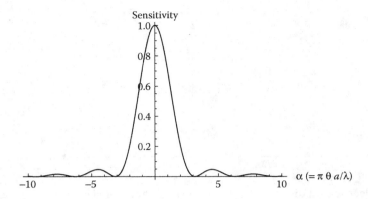

FIGURE 3.13 Power pattern for a uniformly illuminated one-dimensional aperture. On the y-axis is the relative power, normalized to be unity on-axis, and on the x-axis is the quantity $\alpha = \pi a \theta / \lambda$.

beam pattern for a uniformly illuminated aperture of size a is shown in Figure 3.13. For this idealized one-dimensional case, the angular resolution (defined by the full-width-at-half-maximum of the beam) is

$$\theta_{\mathrm{FWHM}} = 0.89 \, \lambda / a$$

and the first sidelobe level is at 5% of the peak response.

For two-dimensional apertures, even circular apertures with uniform illumination, the Fourier transform calculations are much more complicated and involve Bessel functions. Remember, however, that for radio telescopes, the illumination of the primary reflector will never be uniform and the calculation of the far-field radiation pattern is even more complicated. See, for example, the power distribution shown in Figure 3.10. One can approximate the power pattern of the feed horn antenna as a Gaussian, and thus the power pattern on the primary reflector would be a truncated Gaussian, with the power level at truncation depending on the amount of edge taper. The response patterns for such telescopes have been computed numerically (a good book covering this subject is *Quasioptical Systems: Gaussian Beam Quasioptical Propagation and Application* by Paul Goldsmith[*]). It is from these numerical calculations that the optimum edge taper, discussed in Section 3.1.3, was deduced.

3.2 HETERODYNE RECEIVERS

The purposes of the radio telescope *receiver* are to define the frequency range, or *passband*, over which the received power will be collected, and to produce a signal proportional to the collected power that can be recorded. The components that make up a receiver are often divided between two separate sections referred to as the *front-end* receiver and the *back-end* receiver. These names arise because it is usually the case that the front-end components are located very near the focus of the telescope, while the back-end receiver components are

---

[*] Paul Goldsmith, 1998. *Quasioptical Systems: Gaussian Beam Quasioptical Propagation and Application.* Wiley-IEEE Press, New York.

located off the telescope and could be some distance away. A schematic of the components that make up a typical radio-wavelength receiver system designed to detect broadband astronomical signals is shown in Figure 3.14. There are a number of components shown in Figure 3.14 that affect the radiation as it travels from the feed to the recorder/computer and which we need to explain. First, though, we must consider the means by which the signal travels through this system, from the feed, to the front-end receiver components and then to the back-end receiver components, that is, the *transmission lines*.

3.2.1 Transmission Lines

There are several types of transmission lines, but they have in common the ability to transmit EM waves along their length. Two examples of commonly used transmission lines are *waveguides* and *coaxial cables*, such as those illustrated in Figure 3.15. In all transmission lines the electric and magnetic fields oscillate sinusoidally in time and position, the

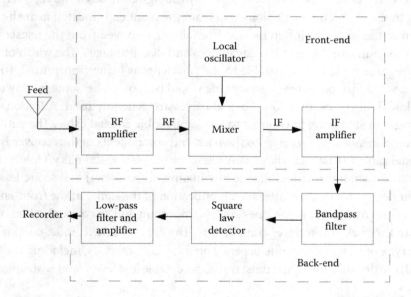

FIGURE 3.14   Schematic diagram showing components of a radio telescope's receiver, front-end and back-end.

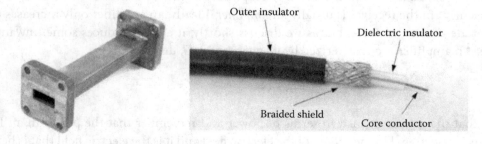

FIGURE 3.15   Examples of two types of transmission lines: a rectangular waveguide on the left and a coaxial cable on the right.

same as in free space. Additionally, all of the spectral information that was present in the waves propagating in free space is preserved in the waves propagating in bound modes (modes confined within the waveguide or coaxial cable). Generally, transmission lines function in what is referred to as single mode. For simplicity we only need to know that this means that just one of the two polarizations is allowed to be transmitted, and so half the power from an unpolarized source is passed through a single transmission line. Two lines must be provided to pass the full received power.

Most feed horns narrow down to a waveguide and so the transition from feed to waveguide occurs naturally. Sometime it is advantageous to transition from waveguide to co-axial cable and this can be accomplished by placing a short monopole antenna inside the waveguide and connected to the core conductor of the coaxial cable, while the metal structure of the waveguide itself is connected to the braided shield.

Transmission lines are needed through the entire trip made by the EM radiation from the feed to the detector. They are needed to carry the signal through the receiver, from one component to the next, and for the long trip from the front end (located near the telescope focus) down to the back end (often located in a lab, far-removed from the telescope).

Each type of transmission line has advantages and disadvantages. The width of the opening in a waveguide must be comparable to the wavelengths being transmitted, so at longer wavelengths, which can be meters, a waveguide would be too cumbersome for any reasonable design. At low frequencies, therefore, coaxial cables are generally used. Additionally, waveguides are rigid in structure, in contrast to the semiflexible coaxial cable. The path from the front-end components of the receiver to the back-end components of the receiver is generally in the co-axial cable so that it can flex as the telescope moves about the sky. However, co-axial cables have substantially more loss (due to attenuation of the signal) and so are less desirable than waveguides. As we will see, after the amplification of the signal in the front-end components, losses from the transmission lines are less important and so co-axial cables often carry the signals to the back-end receiver components. The choice of waveguide, co-axial cable, or some other type of transmission line depends on a variety of factors, including the frequency to be passed, the distance the signal must travel, acceptable loss levels, and cost, among others.

3.2.2 Front-End Receiver Components

The front-end receiver components perform several important signal processing functions. The first task, usually, is to increase the amplitude of the input radio waves, and so the first element in the receiver is usually an *amplifier*. Ideally, an amplifier only increases the amplitude of the EM wave, but as we discuss shortly, it also introduces some unwanted noise. An amplifier is characterized by its *power gain, G*, defined by

$$G \equiv \frac{P_{\text{out}}}{P_{\text{in}}} \tag{3.3}$$

Note that the gain is defined in terms of power and remember that the power in an EM wave is proportional to the *square* of the electric field and it is the electric field that is being amplified. So, if an amplifier has a power gain of 100, both the electric and magnetic fields are amplified by a factor of 10.

The power gain of an amplifier is usually expressed in logarithmic units called *decibels* (dB), which is related to the actual gain by

$$G(\text{dB}) = 10 \log_{10} G$$

This is a convenient way of expressing the total gain when there are multiple amplifiers involved. If we have two amplifiers in succession, for example, the output of the first is amplified by the second. If the gain in the first amplifier is G_1 and that of the second is G_2, then the total amplification of the initial signal is $G_1 G_2$. In terms of decibels, the gains simply add, so that the total amplification is $G_1(\text{dB}) + G_2(\text{dB})$. For example, if we have a 30-dB amplifier followed by a 40-dB amplifier, the total gain is 70 dB, or a gain of 10^7.

After the initial amplification, it is usually essential to convert the signal to a different (and usually much lower) frequency as early in the process as possible. Higher frequency EM waves experience greater loss of power in the transmission line, and components for high frequencies are more difficult and more costly to build. Moreover, if additional amplification is required, it is often convenient to introduce this amplification at a different frequency, to improve the stability of the system. This conversion to another frequency is accomplished by a device called the *mixer*, where another signal of slightly different frequency, created by the *local oscillator* (LO), is used to shift the input signal by a process that we will explain shortly. The term *heterodyne* relates to the idea of combining or mixing two different signals together to accomplish a conversion in frequency.

Throughout the following discussion, keep in mind that a broad range of radio frequencies (RF) enters the antenna and that each frequency is affected by the components similarly, including the frequency conversion. For simplicity, we describe what happens with a single RF input, leading to a single lower frequency output, but, in reality there is a broad range of output frequencies.

The input radio frequencies to the receiver (i.e., the RF from the source being observed) are commonly abbreviated as the *RF*. The output of the mixer is a band at (usually) lower frequencies that we call the intermediate frequencies, or *IF*. (Further mixing downstream often occurs and so the output from the mixer in the front end is considered an intermediate step.) Thus, the mixer combines the RF and LO signals and produces an IF signal that carries the same information as the RF signal, but at a different frequency.

To follow our explanation of the mixing process, you should find the illustration in Figure 3.16 to be helpful. Again for simplicity, we show at the top of this figure a single-frequency RF signal. The LO produces an adjustable (or tune-able), single-frequency sinusoidal wave. For typical mixers, every time the LO waveform changes sign, the mixer changes from a low-resistance state to a high-resistance state. Thus, the mixer acts somewhat like a switch, opening and closing in response to the LO waveform. This behavior, for an ideal switch, is illustrated in the second row of Figure 3.16. The RF waveform is either blocked or passed by the switch.

For RFs that differ slightly from the LO frequency, the output waveform, shown in the third row of Figure 3.16, is very messy and contains many frequencies. Before leaving the mixer, though, the signal is sent through a *low-pass filter* which blocks all the high frequencies, allowing only low frequencies to continue on for further signal processing.

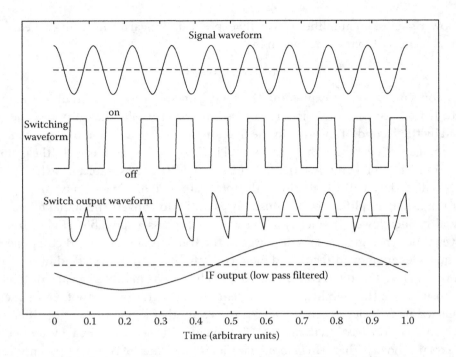

FIGURE 3.16 Model of a mixer as an ideal switch. At the top is shown the RF waveform and below that is the state (on or off) of the switch driven at the LO frequency, which is slightly different from that of the RF. The resulting switched output is shown next and is quite messy. However, after the low-pass filter, the difference frequency or beat frequency remains. (Courtesy of Neal Erickson.)

These low-frequency signals occur at frequencies equal to the difference between the RF and LO frequencies and are represented by the clean waveform shown at the bottom of Figure 3.16. This is the IF that exits the mixer. The IF signal has all of the spectral information that was in the corresponding RF signal, just shifted to a lower frequency range. Real mixers, of course, do not involve such ideal switches, since the change in resistance from high to low cannot occur instantaneously like the step function shown in Figure 3.16. The resistance change, instead, must be continuous, but the desired end result is still achieved— that the output *IF is equal to the difference between the LO and RF.*

A more complete analysis (which we do not present here) shows that in practice, multiple combinations of the RF and LO frequencies are present at the IF. In particular, frequencies that are integer multiples of the RF and LO frequencies and their sums and differences are formed by the mixer. These additional frequencies are undesirable, and in a well-designed mixer they will be quite weak. Furthermore, the bandpass filter at the start of the back-end, discussed in the next section, will block these unwanted frequencies.

The combining of the RF and LO signals to produce an output frequency equal to the difference of the two is the same process that produces a beat frequency. You may have seen an explanation of the beat frequency in a class that discusses waves or you may have some familiarity with it from playing a stringed musical instrument. If you have tuned your instrument relative to another, you probably have heard the beat frequency. When the two instruments are slightly out of tune relative to each other, meaning that the frequencies of their notes differ by a small amount, and they are strummed simultaneously, in addition to the two notes you can hear a very low-frequency amplitude oscillation; this is the beat frequency of the combined sound.

The LO frequency is adjustable and is generally set close to the frequency of radiation that one wants to detect. So, for instance, if we are interested in a 60-MHz wide frequency band, centered at an RF of 1420 MHz, we might set the LO frequency to 1370 MHz, and then this RF band is *mixed down* to a new center frequency of

$$1420 - 1370 \text{ MHz} = 50 \text{ MHz}$$

The IFs output from the mixer that we wish to detect, then, are centered at 50 MHz and range from 20 to 80 MHz.

In simple mixers, the output IF is simply the difference between the LO frequency and the RF, regardless of which is larger. In the example above, for instance, an IF of 50 MHz is the output from combining the LO of 1370 MHz with an RF of 1320 MHz as well as 1420 MHz. Such mixers are called *double-sideband mixers*, and the RF signals above the LO frequency make up the *upper sideband*, and those below are in the *lower sideband*. Both these RF signals will be mixed to the same IF frequency. This is not a problem if the purpose of the observation is to measure the continuum power over a large range of frequencies. In fact, this effect doubles the bandwidth and so increases the sensitivity of the observation. However, this can be a serious problem for spectral line observations because the measured power in the spectral line will include, and be contaminated by, that from the other side-band. *Single-sideband mixers* are designed to eliminate the unwanted sideband and avoid the confusion of which sideband a spectral line comes from and are more commonly used.

In general, for any given IF, the RF that it corresponds to is given by

$$\nu_{RF} = \nu_{LO} \pm \nu_{IF} \qquad\qquad (3.4)$$

where:
+ corresponds to the upper sideband
− to the lower sideband

With a double-sideband mixer, there are two possible RFs for each IF. With a single-sideband mixer, we use the solution for the sideband selected.

Example 3.4:

While making an observation with the LO set to 1350 MHz in a double-sideband mixer, a whopping signal is detected and determined to appear at an IF of 100 MHz. Since the signal is so strong and unexpected, it is suspected that this may be an interference signal. What are the possible frequencies at which the interference source is emitting?

Answer:

The interference might be entering the system late in the path and so the frequency it appears at may be its actual frequency. So, one possibility is 100 MHz. However, more likely, the interference signal can be collected by the reflector and enter the receiver just as an astronomical signal would. In this case, the interference signal is at the RF that gets mixed down with the IF at which it appears. However, since this is a double-sideband mixer, there are two possible RFs (given by Equation 3.4) and these are 1450 and 1250 MHz. In sum, then, the likely suspects for the interference frequency are 100, 1250, and 1450 MHz.

To ultimately make a detection of this signal we still need more amplification and this is more conveniently done at the IF. The mixer is, therefore, followed by an IF amplifier, which is different from the RF amplifier in that it is designed to amplify the lower frequency, IF, EM waves.

As a final statement about the front-end receiver components, note that because the RF amplifier, mixers, and often the transmission lines only work over a finite range of frequencies, different receivers *must* be constructed to detect radio light at different wavelengths. Therefore, a radio telescope that is to be used at different frequency bands will have several different receivers and feeds. When changing observing frequencies, the feed/receiver assemblies must also be changed. In most professional telescopes, several feed/receiver systems are mounted together on the telescope and can be switched in and out in a matter of minutes.

3.2.3 Back-End Receiver Components

The output of the front end is still an EM wave, just amplified and translated to a lower frequency. Now we want to *detect* these waves, which we accomplish by measuring the amount of power they contain. However, before we make any detection, we need to define precisely the range of frequencies that we will detect. This is accomplished with a *bandpass filter*, which is the first element shown in the back-end in Figure 3.14. Such a filter only allows EM waves in a well-defined range of frequencies to pass through and it rejects all waves outside these frequencies. There are two reasons why this filtering is important. First, undesired signals might be present, and we don't want to include them in the power measurement (we mentioned previously, some of these unwanted signals as part of the output to the mixer). Second, because we are measuring a flux *density*, we must know the precise range of frequencies that are included.

In the example given earlier, we were interested in measuring the power from an astronomical source centered at 1420 MHz covering a bandwidth of 60 MHz, and we mixed this RF signal with a range of IFs centered at 50 MHz covering a passband of 20–80 MHz. In this example we would want to use a bandpass filter of width 60 MHz centered on a frequency of 50 MHz. If the range of IFs starts at 0 Hz, this is called the *baseband*. In this example, the baseband would run from 0 to 60 MHz and so we would need to set the LO frequency to 1390 MHz.

Note that by tuning the LO frequency, we can change what RFs get mixed down to the IFs that will be passed by the bandpass filter, and thus change what range of frequencies we detect from the astronomical source. So, if we changed the LO to 1400 MHz, then with the IF centered at 50 MHz we would detect the power from the radio source in a 60 MHz bandwidth now centered at an RF of 1450 MHz. If multiple bandpass filters are available, then the observer can also adjust the total bandwidth.

Example 3.5:

Imagine that we want to measure the continuum emission from a source at a frequency of 1750.0 MHz, but there is also an emission-line source in the foreground in the line of sight. The foreground source emits no continuum and has two narrow emission lines that appear at 1725.0 and 1765.0 MHz. And, imagine that we have access to a telescope that observes at this frequency that has a single-sideband mixer and has a number of IF filters so that we can choose any of the following bandwidths: 7.50, 15.0, 30.0, or 60.0 MHz and the central IF is set at 30 MHz. What would be a good LO frequency and bandwidth to optimize this observation?

Answers:

Our goal is to set our LO and bandwidth so that our measured power is due solely to our source of interest. Therefore, we want to avoid both emission lines. Since the emission lines bracket the frequency we wish to observe at and are separated by 40 MHz, we can set our bandpass to occur between the lines, but our bandwidth must be less than the line separation frequency. But, otherwise, we want a large bandwidth to maximize the sensitivity of the observation (Recall from Equation 3.1 that the measured power depends on bandwidth). Of the possible bandwidths, the largest is too large, while the second largest, 30 MHz is narrow enough to avoid the lines and so we'll go with that one. An RF bandpass that does the job for us, then, is from 1730 to 1760 MHz. So, we want the central RF in this band, 1745 MHz, to be mixed down to the central IF. An LO frequency of 1745 MHz − 30 MHz = 1715 MHz will produce the IF range that we want in the upper sideband. In sum, then, we can ask for

$$\Delta v = 30 \text{ MHz and LO} = 1715 \text{ MHz}$$

Now we can explain *how* the detection of the power occurs. This is normally accomplished with a device called a *square-law detector*, which is composed of either a crystal diode or a semiconductor diode. For signals that are not too big or too small, these diodes produce a current proportional to the square of the electric field of the input EM wave and therefore proportional to the power in the incoming waves. This current then passes through a given resistance, and by Ohm's law then the resulting voltage out of the detector is also proportional to the power.

Imagine, for illustrative purposes, an EM wave of a single frequency entering a square-law detector. The output voltage is proportional to the square of the electric field, so

$$V \propto E^2 \propto \left(E_0 \cos(\omega t) \right)^2 \propto E_0^2 \cos^2(\omega t)$$

Making use of the trigonometric identity mentioned in Chapter 2 (Section 2.5), we can rewrite this as

$$V \propto E_0^2 \frac{1}{2} \left[1 + \cos(2\omega t) \right]$$

The output of the detector is then sent into a *low-pass filter*, which removes the high-frequency variations (at frequency 2ω) while preserving the DC signal, thereby leaving just the factor in front of the brackets, which, we can see, is proportional to E_0^2 and hence to the power.

For a square-law detector, then, the output voltage is given by

$$V = \alpha P$$

where:
  P is the input power
  α is the *responsivity* of the detector (the constant for the conversion of power into voltage)

There is a minimum power required for these detectors to function that typically is of order 10^{-6} W. The power received from an astronomical source is many orders of magnitude smaller than this, which explains the need for all of the RF and IF amplification. A total amplification of order 90 dB (or a factor of a 10^9!) is typically needed ahead of the detector.

The responsivity of a typical detector is of order 100 V W^{-1}, so for a typical input of 10^{-6} W, the output voltage is only 0.1 mV, which is rather small. Therefore, we usually follow the low-pass filter with a DC amplifier that boosts the voltage so that it can be readily digitized by an A/D (analog to digital) converter and recorded by a computer.

3.2.4 High-Frequency Heterodyne Receivers

As we will explain in the next section, the receiver components, especially the amplifiers, also produce unwanted power signals, called *noise*. We will discuss how we can minimize it. We will show that the amount of noise power produced by the first element in the light path, which in the heterodyne receiver outlined in Figure 3.14 is the RF amplifier, is the most crucial. The RF amplifier, therefore, needs to be of relatively low noise output. At frequencies above about 300 GHz (or wavelengths shorter than about 1 mm, often called the *sub-mm regime*), though, it is difficult to make low-noise amplifiers, so the classical design of a heterodyne receiver sketched in Figure 3.14 will not suffice for these observations. Receivers designed to operate at these frequencies mix the input RF signal with the IF first and *then* amplify the IF signal; the schematic for these receivers, then, would list the mixer as the first element, followed by an IF amplifier. These receivers generally have higher noise outputs than the lower frequency receivers we discussed previously, but still lower than if a noisy, high-frequency RF amplifier was used. For these high-frequency receivers, much of the effort by receiver builders is to reduce the noise, as well as loss, of the mixer and to produce very low-noise IF amplifiers. Mixers in use at these frequencies today often use superconducting materials and are cooled to about 4 K.

3.3 NOISE, NOISE TEMPERATURE, AND ANTENNA TEMPERATURE

All the components in the receiver, especially the amplifiers, generate their own electrical signals that propagate through the receiver and are unrelated to the signal from the astronomical source. The power measured coming out of the detector, then, includes these extra signals, which interferes with our ability to detect and measure the power of the radiation from the astronomical source. These extra signals are undesirable, but cannot be avoided. We call this unwanted signal *noise*. We need to be aware of the noise and know how to account for it. It is also desirable for the receiver to be designed so as to minimize the noise as much as possible.

Characterizing noise signals generated in electrical circuits has, of course, always been of great interest in electronics. Nyquist in 1928, for example, found that a resistor in the circuit will add electrical noise with a power per Hz that depends solely on the resistor's temperature. For this reason, the electronic power in a circuit, in general, can be described in terms of an *equivalent temperature*, T_{equiv}, which is equal to the temperature of a resistor that would produce the same amount of power as the resistor. Following this convention, radio astronomers also describe the power traveling in the transmission lines and receiver in terms of an equivalent temperature given by

$$T_{equiv} = \frac{P}{k \, \Delta v} \tag{3.5}$$

where:
  k is Boltzmann's constant
  Δv is the bandwidth of the radiation with power P

Equation 3.5 can also be viewed as using the Boltzmann constant (which has units of energy per Kelvin) as a conversion factor, to express energy (or power per Hz) in units of temperature.

Some of the detected power is due to the astronomical source, which was converted by the antenna to electronic power in the transmission line. We call the equivalent temperature of the power that the antenna delivers to the transmission line, the *antenna temperature*, T_A. The far majority of the detected power, though, is due to noise from the receiver components. We describe the total noise power by the *noise temperature, T_N*, and each component in the receiver is characterized by its own noise temperature. We need, now, to discuss how the final power measured in the output of the receiver relates to the antenna temperature and the noise temperature of each of the components. Keep in mind that both the source signals and the noise are affected by the amplification and losses that occur along the path through the receiver. The equivalent temperature of the final power output, then, is not simply the sum of the equivalent temperatures of all the sources in the path.

Let us first focus on the signal from the astronomical source and see how its power is affected by the processes in the receiver. At each stage, the source signal is either amplified (when passing through an amplifier) or reduced by a loss (such as in a transmission line or in the mixer). We can use the gain, G, defined by Equation 3.3, for each step; when passing through an amplifier, $G > 1$, and when there is a loss, $G < 1$. For example, if we assign a gain of G_1 to the first element, which is the RF amplifier, then the power in the source signal after this stage is

$$P = G_1 k \, \Delta v \, T_A$$

By the time the signal enters the detector, the power due to the radiation entering the antenna has been amplified by a net gain, G, and so

$$P = Gk\,\Delta\nu\,T_A$$

We need, then, to determine how to calculate the total gain from each of the component gains. Also, the power due to the noise is a bit more complicated, because the noise generated later in the path does not undergo the amplifications and losses that occur earlier in the path. So we also need to show how to calculate the total noise power from the component noise temperatures.

Note that the antenna temperature describes the power in the input radiation *before* any amplification. Even though the amount of power increases when the signal is passed through an amplifier, the radiation is still described by the same equivalent temperature. Therefore, regardless of the amount of amplification in the system, the input radiation power will still be described by the same antenna temperature. Now, since we need to compare the noise power with the power from the astronomical source, we must describe both powers in the same way. In particular, an amplifier's noise temperature is defined by the equivalent temperature of the noise power as if it was introduced at the *input* to the amplifier, and hence it is amplified *along with* the astronomical signal. Imagine, for example, that the receiver contained only an amplifier of gain G_1 and noise temperature T_{N1}. Then, the power output would be

$$P = G_1\,k\,\Delta\nu\,T_A + G_1\,k\,\Delta\nu\,T_{N1} = G_1\,k\,\Delta\nu\bigl(T_A + T_{N1}\bigr)$$

and so the antenna temperature and noise temperature, with this convention of defining T_{N1}, are directly comparable.

Now, if we have two amplifiers in succession, the first characterized by G_1 and T_{N1}, and the second by G_2 and T_{N2}, then the noise power due to the first is amplified by a factor of G_2 *along with* the noise produced by the second amplifier. So the total noise power coming out of the second amplifier is

$$P_N = G_2 G_1\,k\,\Delta\nu\,T_{N1} + G_2\,k\,\Delta\nu\,T_{N2}$$

As we stated in Section 3.2.2, the total gain in a succession of devices is just the product of the individual gains, so the total gain here is $G = G_1 G_2$. Hence, we can define the total noise temperature (T_N) in this sequence of devices by

$$P_N = Gk\,\Delta\nu\,T_N$$

and so the total noise temperature relates to the individual noise temperatures as

$$T_N = T_{N1} + \frac{T_{N2}}{G_1} \tag{3.6}$$

Note that the noise contribution of the second amplifier is *reduced* by the gain of the first amplifier. The total noise power, including all elements in the signal path, then, is obtained by considering the G and T_N of all the elements in succession and applying the idea behind Equation 3.6. In general, for many elements in succession, we have

$$G = G_1 G_2 G_3 \ldots \tag{3.7}$$

and

$$T_N = T_{N1} + \frac{T_{N2}}{G_1} + \frac{T_{N3}}{G_1 G_2} + \cdots \tag{3.8}$$

The total noise produced by all the components in the receiver is called the *receiver noise temperature*, T_N.

Note that a lossy element, such as a mixer or transmission line, has $G < 1$, which, in effect, *increases* the contribution to the total noise temperature from all the components that come *after* it. How could a loss cause an increase in the noise temperature? Remember that the noise temperature describes the power before amplification (or loss), which makes it easier to compare the noise temperature and the antenna temperature. The effect of a loss is to decrease the power in the source signal, and therefore any noise generated later in the signal path will appear larger, relative to the source signal, than if there had not been that loss, and so the value T_N of these latter elements must now be larger.

Equation 3.8 also shows that, as we saw in the case of two amplifiers, the noise contribution of each successive device to the total noise is reduced by the product of the gains of the preceding elements. The RF amplifier, the first device the radiation enters into immediately after the feed, therefore, is *the most critical* in determining the total noise temperature. Since RF amplifiers usually have gains of at least a factor of 100 (or 20 dB), the contributions to the noise temperature of all other elements in the receiver and detector are reduced by *at least* a factor of 100. A noisy mixer, for example, does not add that much to the total noise temperature, provided that its loss is not too great. It is, therefore, extremely important that the RF amplifier has as much gain and as little noise as possible. This amplifier should be a state-of-the-art device, and not a device that is mass-produced for commercial use. An enormous effort by receiver builders is directed toward designing and building the lowest noise and highest gain amplifiers. The amplifier noise is largely thermal, so reducing the temperature of the amplifier helps significantly in reducing the noise. Therefore, most receivers on professional telescopes are cooled with cryogenic refrigerators to further reduce their noise.

Example 3.6:

Consider a sequence of three amplifiers with gains of

$$G_1 = 100 \ (20 \text{ dB}), \ G_2 = 20 \ (13 \text{ dB}), \text{ and } G_3 = 200 \ (23 \text{ dB})$$

and noise temperatures

$$T_{N1} = 20 \text{ K}, \ T_{N2} = 100 \text{ K}, \ T_{N3} = 500$$

Calculate the total gain and total noise temperature of this amplifier sequence.

Answers:

We use Equations 3.7 and 3.8 to find the total gain and noise temperature of this cascade of amplifiers. The total gain is given by

$$G_{\text{total}} = 100 \times 20 \times 200 = 4 \times 10^5$$

Note that you can also find the total gain by summing the gains of the amplifiers in decibels, so

$$G(\text{dB}) = 20 + 13 + 23 = 56 \text{ dB}$$

or a linear gain of $10^{56/10} = 4 \times 10^5$. For the total noise temperature we find:

$$T_{\text{N}} = 20 + 100/100 + 500/(100 \times 20) = 20 + 1 + 0.25 = 21.3 \text{ K}$$

As mentioned in the last section, low-noise RF amplifiers are not so easy to build at high radio frequencies. We see, now, that including a noisy RF amplifier as the first element in the receiver would, indeed, be a problem. We obtain a smaller total noise temperature in this case by making the mixer the first element instead. However, since the mixer has some loss, and so its G is less than 1, by Equation 3.8, the noise of all the following elements is increased a bit. Therefore, high-frequency receivers have higher noise temperatures than those at lower frequencies.

Example 3.7:

At high RF the first component of a receiver is the mixer, in which some power is lost. Suppose that we have a mixer with a loss such that $G_{\text{mixer}} = 0.8$ (recall that $G<1$ indicates a loss) and a noise temperature of 50 K. If it is followed by the same three amplifiers as is shown in Example 3.6, calculate the total gain and noise temperature.

Answer:

As is shown in Example 3.6, we use Equations 3.7 and 3.8. The total gain is given by

$$G_{\text{total}} = 0.8 \times 100 \times 20 \times 200 = 3.2 \times 10^5$$

For the noise temperature we find:

$$T_{\text{N}} = 50 + 20/0.8 + 100/(0.8 \times 100) + 500/(0.8 \times 100 \times 20) = 50 + 25 + 1.25 + 0.31$$
$$= 76.6 \text{ K}$$

In comparison to the answers in Example 3.6, we see that adding a lossy mixer in front of the series of amplifiers increases the noise temperature by 65.3 K even though the mixer's noise temperature is only 50 K. This is because the loss in the mixer, effectively, increases the noise temperature of the following components.

With the total gain and noise temperature defined by Equations 3.7 and 3.8, we can write the expression for the total detected power. Recall that the antenna temperature is defined before all the amplification. Therefore, the total power entering the detector is given by

$$P = G\,k\,\Delta\nu\,(T_A + T_N)\tag{3.9}$$

Considering the responsivity of the detector, we have that the voltage out of the detector is

$$V = \alpha\,G\,k\,\Delta\nu\,(T_A + T_N)\tag{3.10}$$

For the majority of astronomical sources, and with most receivers, it is the case that $T_A \ll T_N$. Therefore, even if we observe blank sky, the power out of the receiver is not zero because of the noise power. For these reasons, we cannot readily make total power measurements, but instead, we must make *switched* power measurements in which we measure the difference in voltage between when the telescope is aimed at the astronomical source (called an *on*-source observation) and when it is aimed at blank sky (called an *off*-source observation). The latter observation is made by pointing the telescope more than a beam width away from the source. Switched observations are discussed in more detail in the next chapter. We mention them here to clarify a potential source of confusion about the nature of the noise and the effect it has on our ability to detect a weak astronomical signal. It is tempting to believe that the switched observations completely subtract off the noise power, and hence that noise is not really a concern. However, this is *not* the case. The switched observations remove the *offset* in the measured power caused by the noise, but the *fluctuations* in the noise power still affect our measurement and dominate our uncertainty in the antenna temperature. It is these fluctuations that limit the sensitivity of a radio telescope to detect a faint astronomical source. The statistics of these noise signals, with a focus on the variance of the noise power, therefore, warrants some discussion.

First, recall that the *variance* is the mean square deviation from the average. Following common convention, we will use σ to indicate the standard deviation, and thus σ^2 as the variance. And, therefore, for single-measurement experiments, the uncertainty is equal to the square root of the variance of that type of measurement. Now, when measuring the amount of power in EM radiation, the variance in the measure of that power depends on two effects. Since the amount of power in the radiation is proportional to the number of photons arriving per second, the variance in power must relate to fluctuations in the arrival rate of the photons. The power in the radiation is also proportional to E^2 in the waves, and so the variance also depends on the fluctuations in the waves. The former effect is a consequence of the particle aspect of light while the latter is often called the *wave noise* and is a purely classical effect. When both effects are included in the statistics, the variance is found to have two terms; the variance due to fluctuations in the photon arrival rate depends on the number of photons per mode (with units of photons per second per Hz) and the wave noise depends on the number squared. More specifically,

$$\sigma^2 \propto n + n^2\tag{3.11}$$

where:

$$n = \frac{1}{\exp(h\nu / kT) - 1}$$

which (you hopefully recognized is the last factor in the Planck function) is proportional to the number of photons per mode. Note that the magnitude of n is less than 1 when $h\nu > kT$ and greater than 1 when $h\nu < kT$. Therefore, at higher frequencies, the first term in Equation 3.11 dominates and the uncertainty is proportional to the square root of the number of photons per mode, that is,

$$\sigma \propto \sqrt{n}$$

while at lower frequencies, the second term dominates and the uncertainty is proportional to the number of photons per mode,

$$\sigma \propto n$$

If you are familiar with observations at visible frequencies, you may recall that one generally counts the number of photons—such as when measuring the brightness in the individual CCD pixels—and that the uncertainty in that count is directly proportional to the square root of the count. This is described by Poisson statistics, which describe the data when counting discrete objects. At radio frequencies, though, when the photon energies are so low that the statistics fall into the wave regime, the uncertainty in the number of photons N is directly proportional to N. Similarly, with the measured power we have

$$\sigma_P \propto P_N$$

You probably know that with any measurement involving random errors, the uncertainty in the measure decreases by averaging more values, by a factor equal to the square root of the number of measurements. Similarly, by making the measurement for more seconds or by increasing the bandwidth, we make many independent measurements of the power. In general, the number of independent measurements (or modes) made over a time period of Δt and bandwidth $\Delta \nu$ is given by $\Delta t \Delta \nu$. Therefore, in an observation with a bandwidth $\Delta \nu$ and integration time of Δt, the uncertainty in the power measured, σ_P, is given by

$$\sigma_P = \frac{P_N}{\sqrt{\Delta t \, \Delta \nu}} \tag{3.12}$$

The power measured is that of both the signal from the astronomical source and the total receiver noise power; however, there may also be some additional unwanted noise contributions, and these are discussed further in Chapter 4. Since the receiver noise power is usually much greater than the power from the astronomical source, it is the fluctuations of the noise power that limits the ability to detect a weak astronomical source. For this reason, receiver noise, measured by the noise temperature, is a very important parameter to radio astronomers.

Example 3.8:

1. A radio telescope has a receiver with a total noise temperature of 100 K, a total gain of 1×10^8, and a bandwidth of 1 MHz. What is the total noise power detected?

2. If the observations are averaged over a time period of 1 s, what would be the uncertainty in the measured noise power?

Answers:

1. The power is given by Equation 3.9.

$$P = (1 \times 10^8) \times (1.38 \times 10^{-23} \text{ J K}^{-1}) \times (1 \times 10^6 \text{ Hz}) \times (100 \text{ K}) = 1.38 \times 10^{-7} \text{ W}$$

2. The uncertainty in the measured power is given by Equation 3.12. Note that observing for 1 s in a bandwidth of 1 MHz is equivalent of making 1 million independent measurements of the power. Since the uncertainty in the measured power is equal to the measured power divided by the square root of the number of measurements, the uncertainty is 1.38×10^{-7} W/1000 $= 1.38 \times 10^{-10}$ W.

3.4 BOLOMETER DETECTORS

Another type of detector often used at high radio frequencies is the *bolometer*. These are inherently broadband devices and so are usually only used to detect continuum emission and not spectral lines. At short radio wavelengths (less than a millimeter), bolometers are the most sensitive devices for astronomical observations. Simply put, a bolometer is a thermometer for radiation. It has two components—an absorber and a thermistor. Radiation from the source is incident on material that is a good absorber of radio waves and the power in the EM waves heats the absorber. The increase in temperature of the absorber is proportional to the amount of EM energy incident on it (since $E_{\text{thermal}} \propto kT$), and the rise in temperature is measured by a very sensitive thermistor. At the input to the bolometer is a filter that defines the range of wavelengths to be detected. The absorber is also weakly connected to a cold bath to reset the temperature of the absorber after detection of the incident radiation.

Although bolometers are simple in concept, they can be quite complicated to realize. The amount of radiation power arriving from a distant astronomical source is exceedingly small, and so the temperature change of the absorber is miniscule. Therefore, the thermistor must be extremely sensitive. To prevent thermal fluctuations of the absorber from overwhelming the astronomical signal, the absorber must be cooled to a temperature close to absolute zero. Early bolometers used semiconductor thermistors to measure the temperature change, while more recent bolometers use superconducting materials that are very sensitive to small temperature changes. This latter type is often called a *transition edge sensor*.

An important distinction between bolometers and the heterodyne systems described in Section 3.2 is that the former detects only the energy of the wave and disregards its phase. Such a receiver is called *incoherent* and cannot be used for radio interferometry, a technique, which we describe in Chapters 5 and 6, that requires maintaining the phase information of the wave.

Bolometers to be used at high radio frequencies can be very small and so many (>10,000) can fit in the focal plane. With today's bolometers, radio astronomers can easily obtain large-area maps (of many arcminutes on a side) of the broadband radio emission from astronomical sources. Other technologies are being developed, such as MKIDs (microwave kinetic inductance detectors), that rely on properties other than temperature. The ultimate goal of such technologies is to provide the radio-equivalent of the CCDs that are used for imaging at optical wavelengths.

3.5 SPECTROMETERS

There are many radio-wavelength spectral lines that can reveal valuable information about the physical conditions of astronomical sources and so another important component of a radio telescope is the spectrometer. Radio-wavelength spectrometers are fundamentally different from spectrometers at visible wavelengths. At visible wavelengths the light can be dispersed by a grating via the wavelength dependence of diffraction of visible light. The dispersed light is then detected with a CCD and the spectrum of the light recorded. At typical radio wavelengths, though, the radiation is not diffracted by reasonable size gratings and so radio-wavelength spectra are not obtained by optical means, but rather by electronically processing the signal that is output from the front end of the receiver. A number of techniques are used to effect this processing; we describe two of the more common ones: (1) *analog filter-bank spectrometers* and (2) *digital spectrometers*. In either case, two important parameters characterize the spectrometer—bandwidth, Δv, and spectral resolution, δv. The bandwidth determines the total frequency range over which the spectrum is measured; generally one wants this to be as wide as possible. The spectral resolution determines the ability of the spectrograph to distinguish closely spaced spectral lines and to reveal the details of the shapes of each line; generally, one wants this to be as narrow as possible.

3.5.1 Filter Bank Spectrometer

One of the simplest spectrometer concepts is the filter bank; a schematic of this is shown in Figure 3.17. At the heart of this spectrometer is a series of *bandpass filters*, with each centered at a slightly different IF, but with identical narrow bandwidths. The IF signal output from the heterodyne receiver is amplified further and then sent into a *power splitter*, which disseminates the signal into hundreds of separate *channels*, each containing one of the filters in the filter bank. Each filter is followed by a separate square-law detector (described in Section 3.2.3), which outputs a voltage to be sent to the computer, which assigns the detected powers to the appropriate channels. The bandwidth of the filters determines the frequency resolution of the spectrometer,

$$\delta v = \Delta v_{ch}$$

and the number of channels multiplied by the channel width determines the spectrum bandwidth,

$$\Delta v = N_{ch} \Delta v_{ch} \tag{3.13}$$

Splitting the signal into channels, of course, reduces the power in the signal from the source in each channel. But, as long as there is sufficient gain in each channel so that the noise of the power splitter does not noticeably increase the overall noise temperature, then there is no additional noise added by this process. The sensitivity to weak astronomical sources, then, is not reduced.

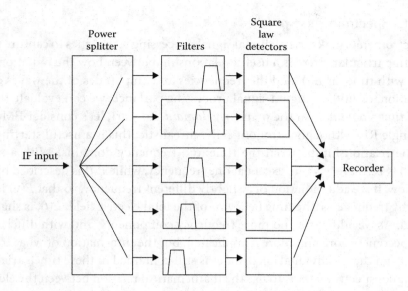

FIGURE 3.17 Schematic of a filter bank spectrometer. For display purposes, we only illustrate five filters here. In general, the signal is split into hundreds of channels, each with its own filter and detector.

Most splitters accomplish the division into many channels by continually dividing signals in two, and so the number of channels in the final output will be a power of 2, that is, 2^n where n is the number of division steps (e.g., 128, 256, 512, or 1024). For instance, we might have a filter bank with 256 spectral channels, each with a width of 100 kHz or 0.1 MHz, yielding a total frequency coverage of

$$256 \text{ channels} \times 0.1 \text{ MHz/channel} = 25.6 \text{ MHz}$$

In Section 3.2.2, we considered mixing an RF signal centered at 1420 MHz to an IF signal centered at 50 MHz. The spectrometer, then, would be designed to cover a frequency range of 37.2–62.8 MHz. The first filter might be centered at 37.25 MHz, the second at 37.35 MHz, the third at 37.45 MHz, and so on until the last filter, which would be at 62.75 MHz. The 21-cm line of neutral atomic hydrogen, for example, which has a frequency of 1420.406 MHz, would be mixed down to a frequency of 50.406 MHz and the spectral line would pass through one of the filters near the center of the filter bank.

Note that by changing the LO frequency we can tune our receiver to detect spectral lines of different frequencies. *In general, the LO determines the observing frequency.* For instance, one of the radio spectral lines of the molecule OH occurs at a frequency of 1665.40 MHz. By changing the LO from 1370 to 1615 MHz, we now mix the RF centered on 1665 MHz to an IF of 50 MHz. The OH spectral line will now be roughly centered in the filter bank. Although the filter bank operates at a fixed IF frequency, by varying the LO frequency we can tune over a range in frequencies and hence detect different spectral lines.

3.5.2 Digital Spectrometers*

Digital spectrometers make use of digital signal processing techniques to capture the power spectrum. In particular, there is a useful relationship between how the radiation's electric field varies with time, or $E(t)$, and the frequencies and amplitudes of the waves contained in the radiation, or $E(\nu)$. In digital signal processing parlance, we can evaluate the electric field in the *time domain* or in the *frequency domain.* To start, let's consider EM radiation of only a single RF. Although astronomically unrealistic, this is a useful starting point to illustrate the relationship between the time and frequency domains. With a single frequency, $E(\nu)$ is zero everywhere except at one frequency, while $E(t)$ is described by a simple sinusoid. Now, if we add radiation of a slightly different frequency, so that $E(\nu)$ is nonzero at two nearby frequencies, the time behavior of the total electric field, $E(t)$, is slightly more complicated. As we add more and more frequency components, and with differing amplitudes, $E(t)$ becomes more and more complicated, but the information of what frequencies are involved and their individual amplitudes is still contained in these time variations. The basis of a digital spectrometer is to use the mathematical relation between the electric field expressed as a function of time (the time domain, which we observe) and the electric field as a function of frequency (the frequency domain, which is the power spectrum we want to measure).

The two domains are related by a *Fourier transform,* which is a powerful mathematical tool used in many branches of science; it is vitally important for digital signal processing. Mathematically, we express this relation by $E(\nu) = \mathcal{F}[E(t)]$, where $\mathcal{F}$ represents the Fourier transform operation. The equation defining this operation is provided in Appendix V. In essence, a Fourier transform deconstructs the time variations of the electric field as the sum of many sinusoidal functions, each with a specific frequency and amplitude. The individual components going into this sum have specific frequencies and amplitudes, and this is the function $E(\nu)$. For our example of light at a single frequency, the time variation is a single sinusoidal function and its Fourier transform contains a single component defining the electric field amplitude at the frequency of that one sinusoidal function. In this simple case, $E(\nu)$ is zero at all frequencies other than that of the single sinusoid.

To obtain the power spectrum, the amplitude of the electric field is sampled and digitized as it enters the spectrometer to measure $E(t)$. This time series of digital data is Fourier transformed, to yield $E(\nu)$, and then squared, to produce the power spectrum, $P(\nu) \propto [E(\nu)]^2$.

There are several important differences between digital and analog spectrometers, that arise from the *digitization* of the IF signal; dealing with them requires additional components in the spectrometer. We show a block diagram of a generic digital spectrometer in Figure 3.18. A key element is the digitization (or analog-to-digital conversion) block.

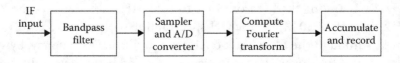

FIGURE 3.18 Schematic of a simple digital spectrometer.

This conversion consists of two steps: *sampling* of the signal at periodic intervals and *quantization* of the sampled signal. The sampled signal is still an analog value (and hence continuous, taking on any value in the acceptable voltage range). We digitize this value by representing it in terms of bits. If n bits are available to represent the value, then the allowed voltage range will be divided up into 2^n levels. Quantization of the signal is the process of selecting the level closest to the sampled value. The small difference between the sampled valued and the assigned value is a source of noise in the measurement, called *quantization noise*. The output of this digitizer is a stream of digital data that is sent to a computer where it is collected into blocks of time intervals that are many orders of magnitude longer than the period of a wave; each of these time intervals is then Fourier transformed and squared to produce the power spectrum.

The sampling process can lead to confusion of the frequency content of the signal with undesired higher frequency components, due to an effect called *aliasing*. Aliasing is an ambiguity that arises when we sample a continuous signal at discrete times. The practical result is that higher frequency signals can *masquerade* as lower frequency, which we might wish to detect. This problem can be eliminated by the use of filters to remove the higher frequency components. If the RF signal is mixed down to the baseband, so that the IF spans a frequency range from 0 to Δv Hz, then a low-pass filter is placed before the digitization block to remove the aliasing frequencies. When the IF is at a higher range, well above DC (or 0 MHz) then a bandpass filter is usually employed, as shown in Figure 3.18.

There are two important parameters that characterize a digital spectrometer: the *sampling rate* and the length of the *time interval* of the data blocks used in the Fourier transform. To recover all of the information in a bandwidth Δv, the electric field must be sampled at a rate at least twice Δv; this is called *Nyquist sampling*. For example, if we wanted a bandwidth of 25 MHz, comparable to that of the example filter bank spectrometer described in Section 3.5.1, the electric field must be sampled at a rate of 50 MHz, or 50 Mega samples per second. Thus, we see that *the sampling rate determines the bandwidth of the spectrometer*.

The resulting data stream is blocked into time intervals much longer than the period of the waves, and each time interval is Fourier transformed to produce a spectrum, and then the spectra from all the different time intervals are averaged together to produce the final spectrum.

Longer time intervals contain more samples, which makes the Fourier transform more computer intensive, but also provides better spectral resolution. In general, if the data are blocked into time intervals of $\Delta t_{interval}$, the spectral resolution, δv, is approximately.

$$\delta v_{resolution} \sim \frac{1}{\Delta t_{interval}} \tag{3.14}$$

and so *the time interval determines the spectral resolution of the spectrometer*.

If we want to achieve the same resolution as the filter bank spectrometer in Section 3.5.1, for example, which had a resolution of 0.1 MHz, or 10^5 Hz, we would use a time interval of 10^{-5} s. And, with the sampling rate of 50 MHz each time block consists of 500 samples.

Example 3.9:

Using a digital spectrometer we want to obtain a spectrum covering a frequency range of 100 MHz with a spectral resolution of 0.01 MHz, what sampling rate must be used and what time interval must the data be blocked into?

Answer:

The sampling rate must be twice the bandwidth of 100 MHz to meet the Nyquist sampling criterion. Thus, one must measure the electric field 200 million times per second, that is, a sampling rate of 200 Mega samples per second (MSPS). The resolution is given by Equation 3.14. To achieve a resolution of 0.01 MHz, or 10^4 Hz, requires a time interval of 10^{-4} s. Note that each block will contain 20,000 samples.

Since each time interval is Fourier transformed, and many intervals are needed to produce a reasonable spectrum with sufficiently low noise level, the processing power available is often a limiting factor for the resolution of the spectrometer.

The computer time can be reduced with an alternative and less computationally intensive implementation of digital spectrometers. Principal among these is the *autocorrelation spectrometer*. This design makes use of the autocorrelation function (ACF) of the electric field and its relationship with the power spectrum. The *correlation* of two functions is a measure of how similar they are to one another. The *auto* (or *self*) correlation is how similar a function is to *itself* when we translate it in time. For example, a square wave, shifted by half a period, is the *inverse* of itself. But when shifted by an integer number of periods, it is *identical* to itself. Mathematically, we can quantify this self-similarity by the ACF:

$$h(\tau) = \frac{1}{t_{\text{interval}}} \int a(t)a(t-\tau)dt$$

The integral is divided by t_{interval}, the time interval over which the correlation is performed, so that the result is an average.

Conceptually, what this equation does is shift the original function by τ time units and then multiply the new shifted function by the original function. The product is then averaged over a long time interval to produce a numeric measure of how similar the two functions are for the corresponding shift τ. For values of τ that produce highly self-similar functions, the integral will be large; for values of τ where the two functions are highly dissimilar, the integral will be small. The independent variable, τ, is called the *delay*.

An important theorem (the Wiener-Khinchin Theorem) from signal processing (see Appendix VI) states that the power spectrum, $P(\nu)$, is equal to the Fourier transform of the ACF. This equivalence is the basis for the autocorrelation spectrometer. As before, the RF signal is mixed down to an IF, then filtered and digitized. The resulting digital data stream is then split into many different paths, each experiencing a different time delay, and then the autocorrelation is performed by simply multiplying the original signal by each delayed signal, to yield $h(\tau)$. A Fourier transform is then performed on $h(\tau)$, thus producing the desired power spectrum.

The ACF is calculated easily using digital circuits. This is accomplished in real time and can be averaged over a relatively long integration time to produce a single low-noise $h(\tau)$, which is then converted to a power spectrum $P(\nu)$ with a single Fourier transform. The entire process can be repeated to yield a more sensitive spectrum, but the point is that only one Fourier transform must be performed for each integration time, which can be seconds long, whereas the Fourier transform spectrometer requires a Fourier transform to be performed much more often, requiring significantly more computer power.

Because we are performing this process *digitally* on sampled data (not analytically, on continuous data, as the equation mentioned above suggests), we must calculate the *discrete ACF* of E(t). This is accomplished by using a finite number of time delays, τ. The discrete ACF is computed from the digitized time stream of electric field amplitudes by using a sum instead of an integral, that is,

$$\mathrm{ACF}\big[E(\tau)\big] = \sum E(t)E(\tau-t)$$

The hardware that produces the time delays, τ, multiplies the two signals, and then sums (or accumulates) the multiplied result, as outlined in the schematic shown in Figure 3.19.

The number of time delays, or lags, determines the number of terms in the ACF. The Fourier transform of the ACF then provides the power spectrum. The total frequency coverage or bandwidth is still determined by the sampling rate; remember that the bandwidth is one-half of the sampling rate in accord with the Nyquist theorem. The spectral resolution is approximately the bandwidth divided by the number of time lags. So, if the spectrometer has a bandwidth 25 MHz and 256 lags, then the spectral resolution would be about 100 kHz. Note that by changing the sampling rate and filters, we can increase or decrease the bandwidth and consequently decrease or increase the spectral resolution. Digital spectrometers are much more flexible than filter bank spectrometers and for this reason most new spectrometers are digital.

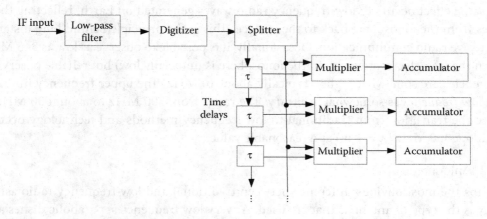

FIGURE 3.19 Schematic of a digital autocorrelation spectrometer.

Apart from filter banks and digital spectrometers, many other spectrometer designs exist, which we do not discuss. The decision of what sort of spectrometer to use depends on the details of the telescope and observational goals.

3.6 VERY LOW-FREQUENCY RADIO ASTRONOMY

Since its inception, radio astronomy has pushed toward observing at higher frequencies. Early on, this was an obvious way to achieve higher angular resolution, because $\theta_{resolution} \propto \lambda/D$. As the field evolved, other reasons for observing at high frequencies also became apparent; particularly important was the fact that many molecules and interstellar dust emit radiation at shorter radio wavelengths. These and other factors pushed radio astronomers to develop telescopes and receivers working at ever higher frequencies, and the trend is likely to continue. As technology improves, the distinction between radio and the far-infrared regimes is becoming increasingly blurred, as radio techniques (in particular, coherent receivers) are being applied to even shorter wavelengths.

Despite this trend toward higher frequencies, the beginning of the twenty-first century witnessed an important resurgence of *very low-frequency* radio astronomy, in the 10–300 MHz range. This renaissance is mostly driven by scientific motivations, but as we shall see, there are important technological developments involved as well, with advances in computer technology being even more important than those in radio-frequency electronics. For example, the first major radio-telescope initiative of this century—the Square Kilometer Array—is based on the design concept known as *Small-D, Large-N*, in which many antennas (large N), each one of relatively small size (small D), work in concert, their individual signals being combined and processed by computers, which currently enjoy an exponential growth in capacity.

3.6.1 Low-Frequency Window

The lower frequency limit for terrestrial radio astronomy is about 10 MHz. This limit is determined by the plasma frequency (see Volume II) of the ionosphere, which is usually in the range of 3–5 MHz. Below this cut-off frequency, radio waves arriving at Earth are reflected outward by the ionosphere, thus never reaching radio telescopes on the surface. The same effect occurs to low-frequency radio waves generated *on* Earth, reflecting these waves from the ionosphere back to the surface. This is the principle behind long distance shortwave radio communication. Occasionally, it is possible to observe as low as 3–5 MHz (when the free-electron density in the ionosphere is unusually low) but reliable observing frequencies are somewhat higher, typically about 10 MHz. The upper frequency limit for *very low frequency* is somewhat arbitrary. The range from 300 MHz to about 600 MHz is where the transition from traditional to low frequency methods and technology occurs. We adopt 300 MHz ($\lambda = 1$ m) as a reasonable value.

3.6.2 Antennas

Perhaps the most obvious difference between traditional and low-frequency radio astronomy is the type of antennas that are used. At very low frequencies, parabolic dishes and their associated feed horns become unwieldy because of the large size needed to couple with the free-space wave. A more practical antenna design is based on an arrangement of metal wires or tubes; this can take a wide variety of forms. Especially common are the *dipole* and the *Yagi-Uda* designs (see Figures 3.20 and 3.21). Some low-frequency antennas, particularly Yagi antennas, operating at a few hundreds of MHz, are physically steerable. But low-frequency antennas are usually located in fixed positions and do not move. As we explain below, such antennas can still be pointed or steered, but this pointing is achieved by electronic circuits rather than physically moving the antennas.

For wire antennas, such as the dipole, the electric field of the radio wave induces small electrical currents in the wire, and the small potential difference between the wires is

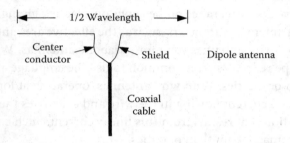

FIGURE 3.20   Half-wave dipole antenna. Each horizontal piece is one-quarter of a wavelength long and is typically made from metal wire or tubing. There is a small gap between the two conductors, where the transmission line is connected. A full-wave dipole antenna has a total length of 1λ, with each branch of the dipole having length 0.5λ.

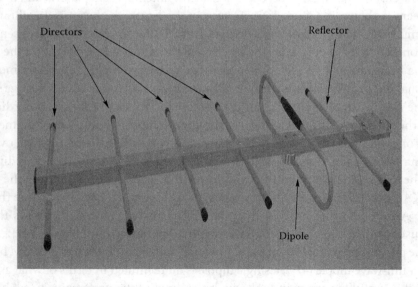

FIGURE 3.21   Yagi antenna. All Yagi antennas have three types of conductor elements: the driven or active element, where we take the signal off the antenna to pass to the transmission line; multiple director elements, which are placed in front of the active element; and one or more reflector elements that are placed behind the active element. The directors and reflector(s) are often referred to as passive elements.

passed directly to the transmission line without the need for a feed horn. The situation is similar for Yagi antennas. These consist of a dipole element (to which the transmission line is attached) but there is also a reflector element (behind the dipole) and multiple director elements (in front of the dipole). Induced currents in these elements act so as to reinforce the current in the dipole element and thus increase the gain and directivity of the antenna. As a result, Yagi antennas provide better sensitivity and angular resolution than dipole antennas, although they are more costly and complicated to construct.

Coaxial cables are usually employed as the transmission line for dipole and Yagi antennas. Other possibilities exist; one of the more common designs is the *ladder line* design. This transmission line is made of two wires, kept at a uniform separation. The spacers placed between the wires have the appearance of the rungs on a ladder, giving rise to the name.

The calculations of antenna parameters, such as the effective area and the beam FWHM, are rather different for wire antennas than for parabolic antennas. With a parabolic dish (or other aperture-type reflectors), it is common to use the language and tools of optics to calculate the antenna properties. With wire antennas, operating at long wavelengths, one imagines a hypothetical current flowing in the wire and calculates the angular power distribution of the radiation that results from this linear current. Such calculations are more in the realm of electromagnetism than of optics.

Even the simple dipole antenna can appear in a wide variety of forms. Two of the most common types are the *half-wave* and *full-wave* dipoles. The names indicate the length of the dipole elements: the half-wave has a total length of $\lambda/2$ (in two sections of $\lambda/4$ each), while the full-wave has a total length of λ (in two sections of $\lambda/2$ each). A schematic of the half-wave dipole, including the current distribution used to calculate the antenna beam pattern, and the resulting antenna beam is shown in Figure 3.22. As seen in the figure, the beam is doughnut-shaped, with maximum sensitivity in directions perpendicular to the dipole axis and zero sensitivity in the directions along the dipole axis. The FWHM of the half-wave dipole is 78°; the full-wave dipole has a beamwidth of 47°. Dipoles longer than one wavelength are not commonly used; the additional length does not provide for a narrower beam, but rather results in large sidelobes.

Even the simple dipole antenna can take on a wide variety of forms, according to the diameter, the shape, and the orientation of the conducting elements. Large diameter wires or tubes—or even flat metal plates—can substantially increase the bandwidth over which the antenna is resonant. Bending the elements downward (an inverted-Vee dipole) can increase the size of the beam pattern and lower the space requirement for the antenna. Several *SKA-pathfinder* projects—prototypes for what may eventually become the Square Kilometer Array—use this style of antenna. For example, Figure 3.23 shows the dipole antennas used by the Long Wavelength Array that operates from 10 to 88 MHz.

With the poor angular resolution of an individual half-wave dipole (78° FWHM beam size) or even of a full-wave dipole (47°), a single dipole is quite limited in terms of what astronomical objects it is useful for observing. Although the dipole antenna remains fixed, because its beam is so large, the astronomical object will be within the beam for most of the time that it is above the horizon. The orientation of a dipole, whether aligned east–west or north–south, is usually not very critical; although depending on the latitude of the telescope and the source being observed, there may be advantages to one orientation compared to another.

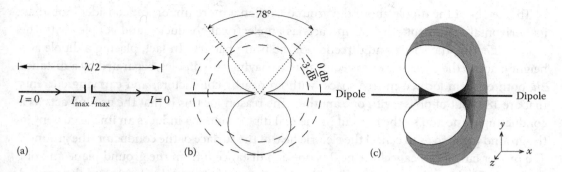

FIGURE 3.22 (a) Half-wave dipole, showing the currents flowing in the two arms. The currents must be zero at the end-points of the dipole arms and are maximum at the midpoint, where the transmission line is connected. (b) A cross-sectional view of the antenna beam pattern, shown in polar coordinates, of a dipole antenna in free space (far from the ground or other conducting surfaces). Because the beam pattern is normalized, its maximum value is 1 or 0 dB. Also shown is the locus of points at half the maximum (–3 dB). Where the antenna pattern crosses the –3 dB line defines the FWHM of the beam; in this case of size 78°. (c) In a three-dimensional view, the antenna pattern of a free-space dipole is torus-shaped. The direction of maximum sensitivity is perpendicular to the dipole axis and there is no sensitivity in the direction of the dipole axis. In the coordinate system shown, this dipole has no sensitivity in the $\pm x$ directions, and maximum sensitivity in the yz-plane.

FIGURE 3.23 (**See color insert.**) LWA dipole antenna pair. This inverted dipole is actually two orthogonal dipole pairs. The construction is from aluminum tubing, and it is designed to work from 10 to 88 MHz. For comparison, a traditional parabolic reflector can be seen in the background.

The height of the dipole above the ground is often a more important consideration than its horizontal alignment. The ground acts as an electrical conductor and couples with the dipole, affecting the shape and direction of the beam pattern. In fact, placing a dipole at a height h above the *ground plane* is equivalent to having two dipoles, separated by a distance $2h$, both of them located in free space (with no nearby conductors) and carrying currents that are 180° out of phase with one another. The reason for this is that the ground acts as a conducting plane and so the current in the real dipole antenna induces an image current in the ground plane so as to cancel the electric field at the surface of the conductor (the ground). The proper current to cancel the field is one at a distance h from the ground plane and out of phase with the real current. By changing the distance between the dipole and the ground plane, one can effectively shape and redirect the antenna beam pattern in a variety of ways.

Because individual dipole and Yagi antennas have such broad beam patterns, it is common to connect multiple antennas in *arrays*. This serves both to reduce the size of the main beam and to control the direction in which it points. In this case, each individual antenna is referred to as an *element* of the array. The physical locations of the elements with respect to one another and the lengths of the transmission lines connecting the elements to the receiver are important factors for determining the shape and direction of the array beam.

Recall in our discussion in Section 3.1.2 that the width of the beam of a classical radio telescope is determined by the interference of the incoming waves; at some angles of approach, there are extra distances that waves need to travel to some parts of the reflector than to others, and these extra path lengths produce phase differences. The direction angle of the first null of the beam occurs when the sum of all the waves, with their assorted phases, yields complete destructive interference. The beam of an array of dipole or Yagi antennas is made narrower by a similar process. Before the signals from all the elements are combined, each of their phases can be altered in a clever way so that when they are added together, destructive interference occurs for signals from some directions, hence yielding a narrower beam. The phases can be altered either by adjusting the cable lengths connecting the antennas or by introducing electronic phase shifts into the signals. As an example of an electronic phase shift, we can multiply a sine wave by -1; this has the same effect as shifting the phase of the wave by 180°. The optical analog of this is when a wave reflects off the surface of a material of higher index of refraction (compared to the index of refraction where it is traveling). The reflected wave will be phase shifted by 180° even though the difference in path length (caused by the reflection) is zero.

The point of all this is to realize that we can control the interference pattern—obtaining constructive or destructive interference as needed—by adjusting the cable lengths connecting the antennas or by introducing an electronic phase shift into the signals. Thus, an array of antenna elements can be made to have sensitivity in any direction that we wish.

For an antenna array, the resulting, composite beam pattern is obtained by forming the vector sum of the signals arriving at each antenna. In this context, a *vector sum* means that we take the phase of the signal into account when summing the signals. Thus, two equal amplitude signals, but 180° out-of-phase, will cancel one another. For example, if two dipole antennas are spaced half a wavelength apart—as seen from a particular position on the sky—then radiation from that position arriving at the more distant dipole will have a 180° phase lag with respect to the first dipole. These two signals will cancel and the pair of dipoles will have no sensitivity to sources at that particular sky position. In contrast, if some other sky position has the same path length to both dipoles, then the signals will add in phase, and the combination of antennas will be twice as sensitive as either one individually.

The most common way to design an antenna array is to begin with N identical antennas, all with the same orientation and alignment, and to place them in some regular geometric pattern. Moreover, the electronics of the array are designed to allow control of the amplitude and phase of the signal arriving from each antenna. The array beam pattern is the vector sum of the N individual signals.

The contribution of each antenna to the vector sum is the product of two terms: the sensitivity pattern of the individual antenna element, which we introduced in Section 3.1.2 (see also Sections 3.1.5 and 4.2.1), and an *array term*, which depends on the location of the antenna and the relative amplitude and phase adjustment that are applied to its signal. Because we assume all antennas to be identical and to have the same orientation, the sensitivity pattern term will be the same for all elements of the array. The shape and direction of the array pattern, then, are largely determined by the array term, which reflects the geometric location of each element within the array and its relative amplitude and phase.

The problem is essentially one of "creative interference." The idea is to use the geometric phase lag (caused by the wave reaching some antennas later than others) in combination with artificially induced phase lags (and possibly amplitude corrections) to control the location in the sky where constructive and destructive interference will occur. Although the individual antenna beam patterns may be quite large (e.g., the torus of a dipole has a FWHM of 78°), a large number of dipoles, judiciously placed and with their signals properly combined, can easily achieve beam sizes smaller than 1°.

Moreover, if the instrumental phase shifts are introduced electronically (and not by changing cable lengths, as was done many years ago) then the direction of the beam can be made to move on very short (tens of millisecond) timescales. Thus, although a dipole array may be physically fixed to the ground, it can be made to track a source as it moves through the sky, or even to re-point the entire array to a different region of the sky almost instantaneously, in response to some astronomical event, such as a gamma ray burst.

3.6.3 Receivers

As mentioned in Section 3.2.2, sometimes the mixing process converts the RF directly to baseband. Such receivers are called *direct conversion* and are common at lower frequencies. Indeed, the IFs of many heterodyne receivers are *higher* than the low-frequency observing range of 10–300 MHz; one would need an up-conversion of frequencies in this case, rather than the more normal down-conversion.

An alternative to direct-conversion heterodyne receivers is *direct sampling* of the low-frequency radio signal. Modern analog-to-digital converters can easily sample signals in the 10–300 MHz range, so there is no need to convert them to an IF. The auto-correlation of this digitized signal (as described in Section 3.5.2) provides its power spectrum, so we can measure the power in the observed passband directly from the sampled voltage values. A receiver of this type might properly be called a *digital receiver*. There will always be a need for an analog RF amplifier following the antenna, but in a fully digital receiver, the signal is digitized immediately after this amplification, and *all* subsequent processing is done in the digital domain. In this case, it is the speed of the digital electronics (or the computer) that determines receiver parameters such as bandwidth and number of spectral channels.

We show a block diagram of a digital receiver in Figure 3.24. Compared to the heterodyne receiver shown in Figure 3.14, the digital receiver looks simpler, owing to the lack of the analog electronics needed for the frequency conversion stage. There are several other important differences worth noting. First is the presence of an RF filter between the antenna and the RF amplifier. As we discuss below, interfering radio signals are a big problem at low frequencies. Although the antenna may not be optimized (i.e., have the best sensitivity) for the frequencies of these interfering signals, these signals can easily be millions of times stronger than the astronomical signal we wish to study. If such signals are allowed to reach the RF amplifier, they can cause it to saturate and we will not obtain a reliable amplification of the astronomical signal. The solution to this problem is to insert a bandpass filter before the RF amplifier.

The second difference worth noting is that the receiver shown in Figure 3.24 can be used for *both* continuum and spectral line observations. Processing the signal as broadband or in narrow spectral channels depends largely on the software being run on the computer. The analog-to-digital (A/D) conversion stage will place some constraints on the bandwidth, but the digital signal processing that occurs on the computer is what determines if we are making a continuum or spectral line observation.

Another important simplification of low-frequency receivers is that their RF amplifiers do not need to be cooled to cryogenic temperatures as high-frequency amplifiers do. This greatly reduces the complexity and cost of the receiver. The reason that room temperature amplifiers are adequate deserves some attention. Recall Equation 3.9, in which $P = G\,k\,\Delta\nu\,(T_A + T_N)$. In our discussion in Section 3.3, we omitted a term from this equation that can be quite important at low frequencies: the brightness of the sky background. This power contribution may be represented by T_{sky}, and it adds directly to T_A and T_N,

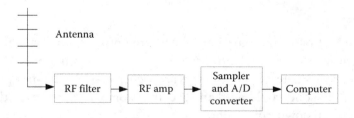

FIGURE 3.24 Block diagram of a directly sampled radio receiver. Compared to the heterodyne receiver shown in Figure 3.14, this receiver is much simpler. Because there are fewer analog electronics involved, it tends to be a more robust system.

$$P = G\, k\, \Delta v \left(T_A + T_N + T_{sky} \right)$$

The sky brightness is strongly dependent on frequency and is much stronger at low frequencies. For example, at 10 GHz the primary contribution to the sky background is the cosmic microwave background, with a temperature of 2.7 K. At this frequency, a cryogenically cooled amplifier might have a T_N of 20 K, and so the sky brightness noise can make a noticeable addition to the amplifier noise. Below 300 MHz, however, galactic synchrotron emission is much brighter than the microwave background (see Volume II); at 100 MHz it contributes a noise temperature, T_{sky}, of 1000 K and this rises to over 20,000 K at 30 MHz! Meanwhile, an uncooled amplifier might have a noise temperature, T_N, of several hundred K. In the sum of all these sources of power, that is, $T_A + T_N + T_{sky}$, there is little to gain by lowering T_N from several hundred K to several tens of K via cryogenic cooling of the receiver; the sum is dominated by T_{sky}, which we cannot reduce by any means. Despite the intense galactic synchrotron background, there are still many other astronomical sources that can be detected at these frequencies.

3.6.4 Radio Frequency Interference

A significant complication at frequencies below 300 MHz is radio frequency interference (RFI). RFI is any radio signal, made by nature or by humans, that interferes with the radio waves we wish to detect. Such interference also occurs at higher frequencies. Notable examples are cell phones, GPS, and Wi-Fi signals, occurring from about 0.8 to about 2.5 GHz. Radar systems and telephone microwave links commonly operate in the 4–8 GHz range, and geostationary satellites are problematic from about 11 to 15 GHz. Above these frequencies, however, the radio spectrum becomes much *cleaner*, with relatively little RFI being present.

The situation gets worse when we go to lower frequencies. In general, the lower in frequency one wishes to observe, the more cluttered the radio environment becomes. Although some of this interference occurs naturally, most of it results from human activity; in this way, RFI is to the low-frequency radio astronomer as light pollution is to visible-light astronomers. There are a myriad of possible sources for this interference. The most common form of natural emission is electrical storms (lighting). Humans produce both unintentional and intentional emission. Examples of the former include spark plugs in gasoline motors, faulty insulators on power lines, and poorly shielded household electrical appliances. The latter category is chiefly commercial and private radio transmissions and commercial television, but there are many other possible sources, ranging from baby monitors to garage door openers.

A number of methods are available to mitigate the effects of this interference. The most obvious methods are to locate the telescope as far away from human activity as is reasonably possible and to install filters to keep interfering signals out of the receiver path. Although both these techniques are very useful, they are often limited by factors beyond our control. Another technique, less obvious than these, is to divide the observed passband into many spectral channels. Even if the goal is a continuum observation, observing with spectral channels (in a *pseudo-continuum* mode) can be very helpful. For example, below 300 MHz it is unlikely that we would find a 20-MHz-wide window that is completely free from RFI. If we treat a 20-MHz passband as a *single* channel, then all frequencies are lumped together, and RFI occurring anywhere in the passband would ruin our observation. But if we process the 20-MHz passband as 2000 channels, each one 10-kHz wide, then we might find that the RFI is restricted to a few dozen channels. These channels can be edited out of the data, and the remaining data will serve for our continuum measurement.

Because RFI can present itself in so many forms, some ingenuity and flexibility is needed to deal with it. As an example, consider interference from pulsed radar systems. Such radar signals are commonly used for meteorological purposes and for tracking aircraft. These radars typically operate at frequencies somewhat higher than 1 GHz, and hence do not fall into the *very low frequency* category; they provide an illustrative example, nonetheless. A typical radar of this type will emit several hundred pulses of radiation per second, with each pulse lasting about 2 ms. Thus, a simple first approach for astronomical observations is to sample the data on very short time scales and edit out the data that are corrupted by the radar pulses, leaving the inter-pulse data to serve for the astronomical observation. In some cases, this may be an adequate solution. For sufficiently sensitive telescopes, however, complications will soon reveal themselves. In particular, the *reflected* waves, coming from clouds or aircraft, although much weaker, will arrive over a much longer time period than the pulse duration. To further complicate matters, in the case of aircraft, the pulses will be Doppler shifted to a different frequency. One technique for dealing with this form of RFI is by means of digital filters. These are mathematical functions that can be applied to the digital data obtained from the telescope. Such filters can identify data affected by RFI and automatically remove it. RFI identification and excision are active areas of research, and will continue to be so, as new forms of RFI continually present themselves.

QUESTIONS AND PROBLEMS

1. One of the brightest radio sources in the sky is the supernova remnant Cas A. Cas A has a flux density at a frequency of 178 MHz of 11,000 Jy. If you had a radio telescope with an effective area of 100 m^2, Cas A would be a point source. How much power would it collect in a bandwidth of 5.00 MHz at a frequency of 178 MHz?

2. A telescope with an effective collecting area of 40.0 m^2 is used to observe a radio source at a frequency of 15.4 GHz, with a bandwidth of 750 kHz. The amount of power in the radiation entering the receiver is 8.84×10^{-20} W. Assuming that the solid angle of the radio source is much smaller than the FWHM of the telescope's beam, what is the flux density of the source at this frequency?

3. The angular resolution of ground-based optical telescopes is limited by the effects of the Earth's atmosphere and is typically of order 1 arcsec. Radio telescopes, on the

other hand, are usually diffraction limited. What size radio telescope operating at a wavelength of 1 cm is needed to achieve the same angular resolution as a typical ground-based optical telescope?

4. Explain why it is important to minimize the level of the sidelobes of a radio telescope.

5. List three reasons why a telescope's effective collecting area is less than the geometric area of the plane of the front of the reflector.

6. The Green Bank Telescope is one of the largest fully steerable radio telescopes in the world with a diameter of ~100 m. If the telescope had the optimum illumination taper, as discussed in Section 3.1.3, what would be the angular resolution (θ_{FWHM}) at a wavelength of 20 cm?

7. The *rms* surface accuracy of the Green Bank Telescope is 250 μm (0.025 cm).

   a. If we require at least 1/20th of wave surface accuracy for operation, what is the shortest wavelength at which this telescope can be operated?

   b. To put the *rms* surface accuracy of the Green Bank Telescope into perspective, estimate how many sheets of paper piled together would equal 250 microns?

8. Explain the following terms: sidelobes, edge taper, illumination efficiency, and spillover.

9. Why is it important to minimize the receiver noise in order to detect weak astronomical sources?

10. A radio telescope receiver is composed of a cascade of three amplifiers. The first amplifier has a gain of 10 dB and a noise temperature of 20 K, the second amplifier has a gain of 20 dB and a noise temperature of 40 K, and the last amplifier has a gain of 30 dB and a noise temperature of 100 K.

    a. What is the overall gain and noise temperature of the receiver?

    b. If the amplifiers were placed in the opposite order, what would be the new gain and noise temperature of this receiver?

11. Assume an observed astronomical source produced an antenna temperature of 0.1 K and this signal was input into a receiver that had an overall noise temperature of 30.0 K, a gain of 60 dB and a bandwidth of 5.00 MHz.

    a. What would be the voltage out of the detector if the detector had a responsivity of 100 V W^{-1}?

    b. What fraction of the voltage out of the detector would be due to the astronomical source?

12. Consider a telescope of 2 m designed to be operated at 21 cm (like the SRT) and trying to use it to observe at 8.40-cm wavelength: Of course, you would have to install a new feed and receiver to match the new observing wavelength.

   a. The telescope is determined to have an effective collecting area at 21 cm equal to one-half of the geometrical area. The blockage due to the feed/receiver assembly and feed legs is 10% of the area. The 21-cm feed has optimum edge taper. What percentage of the incident power is lost due to surface irregularities?

   b. What is the *rms* size of the surface irregularities?

   c. Assume that your new feed has optimum edge taper for this dish when observing at 8.40 cm. Calculate the effective collecting area of this telescope at 8.40 cm using your new feed/receiver.

CHAPTER 4

Single-Dish Radio Telescope Observations

T HERE ARE TWO PRINCIPAL—but greatly different—methods for making radio astronomy observations: single-dish observations and observations made with multiple telescopes that are combined together using interferometric techniques. In this chapter, we present a thorough discussion of the basic considerations of radio observations made with a single telescope, generally referred to as *single-dish observations*. Interferometric observing techniques will be discussed in Chapters 5 and 6.

The simplest type of observation you can make is to measure the power of the radiation coming from a source and entering your antenna. In Chapters 2 and 3, we presented an equation (Equation 2.2 and 3.1, respectively) that relates the power collected by the telescope to the flux density, F_v, of the source radiation, which we repeat here to facilitate our discussion.

$$P = F_v A_{eff} \Delta v \tag{4.1}$$

where:
F_v is the flux density of the source
A_{eff} is the effective area of the telescope
Δv is the bandwidth over which the detection occurs, as determined by the receiver

This seems like a straightforward equation, suggesting a straightforward calculation of the source flux density, that is, point the telescope at a source, note the power measured by the receiver, and use Equation 4.1 to calculate to flux density. However, as you may have guessed, the actual process is not so simple. There are many details and complicating factors, which we will use much of this chapter to address. The most fundamental issue involves the understanding and proper treatment of the power measurement.

4.1 BASIC MEASUREMENTS WITH A SINGLE-DISH TELESCOPE

4.1.1 Switched Observations

Imagine that you point a single-dish radio telescope at a known astronomical radio source and read the amount of power measured by the detector. Is all of this measured power due to the astronomical radio source? No. As explained in Chapter 3, even when the telescope is pointed at a bright radio source, the power measured by the detector is dominated by noise—primarily due to the electrical components in the receiver, but also, depending on the wavelength and the particulars of the observation, to background radiation from the sky and/or thermal radiation from the ground and the dish. Therefore, the total detected power is actually a very poor measure of the power emitted by the radio source. In fact, the measure of a nonzero power, alone, does not prove that you have *even detected* the astronomical source.

To determine the amount of detected power due solely to the astronomical source, one must subtract off the power due to receiver noise and other sources of unwanted radiation. This is a fundamental principle of single-dish radio observations—all single-dish observations require the subtraction of the noise signal. This is accomplished by a process known as *switching*, in which the power is also measured when observing a nearby patch of sky that does not contain the astronomical source. The power measurement when the telescope is pointed *at* the source is generally called the *on* observation, and the measured voltage is denoted as V_{on}, and the measurement when pointed away from the source is the *off* observation, with voltage equal to V_{off}. Another form of switching, called *frequency switching*, which can be used for spectral-line observations, is discussed in Section 4.4.

As explained in Chapter 3, radio astronomers usually express the power measured from an astronomical source by its equivalent temperature called *antenna temperature* (T_A), while the noise power produced by the telescope receiver components is referred to as the *noise temperature* (T_N). We lump all of the unwanted power together, and call it the *system temperature* (T_{sys}). The voltage measured by the detector when the telescope is pointed at the source, or V_{on}, follows from Equation 3.10 as

$$V_{on} = \alpha G k \, \Delta\nu \left(T_A + T_{sys} \right) \tag{4.2}$$

where:
   α is the responsivity of the detector (and has units of V/W)
   G is the dimensionless total receiver gain (discussed in Section 3.3)
   k is Boltzmann's constant
   $\Delta\nu$ is the observing bandwidth (set by the bandpass filter)

Similarly, the voltage when pointed off-source, V_{off}, is given by

$$V_{off} = \alpha G k \, \Delta\nu \, T_{sys} \tag{4.3}$$

Subtracting Equation 4.3 from 4.2 gives us

$$V_{on} - V_{off} = \alpha G k \, \Delta v \, T_A \tag{4.4}$$

which is independent of the noise. So to find T_A, one then only needs to determine the conversion between temperature and voltage. Instead of substituting in all the factors involved ($\alpha G k \Delta v$), this is accomplished by a calibration of the system temperature, which we explain in the next subsection.

There is another important issue about switching that we must address first. The G in Equations 4.2 through 4.4 is the total gain of the radio receiver, which varies over time and can cause significant errors if not accounted for. A variation in G of a few percent causes a deviation in the conversion from detected power, in volts, to antenna temperature, in Kelvins, in Equation 4.2 by a few percent. However, since the detected power from the source, V_{on}–V_{off}, is a tiny fraction of the total detected power V_{on}, a small variation in G, causing a small change in V_{on}, translates into a tremendous error in V_{on}–V_{off}, and could even result in a negative value for the calculated antenna temperature. To prevent gain variations from producing such enormous errors, the *on* and *off* observations must *both* be made regularly throughout the observation. This technique was developed in the mid-twentieth century, and is known as *Dicke switching*, after its inventor Robert Dicke.

To be effective, switching must occur at a rate faster than any power variations. Because there are different sources of variations, often with very different timescales, different forms of switching have been developed to deal with them. Broadly speaking, the variations arise either from receiver electronics or from changes in the antenna and the atmosphere. The former can occur on timescales as short as seconds, while the latter usually is on timescales of minutes to hours. Therefore, the switching rate is usually dictated by the electronics.

The switching can take a number of forms; here we mention two of the most common types: *position-switching* and *beam-switching*. In both cases, the telescope is made to look at a different, but nearby, sky position, which is taken as the *off* position. In the first technique, the entire telescope is moved, to repoint to a different position on the sky. In the second technique, the switching may be accomplished by moving the secondary reflector of the telescope slightly, a procedure also called *nodding* of the secondary reflector. Because the secondary reflector is smaller and lighter, it can be moved more quickly and with less disturbance to the telescope. Although the switching times are highly dependent on frequency and observing conditions, position-switching is usually done on 10 to 30 s timescales, while beam-switching is often done on sub-second timescales, with switching rates of a few hertz.

How large an angular distance one must use for the *off* position depends on several factors, such as beam size, source size, and the presence of other sources. The offset position should be at least one main beam width away (to be completely off the source), but beyond that restriction, the *off* should be as nearby as possible so that all other factors are the same between the *on* and *off* positions. For larger telescopes, with beam sizes of about 10 arcmin or less, the *off* is usually less than 1° away. Especially at higher frequencies (millimeter wavelengths), the atmosphere can be noticeably different (particularly if the offset is in elevation) when pointing even a few degrees away, and hence will not provide a good sky subtraction.

4.1.2 Determination of System Temperature

The additional sources of unwanted power include radiation from the cosmic microwave background, background radiation from the Galaxy, radiation emitted by the Earth's atmosphere, radio frequency interference (RFI), radiation emitted by the telescope's primary reflector, and even radiation emitted by the ground. The cosmic microwave background (relic radiation from the Big Bang at a temperature of 2.73 K) is present in all directions and so is present in every observation, regardless of what source we observe. At low frequencies, there is a strong (and direction-dependent) background of synchrotron radiation from the Galaxy, due to high-energy electrons interacting with the galactic magnetic field. The Earth's atmosphere, although largely transparent, can be a substantial source of emission (and absorption) at wavelengths shorter than about 2 cm, due to molecules of water and oxygen, and longer than half a meter due to the Earth's ionosphere.

The primary reflector also radiates thermally, but is usually not a very important source of noise power. Finally, spillover, discussed in Chapter 3, may allow thermal radiation emitted by the ground to enter the telescope.

The determination of the system temperature is part of the standard set of *calibration* steps that are performed when observing. Most radio observatories use a device called a noise source or *noise diode* for the T_{sys} calibration. A noise diode is a device that produces a known amount of radio-frequency power per unit bandwidth. Therefore, these noise diodes have a known equivalent temperature, which is called the *calibration temperature,* or T_{cal}. The noise diode is configured to inject its power into the telescope's receiver. If we look at blank sky and measure the voltages with the noise diode turned on and off, we can make another type of switched measurement. When the noise diode is on, the measured voltage is V_{cal} and when it is off, we measure V_{off}. Since V_{cal} includes the total noise power (T_{sys}) in addition to the calibrator power (T_{cal}), we can write analogous equations to Equations 4.2 and 4.3, and then divide one by the other to obtain the voltage and temperature ratio relation

$$\frac{V_{cal}}{V_{off}} = \frac{T_{cal} + T_{sys}}{T_{sys}}$$

Since T_{cal} is known (from the specification of the noise diode being used) this switched power measurement permits the determination of T_{sys} via

$$T_{sys} = \frac{V_{off}}{V_{cal} - V_{off}} T_{cal} \tag{4.5}$$

This procedure is normally set up to occur almost automatically. The observer need only input the command for a system temperature calibration, and the computer turns the noise diode on, measures V_{cal}, turns the noise diode off, measures V_{off}, performs the arithmetic in Equation 4.5, and records (or reports) the inferred system temperature. It is important to understand that Equation 4.5 assumes that the power measurement is made while pointing at empty sky, with no radio source in the telescope beam, other than the astronomical background discussed earlier.

This calibration procedure also provides the conversion from voltage to temperature. Keep in mind that the detection of an astronomical source is described by the antenna temperature (in Kelvins), while the telescope detects a power, measured as a voltage, which depends on the amount of gain in the receiver. Radio telescopes can have varying amounts of amplification, though, so the voltage that is measured, for the same astronomical source, will differ between telescopes. For a given observation, then, we need to determine the conversion from measured voltage to an equivalent temperature before amplification. Note that in Equation 4.5, T_{cal} describes the power, in Kelvins, of the noise diode before amplification, and therefore we get the conversion factor that we need, given by $T_{cal}/(V_{cal} - V_{off})$. This also means that the total multiplicative factor in front of the temperatures in Equations 4.2–4.4 is determined by calibration and does not need to be calculated explicitly; an observer does not need to know the responsivity, α, of the detector or the total gain, G, in the receiver.

4.1.3 Measurement of Antenna Temperature

The conversion between T_{sys} and V_{off} enables the calculation of the antenna temperature. As with the determination of system temperature, we can set up a ratio of the temperatures from Equations 4.2 and 4.3, yielding

$$\frac{V_{on}}{V_{off}} = \frac{T_A + T_{sys}}{T_{sys}}$$

This can be rewritten to provide an expression for the antenna temperature as

$$T_A = \frac{V_{on} - V_{off}}{V_{off}} T_{sys} \tag{4.6}$$

The dimensionless quantity $(V_{on} - V_{off})/V_{off}$ is, effectively, a measure of the antenna temperature as a fraction of the system temperature. Similarly, in Equation 4.5, $V_{off}/(V_{cal} - V_{off})$ is a measure of the T_{sys} as a fraction of T_{cal}.

4.1.4 Uncertainty in the Measured Antenna Temperature

Since it is almost always the case that the antenna temperature of the astronomical source is much smaller than the system temperature, the uncertainty in the antenna temperature will be dominated by the inherent fluctuations of the noise power. A quantitative, statistical measure of the amount of variance due to random fluctuations is usually calculated from the root mean square (*rms*) of the individual measurements. The mathematical definition of rms is

$$\text{rms} = \sqrt{\frac{\sum_i (x_i - \bar{x})^2}{N}}$$

where:

x_i is the value of the ith measurement

$\bar{x}$ is the average measured value

N is the number of measurements

With regard to a radio telescope measurement, the *rms* of the fluctuations in the noise (discussed in Section 3.3) is given by the *radiometer equation*, which says that in a measurement of duration Δt, made over a bandwidth Δv, the *rms* fluctuation of the antenna temperature, $\sigma(T_A)$, is given by

$$\sigma(T_A) = \frac{T_{sys}}{\sqrt{\Delta t \Delta v}} \tag{4.7}$$

It should be noted, however, that this equation does not take into account other systematic factors (such as variations of the gain in the receiver) which could occur and increase the uncertainties. As we explain in Section 3.3, in a single independent measurement the uncertainty in power is proportional to power (or the uncertainty in temperature is proportional to temperature), and by making many independent measurements, the uncertainty is reduced by a factor of $N^{1/2}$. Since noise is random, sometimes it will have a value larger than the average and other times a smaller value, while the signal from the source is consistent. In the average, some of the positive deviations cancel with some of the negative deviations. For radio observations, the number of independent measurements is given by $\Delta t \, \Delta v$, resulting in the reduction in the temperature uncertainty reflected in Equation 4.7.

Observing with a larger bandwidth will increase the precision of the measurement. However, often the bandwidth may be fixed or dictated by the science, or the receiver configuration, so it is commonly the case that we reduce the uncertainty in the antenna temperature by increasing the duration of the observation.

A demonstration of the dependence of the noise on observing duration is provided in Figure 4.1a and b. This example is for a spectral-line observation and so Δv in Equation 4.7 is the width of each spectrometer channel, which, in Figure 4.1, is only 0.007813 MHz. The emission line in Figure 4.1a is strong and clearly apparent, but because there are substantial noise fluctuations present throughout the spectrum, measurements of the strength and width of the emission line will suffer greater uncertainty. If the duration of the observation was increased, the source signal reinforces itself, while the random noise fluctuations will be reduced. Thus, the average value of the source signal remains the same, while the *rms* fluctuations tend toward zero, as demonstrated in Figure 4.1b. By increasing the time of the observation by a factor of 42, the noise is reduced by a factor of the square root of 42, or about 6.5, and the resulting *signal-to-noise ratio* (SNR) increases by the same factor.

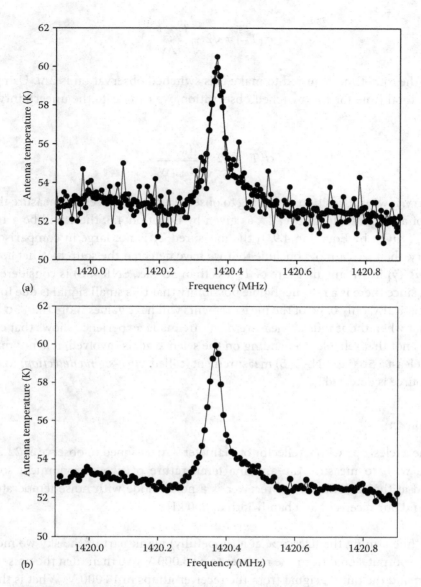

FIGURE 4.1 (a) Measured spectrum in a short observation. (b) An observation with an integration time 42 times longer than that shown in (a).

Since we need to switch between *on* and *off* source measurements, our observing time must include time *off* the source as well. Furthermore, the *on* and *off* source measurements are subtracted, and so both of their uncertainties will contribute to the total uncertainty. Assume we observe *on* source for Δt seconds and *off* source for the same amount of time, so that each *on* and *off* measurement has the same uncertainty as given by Equation 4.7. When we subtract the two measurements, the uncertainties of each value are random and unrelated so they add in quadrature. In this case, that means that the *rms* in the difference will be the square root of two times larger. Therefore, in a position-switched observation the uncertainty in the measured antenna temperature is

$$\sigma(T_A) = \sqrt{2}\,\frac{T_{sys}}{\sqrt{\Delta t \Delta \nu}} \tag{4.8}$$

Note that the total time required to make this switched observation is $2\Delta t$. Using $\Delta t_{obs} = 2\Delta t$ as the total time for the switched observation, we can write the uncertainty expression as

$$\sigma(T_A) = 2\,\frac{T_{sys}}{\sqrt{\Delta t_{obs} \Delta \nu}} \tag{4.9}$$

In summary, if we use position switching to observe a radio source and measure the power in terms of antenna temperature, T_A, as given by Equation 4.6, there will be a 1σ uncertainty in T_A given by Equation 4.9. If the measured T_A is not large in comparison to the uncertainty, then we cannot conclude that we have detected the source. What constitutes a *detection*? Typically, any measure of T_A less than 3σ (i.e., SNR < 3) is considered a non-detection, since there is a non-negligible possibility that this small signal is due to random noise. Statistically, only 0.3% of the measurements will have values more than 3σ from the average, but when thousands of measurements are made, experience shows that even a 3σ criterion is not that reliable. Depending on the science goals involved, it is not unusual to demand at least a 5σ (i.e., SNR > 5) measurement (called a *five-sigma detection*) to conclude that the source is detected.

Example 4.1:

Imagine a telescope with a reflector of diameter 7 m designed to observe at 22.2 GHz and we wish to measure the antenna temperature of two astronomical sources. Inserted at the entrance to the receiver is a noise diode with noise temperature of 250 K and our receiver has a bandwidth of 500 kHz.

1. While aiming the telescope at a (hopefully) blank part of the sky we measure the output signal from the receiver to be 3.000 V. We then turn the noise diode on and the output signal from the receiver jumps to 18.000 V. What is the system temperature of this observation?
2. We immediately point the telescope at the first source and the detector output reads 3.002 V, what is the antenna temperature of this source?
3. After being interrupted by a lengthy phone call, we point the telescope at the second source and the detector now reads 2.998 V. What is the antenna temperature of this source? Does this make sense? What must be done differently?
4. A switched observation is performed on the first source for a total *on + off* observing time of 5 min. Calculate the percent uncertainty in this measurement.
5. If we want to measure the antenna temperature of the first source with an SNR of 10, how much longer must we observe?

Answers:

1. To calculate the system temperature, we use Equation 4.5, and so we have

$$T_{sys} = \frac{3.000\ \text{V}}{18.000\ \text{V} - 3.000\ \text{V}} 250\ \text{K} = 50.0\ \text{K}$$

2. The antenna temperature is given by Equation 4.6

$$T_A = \frac{3.002\ \text{V} - 3.000\ \text{V}}{3.000\ \text{V}} 50.0\ \text{K} = 0.0333\ \text{K}$$

3. There are (at least) two mistakes that can be made. To demonstrate, we will do the incorrect calculations first, and then explain the correct process.

   a. *First incorrect approach*: One might be tempted to take the ratio of the two voltage readings, that is,

   $$\frac{P(\text{first source})}{P(\text{second source})} = \frac{3.002}{2.998}$$

   set this equal to the ratio of the antenna temperatures. But, this ignores the fact that both signals are predominantly due to noise. The noise power in the system must be subtracted from both the numerator and denominator, and so the correct ratio of powers will be larger.

   b. *Second incorrect approach*: So, let us subtract off the noise, as we did in (2). In (1) we saw that when pointed at blank sky, we got a voltage of 3.00 V. So, this must be the signal due to the noise, right? But, if we subtract this value, we conclude that the signal due to the second source is negative! 2.998–3.000 V = –0.002 V. How could this be? Recall that the amplifier gains can vary on timescales of minutes, and a fair bit of time passed between our measurement of the blank sky and the second source. In that amount time, the noise signal changed. This might be due to a time variation in the amplifiers, or to a different elevation angle between the two sources, so that one observation looked through a longer atmospheric path than the other.

   The correct approach, then, is to make blank sky measurements frequently and subtract from the on-source measurements made around the same time and elevation. This is what we called a switching observation. So, the data given in Question (3), in fact, are useless.

4. Using Equation 4.9, and the system temperature result in (1), we have that the uncertainty is

$$\sigma(T_A) = 2 \frac{50\ \text{K}}{\sqrt{5\ \text{min} \times (60\ \text{s}\,\text{min}^{-1})\ 5.00 \times 10^5\ \text{Hz}}} = 0.008\ \text{K}$$

The measured antenna temperature is 0.0333 K, so as a percentage, this uncertainty is

$$\frac{0.008}{0.0333}100\% = 25\%$$

5. The SNR in (4) is 0.0333/0.008 = 4.16. If we wish to have a solid measurement by demanding a 10σ result, we can observe for a longer time. Assuming that our preliminary measurement of the source's antenna temperature of 0.0333 K is correct, we wish to get our uncertainty down to 0.00333 K, which means using a total observing time, Δt_{obs} given by

$$0.00333\ \mathrm{K} = 2\frac{50\,\mathrm{K}}{\sqrt{\Delta t_{\mathrm{obs}}\ \mathrm{min} \times (60\,\mathrm{s\,min^{-1}})\ 5.00 \times 10^5\,\mathrm{Hz}}}$$

or $\Delta t_{\mathrm{obs}} = 30$ min. Since we already have 5 min of observation, we need to add another 25 min.

Once the source is detected, you will, of course, want to use your measured value to describe the amount of radiation coming from the source, and this is not what antenna temperature represents. Antenna temperature is strictly a measure of the power detected due to the source, and this depends on the details of the telescope, in particular its effective aperture. For example, if you use a larger telescope, more of the source's radiation will be collected and fed into the receiver, leading to a larger antenna temperature.

So, how do we describe the source's radiation power, independent of the observation? As explained in Chapter 2, we will use *intensity* or *brightness temperature* (see Section 2.4) and/or *flux density*. Before we can make this conversion calculation, there are a number of other factors that we will need to account for. Keep in mind as we go forward that the antenna temperature, alone, is not sufficient to infer physical information about the source.

4.2 ANTENNA BEAM

An important point to bear in mind is that radiation received by the telescope can come from many different directions, but the ability of the telescope to capture the power of that radiation is not the same in all directions. Hence, the antenna temperature we measure will depend on both the intensity of the radiation from the astronomical source, which is a function of position on the sky, *and* on the sensitivity of the telescope—also a function of position on the sky.

The sensitivity of a telescope as a function of angle relative to the pointing direction is known as the telescope's *beam pattern*, or *power pattern*, and is a fundamental characteristic of the telescope. The beam pattern has a significant impact on the observations. We discussed in Section 3.1 the determination of a telescope's beam pattern, and displayed the profile of a typical beam pattern (see Figure 3.8). To aid our discussion here, we show another plot of a typical beam pattern in Figure 4.2.

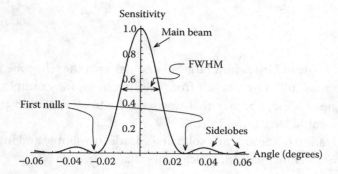

FIGURE 4.2   Sensitivity function, in one dimension for a 1.4-cm observation with a 40-m diameter radio telescope and 10-dB edge taper.

4.2.1 Beam Power Pattern and Antenna Solid Angle

We will denote the normalized beam power pattern in this chapter as $P_{bm}(\theta)$. In principle, the beam pattern is a function of angular position in two orthogonal directions, and is generally represented by $P_{bm}(\theta,\phi)$. The vast majority of radio telescopes have circular primary reflectors, and so their beam patterns are approximately axially symmetric. To simplify calculations in our examples, we will assume axially symmetric beams. A one-dimensional plot of the beam, $P_{bm}(\theta)$, then, is sufficient to describe the telescope's beam pattern. This function is normalized to unity in the direction where the telescope has its maximum sensitivity, which is usually in the direction the telescope is pointed, while in other directions, θ, it is decreased by the factor $P_{bm}(\theta)$.

In Appendix II, as an example, we derive the one-dimensional beam pattern for a uniformly illuminated rectangular horn antenna. The normalized beam function in this case, see Equation II.9, is

$$P_{bm}(\theta)=\frac{\sin^2\left[\left(\pi D/\lambda\right)(\sin\theta)\right]}{\left[\left(\pi D/\lambda\right)(\sin\theta)\right]^2}$$

The value of $P_{bm}(\theta)$ is unity when $\theta = 0$ and decreases for increasing angle θ. For a uniformly illuminated circular aperture, the mathematical form of the beam pattern is a first order Bessel function. A realistic beam pattern includes the effect of the edge taper (discussed in Section 3.1.3) and has a shape that is not a simple analytic function.

The integration of the beam power pattern over all angles is an important parameter called the *antenna solid angle*, often denoted as Ω_A. Including the integration over both angles for completeness, the antenna solid angle is defined by

$$\Omega_A = \int P_{bm}(\theta,\phi)d\Omega \tag{4.10}$$

where:

$$d\Omega = \sin\theta \; d\theta \; d\phi$$

The antenna solid angle indicates how large an angular area the telescope responds to. An antenna that responds fully to radiation from all directions, for example, would have an antenna solid angle of 4π sr, and one that was sensitive to all radiation coming from the forward direction would have $\Omega_A = 2\pi$ sr.

A rigorous derivation of the antenna solid angle leads to a simple and important expression, known as the *antenna theorem*, which states that

$$\Omega_A = \frac{\lambda^2}{A_{\text{eff}}} \tag{4.11}$$

This is a powerful theorem, in part because it is absolutely true, *always*, regardless of how well the reflector dish is made or of the design of the feed horn. We will use the antenna theorem through much of this chapter.

4.2.2 Main Beam and Angular Resolution

The beam pattern shown in Figure 4.2 contains a large central region, which peaks (with a normalized value of 1.0) in the pointing direction of the telescope. As we move further away from the telescope axis in angular distance, the pattern decreases to zero, and then oscillates between zero and ever-decreasing (but nonzero) values. The central, main part of the beam is called the *main beam*. The zero points on either side of the main lobe are called the *first nulls*, and the nonzero responses outside the first nulls are called *sidelobes*.

In practice, the main beam pattern is often approximated as a Gaussian function of the form

$$P_{\text{main}}(\theta) = \exp\left[-4\ln 2\left(\frac{\theta}{\theta_{\text{FWHM}}}\right)^2\right] \tag{4.12}$$

where:

θ_{FWHM} is the full width at half maximum (FWHM) of the main beam (see Figure 4.2)

This function is less cumbersome than the Bessel function for analytical treatment and is a very close approximation to the main beam. The *solid angle of main beam*, Ω_{main}, can then be obtained by integrating Equation 4.12 over angle, which leads to

$$\Omega_{\text{main}} = \frac{\pi}{4\ln 2}\theta_{\text{FWHM}}^2 \tag{4.13}$$

Note, by the way, that the solid angle of the main beam is *not*, simply, $\pi \left(\theta_{\text{FWHM}}/2\right)^2$; this is a very common mistake, which you now know to avoid.

The main beam determines the telescope's *angular resolution* (i.e., the smallest angle at which detail can be discerned). Think of the observing beam as a probe that you use to explore the sky. The map that you make with the probe will not reveal details on scales smaller than the probe itself, which in this case is the beam size. One can infer information on slightly smaller scales by *deconvolving* the observed data from the telescope beam pattern, provided that the signal-to-noise ratio in the data is large and the beam pattern is well known. We will leave a discussion of deconvolution until later and focus at present on the standard angular resolution of the telescope.

The standard convention used in radio astronomy is to define the angular resolution as the FWHM of the main beam of the telescope. As discussed in Section 3.1.3, the FWHM of the main beam depends on several factors, including the diameter of the reflector, the wavelength of the observation, and the *illumination* of the primary reflector by the feed. The best angular resolution occurs when the primary reflector is uniformly illuminated, meaning that the detector is equally sensitive to radiation reflecting off all parts of the dish. In this case, the FWHM of the main beam is

$$\theta_{FWHM} = 1.02\lambda/D$$

where:
    θ is in radians
    λ is the wavelength of the observation
    D is the diameter of the primary reflector

In practice, a radio telescope does not usually have uniform illumination of the primary reflector due to the diffraction that occurs at the feed horn. Moreover, uniform illumination is not necessarily desirable as this will usually imply a large spillover loss (see Section 3.1.3). In general, the center of the reflector is more effective at collecting power than the edge. The ratio of the relative collecting ability per unit area between the edge and the center is called the *edge taper* and is usually expressed in decibels. For the optimum edge taper (the 10-dB taper; see Section 3.1.3), the FWHM of the beam is given by

$$\theta_{FWHM} = 1.15\lambda/D \tag{4.14}$$

For the sake of discussion, we will use Equation 4.14 to represent the resolution in all our examples, but you should keep in mind that any telescope you use may have a different primary reflector illumination, which will alter the resolution slightly. Remember that the optimum edge taper of 10 dB is designed to maximize the effective area of the telescope.

Consider observing at a wavelength of 21 cm (an important wavelength, as you will come to appreciate) with a radio telescope of diameter 2.0 m (such as the Haystack SRT). The angular resolution of this observation is 1.15 λ/D = 1.15 (0.21 m/2 m) = 0.12 radians ~ 6.9°. This is almost 14 times the angular diameter of the Moon. So if there were two radio sources separated by, say, 8 times the Moon's diameter, in a simple observation with this telescope at this wavelength you would not be able to determine that the radio emission was coming from two distinct sources. It should be clear that this antenna/wavelength combination cannot resolve any structure of the Moon, or even determine its angular size;

our *probe* of 6.9° is much too coarse for that. This demonstrates why most telescopes used for studies at 21 cm must be tens to hundreds of meters in diameter.

What is the best resolution obtainable with a single radio telescope? Let us consider a short radio wavelength—say $\lambda = 1$ cm—and a large radio telescope—let us use a diameter of 100 m (the diameter of the Green Bank Telescope). The angular resolution for such an observation is ~ 1.15 (0.01 m/100 m) ~ 0.000115 radians ~ 0.0066° ~ 24 arcsec. This is not so bad, but remember that this is at the high-resolution end and with one of the larger radio telescopes in the world. The same telescope, operating at 21 cm, has an angular resolution 21 times poorer; however, if it could operate at the upper end of the radio window, ~ 1 mm, then the resolution would be 10 times better.

In short, because of the long wavelengths of radio waves, radio telescopes generally have very poor resolution (note that even small visible-wavelength telescopes have angular resolutions of 1 arcsec). Historically, this was a problem for radio astronomy. The invention of aperture synthesis, which we discuss in Chapters 5 and 6, solved this problem. Using a particular form of aperture synthesis called *very long baseline interferometry*, radio astronomers can now observe with micro-arcsecond resolution.

Example 4.2:

Using our 7-m radio telescope introduced in Example 4.1, we observe at a frequency of 1.1 GHz an elliptical radio source that has a major axis of 0.3° and a minor axis of 0.1°. After switching receivers, we repeat the observation at 22.2 GHz. Both feeds have 10-dB edge tapers.

  1. What is the angular resolution in each of these observations?
  2. Will we resolve the elliptical shape of the radio source?

Answers:

  1. With the 10-dB taper, we use Equation 4.14, $\theta_{res} = 1.15\ \lambda/D$. At 1.1 GHz, then,

$$\theta_{res} = 1.15 \frac{3.0 \times 10^8\,\mathrm{m\,s^{-1}}}{(1.1 \times 10^9\,\mathrm{Hz})\,7\,\mathrm{m}} = 0.0448 \text{ radians} = 2.57°$$

And at 22.2 GHz, the resolution is

$$\theta_{res} = 1.15 \frac{3.0 \times 10^8\,\mathrm{m\,s^{-1}}}{(22.2 \times 10^9\,\mathrm{Hz})\,7\,\mathrm{m}} = 0.00222 \text{ radians} = 0.127°$$

  2. At 1.1 GHz, even the major axis is much smaller than the resolution angle and so we cannot detect the ellipticity of this source, or even measure its angular size. At 22.2 GHz, however, the major axis is larger than the resolution angle, and so it can be measured (although, as explained below, a deconvolution will be needed since the major axis is only marginally larger than the beam). The minor axis is smaller than the beam and so is not directly resolved. However, since the minor axis is only a little smaller than the beam, if the SNR of the observation is large enough, the size of the minor axis can be measured via a deconvolution (as we will explain in Section 4.5.2).

4.2.3 Main Beam Efficiency

In Figure 4.2, we saw that in certain directions—well removed from the pointing axis—the antenna beam pattern has a small, but non-negligible sensitivity. These sidelobes are undesirable features of the beam that we must accept. They are problematic because they allow radiation from other parts of the sky to enter the receiver, and we cannot easily distinguish this power contribution from that of the radio source we are pointing directly at. In general, one wants the sensitivity in the sidelobes to be as small as possible. If the sidelobes have a peak sensitivity of 1%, for example, then a source equal in flux density to the target source that happens to fall in the sidelobe will contribute 1% of the power detected. In addition, if the source in the sidelobe were 100 times brighter than the target source, then it would contribute exactly the same power as the target itself.

Not apparent in Figure 4.2, but also a significant concern, is the *error pattern* due to the reflector's surface irregularities (discussed in Section 3.1.4). The existence of the sidelobes and error pattern also means that the telescope, effectively, sees a larger area of the sky than the main beam. Thinking of the telescope as a transmitter (see discussion of reciprocity theorem in Section 3.1.2), we see that some power is radiated into the sidelobes and error pattern. Therefore, with the telescope acting as a detector of radiation, the presence of the sidelobes and error pattern reduces the telescope's sensitivity to radiation in the main beam. In fact, the antenna theorem (Equation 4.11) shows that there is a relation between the effective area of the telescope and the antenna solid angle.

As an example, consider the solid angle of the main beam for a circular telescope with the 10-dB edge taper. Using the expression for the FWHM from Equation 4.14 and the main beam solid angle from Equation 4.13, we obtain

$$\Omega_{main} = \frac{\pi}{4\ln 2}\left(1.15\frac{\lambda}{D}\right)^2 = 1.50\frac{\lambda^2}{D^2}.$$

The geometric area of the reflector dish is

$$A_{geom} = \frac{\pi}{4}D^2 = \frac{D^2}{1.273}$$

and so the relation between the main beam solid angle and the geometric area of the dish in this case is

$$\Omega_{main} = 1.18\frac{\lambda^2}{A_{geom}} \tag{4.15}$$

With the 10-dB edge taper, the maximum effective area is 0.82 of the geometric area (as discussed in Section 3.1.3), or $A_{eff} \le 0.82\,A_{geom}$, and so we can substitute in for the geometric area to give

$$\Omega_{main} \le 0.97\frac{\lambda^2}{A_{eff}}$$

or

$$\Omega_{\text{main}} \leq 0.97 \Omega_A$$

Thus, the antenna theorem says that the main beam solid angle can be up to 97% of the antenna solid angle when using a 10-dB taper.

It is instructive to think of the antenna theorem in the form $\Omega_A A_{\text{eff}} = \lambda^2$. Conceptually, this says that for observations at a given wavelength the product of the antenna solid angle with the effective area is a *constant*. So, if a telescope were modified to increase one of the two quantities, the other must decrease. Because Ω_A represents resolution, while A_{eff} represents sensitivity, an implication of the antenna theorem is that a larger telescope, with more collecting area and hence greater sensitivity, will necessarily see a smaller region of the sky.

Ideally, we would like the difference between the antenna solid angle and main beam solid angle to be as small as possible. The fraction of the antenna solid angle made up by the main beam is called the *main beam efficiency*, which we denote as η_{mb}, and define by

$$\eta_{\text{mb}} = \frac{\Omega_{\text{main}}}{\Omega_A} \tag{4.16}$$

where:
Ω_{main} is given by Equation 4.13
Ω_A is given by Equation 4.11

Similarly, we can describe the ratio of the telescope's effective area to its geometrical area as its *aperture efficiency*,

$$\eta_A = \frac{A_{\text{eff}}}{A_{\text{geom}}} \tag{4.17}$$

From our discussion above, we can show that there is a relation between the main beam efficiency and aperture efficiency. Starting with Equation 4.15, we substitute in for Ω_{main} and A_{geom} in terms of the beam and aperture efficiencies, to give

$$\eta_{\text{mb}} \Omega_A = 1.18 \frac{\eta_A \lambda^2}{A_{\text{eff}}}$$

Then, by the antenna theorem, we can cancel the Ω_A on the left with the $\lambda^2 / A_{\text{eff}}$ on the right and we find that, for the 10-dB edge taper,

$$\frac{\eta_{\text{mb}}}{\eta_A} = 1.18 \tag{4.18}$$

This equation assumes that the FWHM of the main beam equals 1.15 λ/D; however, there are many problems (such as the optics being out of focus or misaligned) that would cause the main beam to be larger and thus increase this ratio.

Example 4.3:

Suppose that our 7-m diameter telescope has an effective aperture at 22.2 GHz of 23.1 m^2.

1. What is the aperture efficiency?
2. What is the antenna solid angle at 22.2 GHz?
3. What is the solid angle of the main beam? (From Example 4.2, we know that at 22.2 GHz, $\theta_{FWHM} = 0.00222$ radians.)
4. What is the main beam efficiency?
5. Do the aperture and main beam efficiencies relate as expected by Equation 4.18?

Answers:

1. Using Equation 4.17, and $A_{geom} = (\pi/4) (7.0 \text{ m})^2 = 38.5 \text{ m}^2$ we have

$$\eta_A = \frac{23.1 \text{ m}^2}{38.5 \text{ m}^2} = 0.60$$

2. By Equation 4.11,

$$\Omega_A = \frac{\lambda^2}{A_{eff}} = \frac{(3.00 \times 10^8 \text{ m s}^{-1})^2}{(22.2 \times 10^9 \text{ Hz})^2 (23.1 \text{ m}^2)} = 7.91 \times 10^{-6} \text{ sr}$$

3. The solid angle of the main beam is given by Equation 4.13.

$$\Omega_{main} = \frac{\pi}{4 \ln 2} (0.00222)^2 = 5.58 \times 10^{-6} \text{ sr}$$

4. The main beam efficiency, then, is

$$\eta_{mb} = \frac{5.58 \times 10^{-6} \text{ sr}}{7.91 \times 10^{-6} \text{ sr}} = 0.705$$

5. The ratio of these efficiencies is

$$\frac{\eta_{mb}}{\eta_A} = \frac{0.705}{0.600} = 1.18$$

which is consistent with Equation 4.18.

4.2.4 Detected Power from Extended Sources

We now discuss how the beam pattern affects the detected power when observing an astronomical source with some significant angular extent on the sky. The radiation from the source can be described by the intensity as a two-dimensional function of angle on the sky, that is, $I_\nu(\theta,\phi)$, with the telescope pointed in some particular direction, which we label (θ_0,ϕ_0). The normalized beam pattern is a function of angle *relative to the direction the telescope is pointed*, so the normalized beam response in direction (θ,ϕ) is given by $P_{bm}(\theta-\theta_0,\phi-\phi_0)$. We know from Equation 4.1 that the amount of power detected by the telescope due to radiation from an astronomical source is given by

$$P = F_\nu A_{eff} \Delta\nu$$

This equation, though, assumes that all of the source's flux density is detected with equal sensitivity, which is not true if the radiation comes from a direction other than the beam center. In essence, a smaller fraction of the flux coming from positions further away from the beam center contributes to the detected power. Consider the radiation coming from direction (θ,ϕ), which has intensity $I_\nu(\theta,\phi)$. If we multiply this intensity by an infinitesimal solid angle, $\delta\Omega$, we get flux density, and the fraction that is passed to the receiver is obtained by multiplying by the normalized beam function at this position, $P_{bm}(\theta-\theta_0,\phi-\phi_0)$. Thus, the detected power from this small piece of sky is

$$\delta P = \left\{ I_\nu(\theta,\phi) P_{bm}(\theta-\theta_0,\phi-\phi_0)\delta\Omega \right\} A_{eff}\Delta\nu$$

By integrating over the entire sky (over angles θ and ϕ), we find the total response of the telescope receiver to the entire distribution of input radiation. The total amount of power detected is given by

$$P = \left\{ \int_{\text{whole sky}} I_\nu(\theta,\phi) P_{bm}(\theta-\theta_0,\phi-\phi_0)d\Omega \right\} A_{eff}\Delta\nu$$

where, as before, $d\Omega = \sin\theta\, d\theta\, d\phi$.

However, this is not quite correct. Remember from our discussion of transmission lines in Section 2.1.1 that most radio receivers only detect one of the two linear (or circular) polarizations of light, so if the light from the astronomical source is unpolarized, then only half of the power is detected. Therefore, the detected power is

$$P = \frac{1}{2}\left\{ \int_{\text{whole sky}} I_\nu(\theta,\phi) P_{bm}(\theta-\theta_0,\phi-\phi_0)d\Omega \right\} A_{eff}\Delta\nu \qquad (4.19)$$

Since the detected power due to the source is generally measured as an antenna temperature, we substitute in for P in Equation 4.19 using Equation 3.5 and obtain for the antenna temperature

$$T_A = \frac{A_{eff}}{2k} \left\{ \int\limits_{\text{whole sky}} I_v(\theta,\phi) P_{bm}(\theta-\theta_0,\phi-\phi_0)d\Omega \right\} \qquad (4.20)$$

Equations 4.19 and 4.20 will be needed in the following discussion to determine the detected power when observing extended astronomical sources.

4.3 OBSERVING RESOLVED VERSUS UNRESOLVED SOURCES

Astronomical sources are often referred to as either being resolved or being unresolved. A *resolved* or *extended* source has an angular size larger than the main beam of the telescope. Observing such a source is intrinsically a very different situation from observing an *unresolved* source (sometimes called a *point source*), that is, one whose angular size is much smaller than the main beam. We examine the antenna temperature, given by Equation 4.20, in each of these two limiting cases.

4.3.1 Unresolved Sources

The solid angle of an unresolved source is much smaller than the main beam solid angle, so when we point the telescope at the source in this limiting case, we can assume that the value of the normalized beam pattern is unity over the entire solid angle of the source. Therefore, we can take the normalized beam pattern term out of the integral in Equation 4.20, set it equal to 1, and integrate only over the solid angle of the source. The measured antenna temperature of the source, therefore, becomes

$$T_A = \frac{A_{eff}}{2k} \left\{ \int\limits_{\text{source}} I_v(\theta,\phi)d\Omega \right\}$$

The integral inside the brackets is just the flux density (see Section 2.1). Therefore, when we observe a point source, the measured antenna temperature is simply given by

$$T_A = \frac{A_{eff}}{2k} F_v$$

Thus, this observation yields a measure of the source flux density, via

$$F_v = \frac{2k}{A_{eff}} T_A \qquad (4.21)$$

This is an important point that we restate for emphasis: *if the astronomical source is a point source, then we obtain a direct measure of the flux density of the source.*

Note that the quantity $2k/A_{eff}$ is the calibration factor that converts antenna temperature to flux density for a point source. In the calibration of a telescope, we actually measure the inverse of this—how much antenna temperature results for each jansky (Jy) of flux density from a point source. This calibration factor is often expressed in units of Kelvins per Jy and is sometimes called the *degrees per flux unit* (DPFU). The DPFU is usually measured directly by observing a source of known flux density and measuring its antenna temperature, that is,

$$DPFU = \frac{T_A}{F_\nu} \qquad (4.22)$$

which should equal $A_{eff}/2k$. We discuss this important calibration in Section 4.6, where we summarize the important calibration steps.

Example 4.4:

We use our 7-m diameter telescope to make the following observations at 22.2 GHz.

1. We do a switched observation of a point source known to have a flux density of 25.0 Jy and obtain an antenna temperature of 0.209 K. What is the effective area of our telescope? What is the DPFU of our telescope at this frequency?
2. We then use the telescope to perform a switched observation of a new source, and get an antenna temperature of 0.05 K. If this new source is unresolved (which still needs to be determined), what is its flux density?

Answers:

1. Using Equations 4.21 and 4.22, we have

$$DPFU = \frac{0.209 \text{ K}}{25.0 \text{ Jy}} = 0.00837 \text{ KJy}^{-1}$$

and

$$A_{eff} = 2k(DPFU) = 2 \times 1.38 \times 10^{-23} \text{ J K}^{-1} \times 0.00837 \text{K Jy}^{-1} \frac{1 \text{ Jy}}{10^{-26} \text{ (W m}^{-2} \text{ Hz}^{-1} \text{ sr}^{-1})}$$

$$= 23.1 \text{ m}^2$$

2. Using our inferred DPFU, we have that the new source's flux density is

$$F_\nu = \frac{T_A}{DPFU} = \frac{0.05 \text{ K}}{0.00837 \text{ K Jy}^{-1}} = 5.97 \text{ Jy}$$

4.3.2 Resolved Sources

For resolved sources, the situation is more complicated. So, consider now a source that is larger than the main beam.

In this case, there are directions where I_v (θ,ϕ) is nonzero and $P_{bm}(\theta,\phi)$ is not unity and so we cannot take the power pattern out of the integral, as in the unresolved source case. Basically, some of the flux density of the source does not get detected because of the reduced sensitivity at angles offset from the central pointing direction.

Let us assume, for discussion sake, that the source has uniform brightness and an angular size larger than the telescope's main beam, but does not extend as far as the sidelobes. In this case, the integrand in Equation 4.20 is zero outside the main beam, and the I_v is constant inside the main beam. The integral, then, becomes the intensity multiplied by the integration over the main beam, or

$$T_A = \frac{A_{eff}}{2k}\Omega_{main}I_v \qquad (4.23)$$

where:

$$\Omega_{main} = \frac{\pi}{4\ln 2}\theta_{FWHM}^{2}$$

If the source does not have uniform brightness, then the I_v in Equation 4.23 is the average intensity of the source over the main beam. The lesson here is that, in contrast to observations of an unresolved source, when the source is *resolved* we do *not* measure the flux density of the source since some of the source's flux density is undetected. Instead, the antenna temperature is a measure of *intensity averaged over the main beam*. Rearranging Equation 4.23 to reflect this result, we have

$$<I_v> = \frac{2k}{A_{eff}\Omega_{main}}T_A \qquad (4.24)$$

This equation is often used in a slightly different form. Using the antenna theorem and the definition of the main beam efficiency, we can rewrite Equation 4.24 as

$$<I_v> = \frac{2k}{\lambda^2\eta_{mb}}T_A$$

Since the measured antenna temperature naturally relates to the flux density of incident radiation, as in Equation 4.21, the average intensity over the main beam from a resolved source is commonly described as the amount flux density per beam. This is usually expressed in units of *Jy per beam* and is an *intensity*, not a flux density. Note that, in this case, we are measuring solid angles in units of *beams*, where 1 beam = Ω_{main}. To convert intensity to Jy/sr we divide by the solid angle of the beam in steradians,

$$I_v(\text{Jy sr}^{-1}) = I_v(\text{Jy beam}^{-1})/\Omega_{main}(\text{sr})$$

The units Jy/beam, which commonly appear in radio astronomy maps, can cause confusion. They give the number of Jy per beam, regardless of whether the source is resolved and regardless of the size of the beam. If the source is larger than the main beam, then the number of Jy per beam describes the average intensity over the solid angle of the beam, but if the source is unresolved, then the number of Jy per beam describes *flux density* of the source, since all the flux density from the source is contained within the beam and is detected. In the former case, we must integrate the brightness over the solid angle of the source in order to obtain the flux density.

Example 4.5:

We now point our 7-m telescope at the center of the elliptical source introduced in Example 4.2, and perform a switched observation at 22.2 GHz for many hours, obtaining an antenna temperature of 0.030 K. (In Examples 4.3 and 4.4, we learned that this telescope has an effective area of 23.1 m², and a main beam solid angle of 5.58×10^{-6} sr. In Example 4.2, we learned that the FWHM of the main beam of the telescope when operating at 22.2 GHz is 0.127°, while the elliptical source has axes of angular lengths 0.1° and 0.3°.) What is the inferred average intensity (in Jy per beam) of the central region of this source?

Answer:

Using Equation 4.24, converting the measured antenna temperature to intensity, we get

$$\langle I_\nu \rangle = \frac{2(1.38 \times 10^{-23} \text{ J K}^{-1})}{(23.1 \text{ m})(5.58 \times 10^{-6} \text{ sr})} 0.030 \text{ K} = 6.42 \times 10^{-21} \text{ W m}^{-2} \text{ Hz}^{-1} \text{ sr}^{-1}$$

In terms of brightness temperature, this is

$$\langle T_B \rangle = \frac{(3 \times 10^8 \text{ m s}^{-1})^2}{2(1.38 \times 10^{-23} \text{ J K}^{-1})(22.2 \times 10^9 \text{ Hz})^2} 6.42 \times 10^{-21} \text{ W m}^{-2} \text{ Hz}^{-1} \text{ sr}^{-1}$$

$$= 0.425 \text{ K}$$

To get the answer in units of Jy/beam, we multiply the intensity by the number of steradians in the main beam and convert to Jy, giving

$$\langle I_\nu \rangle = 6.42 \times 10^{-21} \text{ W m}^{-2} \text{ Hz}^{-1} \text{ sr}^{-1} \frac{5.58 \times 10^{-6} \text{ sr beam}^{-1}}{10^{-26} \text{ W m}^{-2} \text{ Hz}^{-1} \text{ Jy}^{-1}} = 35.8 \text{ Jy/beam}$$

We can also consider how to compare this brightness to the total emission from the source. There are a couple of issues to consider. The first is that the source's minor axis is smaller than the main beam and so some of the main beam is empty. The source's average intensity, therefore, is probably larger than 35.8 Jy/beam. But, the source's major axis is somewhat larger than the main beam, and so some of the source's flux density is not detected. We can infer, then, that the source's total flux density is a little larger than 35.8 Jy, and its average intensity is larger than 6.42×10^{-21} W m^{-2} Hz^{-1} sr^{-1}.

4.3.3 Uniform Source That Fills the Sky

Now imagine a source that fills the entire sky with a uniform intensity. This situation is worth considering because it leads to an insightful understanding about our detection. An example of such a source is the cosmic microwave background. Consider the sky to have a uniform brightness temperature, T_B. For simplicity, assume that there is no spillover in the illumination pattern, so no ground radiation is detected and only sky radiation enters the feed.

We get an interesting result if we first convert intensity to brightness temperature in Equation 4.20. To avoid unnecessary complexity, we use the Rayleigh–Jeans approximation, although our result is true for the Planck function in general. Hence, for the intensity we use

$$I_\nu(\theta,\phi) = \frac{2k}{\lambda^2} T_B(\theta,\phi)$$

which we substitute into Equation 4.20 to get

$$T_A = \frac{A_{eff}}{\lambda^2} \left\{ \int_{whole\ sky} T_B(\theta,\phi) P_{bm}(\theta-\theta_0,\phi-\phi_0)\, d\Omega \right\}$$

Here, the entire sky has uniform brightness, so T_B is a constant and can be taken out of the integral. The measured antenna temperature, then, is

$$T_A = \frac{A_{eff}}{\lambda^2} \left\{ \int_{whole\ sky} P_{bm}(\theta-\theta_0,\phi-\phi_0)\, d\Omega \right\} T_B$$

The integral should look familiar; it is the total solid angle of the beam, Ω_A, given by Equation 4.10. Therefore, we have that the antenna temperature is

$$T_A = \frac{A_{eff}}{\lambda^2} \Omega_A T_B$$

We can use the antenna theorem (Equation 4.11) to substitute in for the antenna solid angle and we find an interesting result. When the sky is filled with radiation of brightness temperature T_B, then the measured antenna temperature is exactly equal to this brightness temperature.

$$T_A = T_B$$

This may enhance your understanding of why the antenna temperature is defined the way it is. If we surround the radio telescope with radiation, then the measured antenna temperature—independent of the specifics of the telescope—will equal the radiation's brightness temperature.

The concept of observing a source that completely fills the beam, and hence having $T_A = T_B$, provides another method of calibrating the system temperature of a telescope. We can cover the opening of the feed horn with absorbing material of known temperature that closely approximates a blackbody at the radio frequency (RF). This can be accomplished by either mounting the absorber on a vane that can be moved in front of the feed or by observing the absorber in reflection by placing a mirror into the beam. The system temperature is calculated as with the noise diode method, that is, by comparing the voltage readings when the absorber *in* and *out* of the way of the feed. When the absorber covers the feed, we know that the antenna temperature should equal the temperature of the absorber. There is a difference from the noise diode method, though, which is that the vane also completely blocks the sky. At higher radio frequencies, attenuation by the Earth's atmosphere can be significant and so needs to be calibrated as well. By using both measurements of the voltage with the absorber in place and that of blank sky, it is possible to determine a system temperature that also corrects for the attenuation of the Earth's atmosphere. We will not show the details here, however. This method is most useful at higher frequencies.

4.3.4 Brightness Temperature versus Antenna Temperature, Beam Dilution, and Beam Filling Factor

Consider again the case of an unresolved source, but this time let the source solid angle be known, but still smaller than that of the main beam. In Equation 4.21, we can replace the source flux density with intensity times the source solid angle, and we can also convert the intensity to brightness temperature as in the previous section. Using the Rayleigh–Jeans approximation, we find that the source brightness temperature, T_B, is related to the measured antenna temperature, T_A, by

$$\frac{2k}{\lambda^2} T_B \Omega_{\text{source}} = \frac{2k}{A_{\text{eff}}} T_A$$

We can cancel the factors of $2k$ on both sides, use the antenna theorem (Equation 4.11) to replace the effective area, and rearrange this equation to show a simple relation between the source brightness temperature and the antenna temperature due to the source. We find that

$$T_A = \frac{\Omega_{\text{source}}}{\Omega_A} T_B \tag{4.25}$$

Equation 4.25 shows that the antenna temperature in an observation is less than the true brightness temperature of the source by a factor of the ratio of the source solid angle to the antenna solid angle. This decrease in T_A relative to T_B is often referred to as *beam dilution*. It occurs because the part of the beam filled with signal from the source is diluted by the part of the beam that is filled by blank sky. The fraction of the telescope beam that contains the source, given by the ratio Ω_{source}/Ω_A in Equation 4.25, is known as the *beam filling factor*.

This can also come into play if the source is larger than the beam but actually contains holes on unresolved scales. Imagine, for example, an observation of a galactic-sized wall of Swiss cheese. There is blank sky in the beam that we are not aware of, and the signal is diluted. In this case, we will underestimate T_B. When analyzing your radio observations, you should always keep this possibility in mind, and remember that the measured T_A is an average over the antenna beam and therefore is a lower limit to the true T_B.

Now consider the situation of Section 4.3.2, where the source was larger than the main beam and of uniform brightness. In this case, the solid angle of the detected radiation is determined by that of the main beam, so the relation between T_B and T_A, now, is

$$T_A = \frac{\Omega_{main}}{\Omega_A} T_B$$

Now the ratio in front of T_B is just the main beam efficiency (defined in Equation 4.16), so this equation can be rewritten as

$$T_B = \frac{T_A}{\eta_{mb}} \tag{4.26}$$

We call this T_B the main beam brightness temperature.

Example 4.6:

We use our 7-m telescope to observe at 22.2 GHz a source known to be of uniform brightness temperature of 55 K. We obtain an antenna temperature of 3.6 K.

1. What is the beam filling factor in this observation?
2. Using the answer to (1) and the fact that have determined that this telescope has a main beam efficiency of 0.705, and a main beam solid angle of 5.58×10^{-6} sr, estimate the angular diameter of the source, under the assumption that it is circular.

Answers:

1. We can calculate the beam filling factor simply by comparing the measured antenna temperature and the source's inherent brightness temperature.

$$T_A/T_B = 3.6 \text{ K}/55 \text{ K} = 0.065$$

2. The source fills only 0.065 of the beam, while the main beam comprises 0.705 of the total beam. Therefore, the source is fully contained within the main beam. Its solid angle, then, is 0.065/0.705 of the main beam, or

$$\Omega_{source} = \frac{0.065}{0.705} 5.58 \times 10^{-6} \text{ sr} = 5.18 \times 10^{-7} \text{ sr}$$

Assuming that the source is circular, its diameter is

$$\theta_{source} = \sqrt{\frac{4}{\pi}(5.18 \times 10^{-7} \text{ sr})} = 8.12 \times 10^{-4} \text{ radians} = 2.79'$$

4.4 SPECTRAL-LINE OBSERVATIONS

So far in this chapter, we have discussed only the simplest type of observation: a single position continuum observation, measuring the total power received over a given bandwidth, with a single pointing of the telescope. A somewhat more complicated measurement is that of a spectral line, and such observations (still at a single position) are the topic of this section.

Section 3.5 explains how a spectrometer can be incorporated in the receiver back-end. Here, we will explain a number of issues important to an observer making a spectral-line observation. Much of our discussion concerning calibration procedures applies equally well to observations using spectrometers; the antenna temperature is computed for each spectral channel identically to the above discussion. With spectral-line observations, though, there is an alternative type of switching that is often used, called frequency switching, which we discuss in Section 4.4.2.

4.4.1 Spectral Parameters

There are three important parameters of the spectrum we must consider for our spectral observation. These are the total bandwidth, Δv, the radio frequency, RF, at the center of the observed bandpass, v_{center}, and the spectral resolution, δv. The central frequency is determined by the combination of local oscillator (LO) and intermediate frequency (IF) (as explained in Chapter 3), while the total bandwidth and spectral resolution are determined by the design of the spectrometer. The RF spectrometer creates a spectrum of the signal by dividing the signal into many different channels (a filter bank spectrometer) or by digitally processing the input signal into a function of frequency (a digital spectrometer). Although digital spectrometers do not physically separate the signal into different channels, the word *channel* is still commonly used to refer to an individual data point in the spectrum. So when we discuss spectral observations in general, we will use channel in this general sense.

With both types of spectrometers, the spectral resolution of the observed spectrum is determined by the frequency width of each channel, which we will write as δv, and the total bandwidth equals the number of channels times the frequency width per channel

$$\Delta v = N_{\text{channels}} \delta v \tag{4.27}$$

With a filter bank spectrometer, the frequency per channel and number of channels are fixed by the design of the filter bank (see Section 3.5.1) and so the spectral resolution and bandwidth are predetermined. In addition, the central IF is also fixed. However, we do need to adjust the LO frequency to place the desired spectral line within the spectrometer frequency coverage. Remember that the central RF from the sky that we detect is related to the LO and central IF (Equation 3.4) by

$$v_{\text{RF}} = v_{\text{LO}} \pm v_{\text{IF}}$$

where the upper sideband RF corresponds to the + and the lower sideband to the −. The observer selects the LO frequency that places the RF of interest at or near the middle of the bandpass, or at the center of one of the sidebands if using a double sideband mixer.

A digital spectrometer is much more flexible in that one can simply change the sampling rate, which changes both the spectral resolution and spectral coverage (see Section 3.5.2). Thus, one is free to adjust all parameters of the observed spectrum and there is more flexibility (and also more decisions) in setting up the observation. The key to spectral-line observations is to ensure that the observed spectrum has sufficient resolution and bandwidth to meet the objectives of the project. Both resolution and bandwidth come at a cost, and often we must trade one off against the other. To obtain greater bandwidth, we must accept lower resolution, or vice versa. Care must be taken that we do not sacrifice so much of either one that our observations become useless. For example, if the bandwidth is too narrow, the spectral line may be too broad to fit into it. On the other hand, if the resolution is too coarse, the line will fall in a single channel, and we will lose information about its width and shape.

As a general guideline, one wants to have at least five channels across a spectral line. With fewer channels, it is difficult, if not impossible, to determine line parameters such as amplitude, center, and width. In some applications, it may be necessary to have dozens of channels across the line, to trace its spectral profile in great detail. For the amount of spectral coverage, a general guideline is that the spectral line should not take up more than about three-fourths of the passband. This allows the mathematical fit to the line profile to determine the zero or continuum level.

As we will see in Volume II, radio astronomers often express frequency resolution and coverage in units of velocity, rather than frequency. This is quite useful, because it allows us to express line properties in terms that relate to the physical properties of the astronomical source. In most cases, the line width is determined either by the temperature of the astronomical source or by large-scale velocities of the gas within the source (we discuss the factors affecting spectral line widths in Volume II). In either case, we can determine the line width in terms of a velocity, which we can then convert to frequency for use in selecting δv and Δv.

When we make a spectral observation, the received signal is divided between N channels and so the uncertainty of T_A in each channel will be a factor of $N^{1/2}$ larger than in a broadband measurement, when the signal is averaged over the entire bandwidth.

In the calculation of the expected *rms* in the spectrum, then, one should use the spectral resolution, $\delta\nu$ in place of the bandwidth, $\Delta\nu$, in Equation 4.9. This has a huge impact on the noise in the measurement: spectral channels typically are 3 to 4 orders of magnitude narrower than the passband, so the uncertainty in T_A can easily be 100× greater than in continuum (broadband) measurements.

Example 4.7:

We wish to measure an emission line at 22.364 GHz in the upper sideband of a filter bank spectrometer, which has 1024 channels of 12 kHz per channel, and a central IF of 8 MHz. We perform a position-switching observation for a total of 30 min. The system temperature is determined to be 50 K.

1. What should the LO frequency be for the emission line to appear in the middle of the bandpass?
2. What is the bandwidth of this observation?
3. What is the expected *rms* in the spectrum we will obtain?

Answers:

1. We want the central RF to be 22.364 GHz and to appear in the upper sideband, so we set the LO to be

$$\nu_{LO} = \nu_{RF} - \nu_{IF} = 22.364 \text{ GHz} - 8 \text{ MHz} = 22.356 \text{ GHz}$$

2. The total bandwidth (Equation 4.27) is 1024 channels × (12 kHz/channel) = 12.288 MHz.

3. Using Equation 4.9 and the $\delta\nu$ per channel, the *rms* in the spectrum is

$$\sigma(T_A)_{\text{spectrum}} = 2\frac{50 \text{ K}}{\sqrt{(30 \times 60 \text{ s})(12,000 \text{ Hz})}} = 0.02 \text{ K}$$

4.4.2 Frequency Switching

With observations of spectral lines, an alternative, and often preferred, type of switching is possible that cannot be used for broadband measurements. Since the signal of a spectral line is only present in a narrow range of frequencies, instead of moving the telescope *on* and *off* the radio source, we can switch in frequency (by changing the LO frequency), slightly shifting the frequency of the observation *on* and *off* the spectral line. Provided that the frequency offset is not too large, all the noise sources (such as that due to the receiver and/or atmosphere) contribute equally to both frequency settings and so Equations 4.5 and 4.6 can still be used. We call this observing technique *frequency switching*. Moving the telescope to change the pointing direction takes time, while frequency switching is very rapid.

Moreover, by careful selection of the passband and the frequency offset, we can double our time on the source. Often (depending on the width of the line, the presence of other lines, and the bandwidth of the receiver) the frequency shift does not actually have to shift the spectral line out of the observing passband, but rather to another part of the passband. In this case, the *on* and *off* observations in one part of the passband are the reverse for the other part of the passband. With the second setting, the channels that had contained the spectral line in the first setting now only contain noise and act as *off* scans, while the channels that now contain the line act as *on* scans to compare with the power in those same channels in the first setting. When this is possible, we gain a significant advantage in sensitivity because we are always obtaining both *on* and *off* scans. In terms of the uncertainty in the measured antenna temperature, then, Equation 4.8, rather than Equation 4.9, is the relevant expression with Δt still representing the total observation time. In other words, frequency-switched observations are often a factor of $\sqrt{2}$ more sensitive than position-switched observations.

4.5 OBTAINING RADIO IMAGES

If the angular size of the astronomical source is larger than the telescope's main beam, one might wish to make an *image* of the source. At visible wavelengths, this is straightforward—one just takes a picture with a CCD camera. Obtaining an image with a single-dish radio telescope is a somewhat more involved process, and requires some explanation.

Most radio telescopes have a relatively small number of detectors—in the form of feed–receiver assemblies—in contrast to the megapixel CCDs used at visible wavelengths. Each feed must have opening dimensions as large as the wavelengths being observed so that the electromagnetic waves propagate through its opening rather than being reflected, and so a large number of feeds cannot fit into the focal plane. Additionally, the complexity and cost of radio astronomy receivers are too prohibitive for a large number to be constructed. Except for some millimeter and submillimeter telescopes that contain a large focal plane array of bolometer detectors, single-dish radio observations generally produce a relatively small number of antenna temperature measurements (sometimes only one) for each pointing direction. For the sake of simplifying the discussion, we will focus on single-receiver telescopes (the Haystack SRT is an example).

To obtain a map with a single-dish radio telescope, one must undertake a process of pointing the telescope in a particular direction, making a switched observation, recording the measured antenna temperature, then moving the telescope to a new position and repeating the procedure, continuing until the desired sky area is covered. The antenna temperatures at all positions can then be converted into a map of the radio source.

Example 4.8:

Using a radio telescope with a resolution of 10 arcmin we obtain mapping data of a new source. We measure T_A at 45 positions and convert to average I_v (following

Equation 4.24) obtaining the data shown in Table 4.1. (Recall that in a map of the sky, west is to the right.) Draw a map of this source.

Answer:

Filling in the map grid with the measured I_ν we get Table 4.2. (Note that we could have mapped T_A directly, without converting to intensity, since the two quantities are proportional and so reveal the same structure. In fact, T_A maps are quite common.)

If you now draw lines between different numbers, with each line representing an equal step in intensity, you will obtain a contour map of this source (which looks, roughly, like a double-lobed radio galaxy, which you can read about in Volume II).

Because the beams of most single-dish radio telescopes are relatively large, the map we obtain is significantly affected by the telescope beam pattern. Understanding how to interpret these maps requires an appreciation of the effect of the beam, which we discuss next.

TABLE 4.1 Measured Average Intensities versus Position Relative to the Center of the Source

Position Relative to Image Center	Measured I_ν	Position Relative to Image Center	Measured I_ν	Position Relative to Image Center	Measured I_ν
(west, north)	(Jy/beam)	(west, north)	(Jy/beam)	(west, north)	(Jy/beam)
(−20', −10')	0.50	(−05, +10')	0.50	(+10', −10')	0.50
(−20', −05')	0.50	(−05', +05')	0.50	(+10', −05')	10.00
(−20', 00')	0.50	(−05', 00')	0.50	(+10', 00')	10.00
(−20', +05')	0.50	(−05', −05')	0.50	(+10', +05')	10.00
(−20', +10')	0.50	(−05', −10')	0.50	(+10', +10')	0.50
(−15', +10')	0.50	(00', −10')	0.50	(+15', +10')	0.50
(−15', +05')	10.00	(00', −05')	0.50	(+15', +05')	10.00
(−15', 00')	15.00	(00', 00')	4.00	(+15', 00')	15.00
(−15', −05')	10.00	(0', +05')	0.50	(+15', −05')	10.00
(−15', −10')	0.50	(0', +10')	0.50	(+15', −10')	0.50
(−10', −10')	0.50	(+05', +10')	0.50	(+20', −10')	0.50
(−10', −05')	10.00	(+05', +05')	0.50	(+20', −05')	0.50
(−10', 00')	10.00	(+05', 00')	0.50	(+20', 00')	0.50
(−10', +05')	10.00	(+05', −05')	0.50	(+20', +05')	0.50
(−10', +10')	0.50	(+05', −10')	0.50	(+20', +10')	0.50

TABLE 4.2 Inferred Average I_ν versus Position on Map Grid

	−20'	−15'	−10'	−05'	00'	+05'	+10'	+15'	+20'
+10'	0.50	0.50	0.50	0.50	0.50	0.50	0.50	0.50	0.50
+05'	0.50	10.00	10.00	0.50	0.50	0.50	10.00	10.00	0.50
+00'	0.50	15.00	10.00	0.50	4.00	0.50	10.00	15.00	0.50
−05'	0.50	10.00	10.00	0.50	0.50	0.50	10.00	10.00	0.50
−10'	0.50	0.50	0.50	0.50	0.50	0.50	0.50	0.50	0.50

4.5.1 Convolution with Beam Pattern

Using Equation 4.20, the measured antenna temperature at each pointing position, (θ_0, ϕ_0), of the telescope is given by

$$T_A = \frac{A_{eff}}{2k}\left\{ \int\limits_{whole\ sky} I_\nu(\theta,\phi) P_{bm}(\theta-\theta_0,\phi-\phi_0) d\Omega \right\}$$

and we now know that the factor in front is the conversion from Jy to K, often called the telescope's *DPFU*.

To make a map, we want to measure T_A for each direction that we point the telescope, so that we get antenna temperature as a function of position, that is, $T_A(\theta, \phi)$. Strictly speaking, the map that we produce is the correlation of the normalized beam pattern and the source intensity. Radio astronomers often say that the observed map is the beam pattern *convolved* with the true sky intensity (or brightness) distribution, even though this is, to be precise, a correlation. However, for a symmetric beam pattern, in which $P_{bm}(\theta - \theta_0, \phi - \phi_0) = P_{bm}(\theta_0 - \theta, \phi_0 - \phi)$, which is most often the case, there is no difference between a correlation and a convolution. We will follow the standard convention in radio astronomy and refer to this as a convolution. The effect of a convolution is to smear out the true intensity distribution over an area the size of the telescope's beam. For a given source size, convolution with a smaller beam has a smaller effect on the map. This is, essentially, a statement about resolution.

We demonstrate the concept of a convolution with a few simple one-dimensional examples.

To start, imagine mapping a double-triangle source with a square beam whose width is equal to the distance between the peaks of the triangles, as shown in Figure 4.3. What will the observed map look like? Note first that the observed antenna temperature in Equation 4.20 involves integration over a solid angle, and the examples below involve only one-dimensional angles. We will, then, get unreal units if we apply Equation 4.20 literally. These examples are still helpful in demonstrating the source convolution with the beam, but it would be obfuscating to worry about the units of the sky angle. We will, therefore, use arbitrary angle units in the following examples. Additionally, to avoid the unit problem when converting to antenna temperature, we instead calculate the average intensity over the beam, which is given

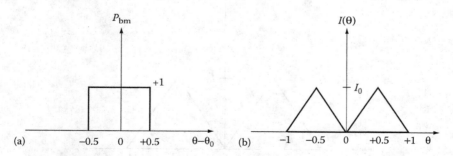

FIGURE 4.3 (a) Square observing beam of width $\theta_{bm} = 1$ and (b) double-triangle source of maximum intensity I_0 and angular distance between peaks $\Delta\theta_{source} = 1$.

by the term in the curly brackets in Equation 4.20 divided by the main beam solid angle. The angle unit, then, occurs in both the numerator and denominator and so cancels; we start with intensity of the source and end up with intensity averaged over the beam.

Consider, first, the detected power when the telescope is pointed at $\theta = -1$. Imagine sliding the beam over the plot of the image so that the center of the beam is at -1. In this case, the point in the beam indicated by $\theta - \theta_0 = 0$ lines up with the $\theta = -1$ point of the image, as shown in Figure 4.4. Then consider the integrated area under the curve of the source's intensity that is inside the beam. The beam extends from $\theta = -1.5$ to $\theta = -0.5$ and so the part of the source that is inside the beam is just the left half of the left triangle. The detected power, then, is proportional to the area of a right triangle with base $= 0.5$ and height $= I_0$. Also, the beam in this example has a constant value of one over an angle of one unit and zero everywhere else, and so $\Omega_{main} = 1$ in these arbitrary angle units. The integral in Equation 4.20 is numerically equal to the area of this right triangle, that is, $\frac{1}{2} bh$, and so the intensity is

$$I(\theta_0 = -1) = \frac{1}{2}(0.5)I_0 = \frac{1}{4} I_0$$

Now consider centering the beam at $\theta_0 = -0.5$. Then, the entire left triangle will be in the beam and so the inferred average intensity will now be

$$I(\theta_0 = -0.5) = 2 \times \frac{1}{2}(0.5)I_0 = \frac{1}{2} I_0$$

When we center the beam at $\theta_0 = 0$, we will get half of each triangle, which is the same as one whole triangle, which is what we had at $\theta_0 = -0.5$. So, at $\theta_0 = 0$ we infer an average intensity, again, equal to $\frac{1}{2} I_0$. Likewise, at $\theta_0 = +0.5$. And at $\theta_0 = +1$, the inferred intensity is $\frac{1}{4} I_0$.

If we analyze the total detected power for each beam position we will infer an intensity as a function of θ that looks like that shown in Figure 4.5.

It is interesting to compare the inferred image, represented in Figure 4.5, to the actual source, shown in Figure 4.3. Remember that, in general, we do not know *a priori* what the

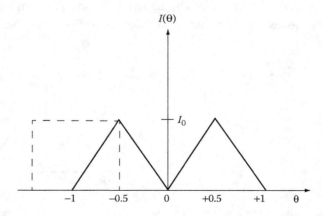

FIGURE 4.4   Square beam centered at $\theta = -1$.

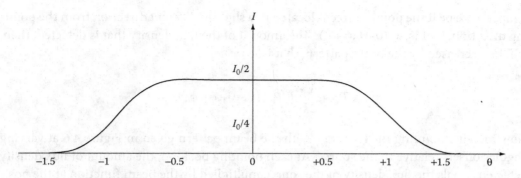

FIGURE 4.5 Resultant map obtained observing the double triangle source with the square beam shown in Figure 4.3.

source actually looks like—that is why we are making a map of it. Yet, if we observe with the square beam of Figure 4.3, whose width is exactly equal to the distance between the peaks of the source, we obtain a map that shows only a single, wide source. The fact that the source is double is not revealed at all. This happens because the width of the beam is just large enough to fail to resolve the double-peaked structure of the source.

Consider, now, using a more complex beam to observe a point source. As we already discussed, a point source is any source whose angular size is very small compared to the telescope's main beam. This is a fairly common situation in radio astronomy, and leads to a very useful result. So, imagine observations of a point source made with a beam pattern as shown in Figure 4.6. What will the convolved image look like?

In Section 4.3.1, we showed that if we point the telescope directly at a point source the measured antenna temperature is given by

$$T_A = \frac{A_{\text{eff}}}{2k} F_v$$

This resulted from the fact that the source's intensity was zero everywhere except at the center of the beam, where the normalized beam pattern equals 1. So, what would the antenna

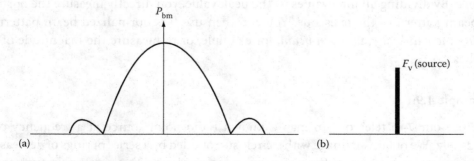

FIGURE 4.6 (a) Complex beam and (b) point source.

temperature be if the point source is located in a slightly *different* direction from the point-ing direction, that is, at $(\theta-\theta_0,\phi-\phi_0)$? The amount of the flux density that is detected, then, will be decreased by the beam pattern, that is,

$$T_A = \frac{A_{eff}}{2k} F_v P_{bm}\left(\theta-\theta_0,\phi-\phi_0\right)$$

Now imagine pointing the telescope with the beam pattern given in Figure 4.6 at varying angular offsets relative to the source. At each pointing position, the amount of flux density detected equals the flux density of the source multiplied by the beam function at the posi-tion of the source relative to the center of the beam. As the beam is moved in angle, then, the measured antenna temperature will trace out the same shape as the beam pattern, scaled by the flux density of the source and by the DPFU. The measured antenna temperature as a function of pointing direction, in one dimension, then, is given by

$$T_A\left(\theta_0,\phi_0\right) = \frac{A_{eff}}{2k} F_v P_{bm}\left(\theta-\theta_0,\phi-\phi_0\right)$$

So, when we convolve the beam pattern with a point source we get a map that looks just like the beam pattern! And, if we convert from T_A to flux density (by dividing by the DPFU), the peak of the map has a flux density equal to the flux density of the source.

As we stated above, this is a common situation, and one that is fruitful to contemplate. When you make a mapping observation of an unknown source and get a map that looks just like your beam pattern, then you immediately learn two important things about your source.

1. The source is unresolved (and hence very small compared to your beam).

2. Its flux density is equal to the peak value in the map.

This is a very useful situation to remember for another reason as well. If you have a new radio telescope, you will want to make initial measurements to learn all the characteristics of your telescope. One of the initial things you can do with it is to *map a known point source* (such as a distant quasar) *and the resultant map will reveal the beam pattern of your telescope.* By dividing all map values by the peak value, you directly measure the normal-ized beam pattern of the telescope. You can then use this normalized beam pattern to *measure the width of your main beam*, for example, or to measure the magnitude of the sidelobes.

Example 4.9:

We use our 7-m telescope to map a known radio point source at a frequency of 22.2 GHz. We obtain an image with a circle surrounded by a series of rings of decreas-ing amplitude. A profile of the power versus angle along a line passing through the center is shown in Figure 4.7. What is the resolution of our telescope? How does it compare with the minimum resolution angle calculated in Example 4.2 (in which we calculated $\theta_{res} = 0.127°$)?

Answer:

The resolution angle is the full width of the profile of a point source at half the maximum intensity, which is at $I/I_{max} = 0.5$. This width is about 0.13°. The true resolution of our telescope, therefore, is about the theoretical value, within the uncertainty of measuring the half-width from this plot.

To more rigorously demonstrate the mathematics of the convolution, let us put in some typical functions for the beam and source profiles. To avoid unnecessary complications, we will still restrict our discussion to one dimension.

Now, the edges of a real source are not likely to be sharp, but rather will fade off as we move away from the source. A reasonable function to describe the intensity versus angle is a Gaussian, which describes the source intensity distribution by

$$I(\theta) = I_0 \exp\left[-4\ln 2\left(\frac{\theta}{\theta_S}\right)^2\right]$$

where:

I_0 is the intensity at the center of the source
$\theta = 0$ corresponds to the center of the source
θ_S is the FWHM of the source profile

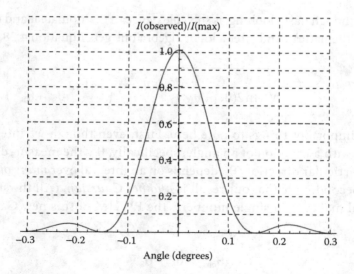

FIGURE 4.7 Profile of intensity versus angle in one dimension through the peak of the observed map of a known point source.

For the beam profile, we will ignore the sidelobes and error pattern and consider just the main beam, a reasonable fit to which is also a Gaussian and given by Equation 4.12. Using the same form as the equation for the source profile, we use θ_B to represent the FWHM of the beam and then we have

$$P_{\text{bm}}(\theta - \theta_0) = \exp\left[-4\ln 2 \left(\frac{\theta - \theta_0}{\theta_B} \right)^2 \right]$$

Now, the convolution in one dimension is given by multiplying these two functions and integrating over all angles on the sky, from $-\pi/2$ to $+\pi/2$ radians, that is,

$$\int_{-\pi/2}^{+\pi/2} I_\nu(\theta) P_{\text{bm}}(\theta - \theta_0) d\theta$$

(Remember that in reality we must integrate over two angles. The one-dimensional integral is still enlightening, though, in terms of what happens with the apparent angular size of the source.) In our observation of this source, the measured antenna temperature as a function of pointing direction of the telescope, (θ_0), is proportional to this convolution integral, and is given by

$$T_A(\theta_0) = \frac{A_{\text{eff}}}{2k} \left\{ \int_{-\pi/2}^{+\pi/2} I_0 \exp\left[-4\ln 2 \left(\frac{\theta}{\theta_S} \right)^2 \right] \exp\left[-4\ln 2 \left(\frac{\theta - \theta_0}{\theta_B} \right)^2 \right] d\theta \right\}$$

This integral, although nasty looking, is solvable (you can manipulate it and use an integral table, or use Mathematica), and after a lot of algebra one gets Equation 4.28.

$$T_A(\theta_0) = \frac{A_{\text{eff}}}{2k} \sqrt{\frac{\pi}{4\ln 2(\theta_S^2 + \theta_B^2)}} \theta_S \theta_B I_0 \exp\left(-4\ln 2 \frac{\theta_0^2}{\theta_S^2 + \theta_B^2} \right) \tag{4.28}$$

There are two important things to note here. First, even though we integrated over θ, we still end up with a function of an angle, specifically, θ_0. The measured antenna temperature in a particular direction, θ_0, depends on an integral over *many other* directions on the sky, θ. Second, note that our result is *another Gaussian*. (All the stuff in front of the exponential makes up a single number.) The FWHM of this new Gaussian, θ_{obs}, is given by

$$\theta_{\text{obs}}^2 = \theta_S^2 + \theta_B^2 \tag{4.29}$$

We find that the widths of the beam and the source add in quadrature (similar to the way the sides of a right triangle relate to the hypotenuse) to give us the observed source size.

If the beam and the source have the same size, for example, then the observed map will show a Gaussian feature with a width that is $\sqrt{2}$ times larger than the true width of the source.

Example 4.10:

Describe what our map at 22.2 GHz of the elliptical source in Example 4.2 will look like. Assume that the source is an elliptical Gaussian, meaning that its profiles along the major and minor axes are Gaussian functions, but with different widths. (The resolution of the telescope at 22.2 GHz is 0.127°, and the source's major and minor axes are 0.3° and 0.1°.)

Answer:

Our map of the point source (in Example 4.9) reveals the beam pattern. In the observed map of the elliptical source, this beam pattern will be convolved with the ellipse. In short, the map in comparison to the beam pattern will be smeared a bit, with more noticeable stretching in the direction of the major axis of the source. The sidelobes are difficult to model mathematically, but are easily recognized as sidelobes and so can be removed fairly easily as long as the source size is not too large. What is left, then, will be just the convolution of the main beam with the elliptical Gaussian source. The main beam of the telescope is approximately Gaussian, and so the resulting map is the convolution of two Gaussians. The width of the major axis of the elliptical source is 0.3°, a little more than double the width of the beam, and the width of the minor axis is 0.1°, slightly less than the beam. Using Equation 4.29, then, we infer that the apparent widths of the major and minor axes of the image of the source in the resultant map are as follows:

$$\theta_{\text{major axis}} = \sqrt{0.3^2 + 0.127^2} = 0.33°$$

and

$$\theta_{\text{minor axis}} = \sqrt{0.1^2 + 0.127^2} = 0.16°$$

The source is still clearly elliptical, but it appears slightly larger in both directions, but with a noticeably greater percent increase along the minor axis, and hence a smaller apparent ellipticity.

4.5.2 Deconvolution

Now that we have seen the effect that convolution with the beam pattern has, you may be able to imagine how to undo the convolution and hence obtain a better representation of the true source structure. This process, as you might expect, is called a *deconvolution*. Consider the case in Example 4.10. From our map of a point source in Example 4.9, we know what the beam pattern looks like. So, when we observe the elliptical source and get the apparent full widths at half-maximum given in the answer to Example 4.10, we can simply reverse the calculation that we made in Example 4.10 and infer that the true widths of the elliptical source are 0.3′ and 0.1′. This demonstrates how important it is to know the shape of the beam pattern, because once you do, you can deconvolve it from the observed data. The effective resolution of an observation, therefore, can be slightly better than the fundamental resolution, given by the FWHM of the main beam (see Problem 11).

The quality of the deconvolution depends on the characteristics of the map. First, a reliable deconvolution of the beam requires that the data points in the map have angular separations no more than half the width of the main beam; this is called *half-beam spacing*. This is the spatial version of the Nyquist sampling theorem that we saw in Section 3.5.2.

Second, the noise in the data will negatively affect our ability to fit and deconvolve the beam pattern. In principle, the ability to reveal structure smaller than the width of the beam via a deconvolution depends on the SNR. Without derivation, the general rule of thumb, assuming perfect knowledge of the beam profile, is that the smallest angle at which one can infer structural information (such as angular sizes or relative positions) is roughly the fundamental resolution angle divided by the square root of the SNR, where the signal is the peak intensity of the feature being measured, and the noise is the rms intensity in the blank areas of the map.

$$\theta_{\text{effective res}} \sim \frac{\theta_{\text{fundamental res}}}{\sqrt{\text{SNR}}} \tag{4.30}$$

This equation should not be used as a precise statement of angular uncertainty, but rather as an approximate guide to whether inferred angular differences are significant.

Example 4.11:

In our map of the elliptical source, we measure the rms in a blank area to be 1/16th of the intensity of the peak in our map. How reliably can we infer the true angular sizes of the elliptical source? ($\theta_{\text{res}} = 0.127°$, and source's widths are $\theta_{\text{major}} = 0.3°$, $\theta_{\text{minor}} = 0.1°$)

Answer:

We have an SNR of 16, so, by Equation 4.30, we can expect to infer information on angles of

$$0.127°/4 = 0.032°$$

This is a smaller angle than the FWHM of both the major and minor axes and so, even though the fundamental resolution angle is larger than the minor axis, we should be capable of extracting the true angular sizes of this elliptical source. In practice, we accomplish this by fitting a Gaussian to the observed profile and making use of Equation 4.28, knowing the true beam width. Note, though, that our uncertainty in the minor axis width will be about one-third of the true width, so this will be considered a rough measurement.

4.6 CALIBRATION OF A RADIO TELESCOPE

The various telescope parameters that we have introduced in this chapter all need to be measured in order for a telescope to be useful. In this section, we list and briefly describe the more important procedures that must be completed to calibrate a telescope.

4.6.1 Pointing Corrections

First and foremost, we want to be sure that the telescope is pointing in the direction that we think it is. With a visible-wavelength telescope, this issue is of no concern because one immediately gets a picture, showing the pattern of stars, which can then be compared to a finder chart. One can visually establish with certainty that the desired object is in the field of view. With a radio telescope, though, the observer does not get any such visual information about the telescope's field of view. We begin, therefore, with a discussion of how to check and improve the *pointing* of the telescope.

The pointing routine generally involves commanding the telescope to move to a known bright and compact (point) source and measuring the antenna temperature at several positions, at and around the source position. A variety of patterns are used; perhaps the most common pattern is the *five-point* scan, in which the central position is observed, and four points surrounding it, each one offset by half a beamwidth in azimuth or elevation. The array of measurements is then fitted in two dimensions, typically with a parabolic function, and the position of the peak in the fitted pattern is taken to be the true direction of the source. The offsets of this peak position relative to the commanded position are the *pointing corrections*. The offset positions are usually made in azimuth and elevation, so there is one pointing correction for each axis. These pointing corrections can then be applied when the telescope is slewed to the source to be observed.

The size of the pointing correction often depends on the part of the sky where the telescope is pointed. The pointing correction in elevation is often the larger one, and varies with elevation because of gravitational deformation of the primary reflector due to its own weight. In addition, variations in the temperature of the dish, due to irregular heating by the Sun, for example, can cause deformations that change with time. Usually the staff at a radio observatory will develop a *pointing model* that will take many known factors into account. This model is applied automatically, and makes an initial correction depending on the Az–El range where the telescope is pointed, serving as a coarse pointing correction. For more accurate pointing, the corrections must be determined for that particular pointing direction and for the conditions that exist at the time of the observations. The easiest and most effective method to accomplish this is to measure the pointing correction on a calibrator source that is near the target source. The corrections should be measured at the beginning of the observation and periodically thereafter. Whenever the observer changes to a new source in a different part of the sky, or when there is a substantial change in conditions (such as sunrise or sunset) a new pointing correction should be measured.

4.6.2 Calibration of the Gain, Effective Area, and Gain Curve

In Section 4.3.1, we introduced the DPFU, which indicates the response of a telescope, in terms of the antenna temperature in Kelvins, to flux density of radiation, in Jy, entering the center of the telescope beam. This value is also sometimes called the *gain* of the telescope, or simply the *Kelvins per Jansky* of the telescope. It characterizes the telescope's sensitivity; a large gain means that the telescope is very sensitive—it has a strong response to the incident radiation.

You may recall from Chapter 3 that we also use gain to describe the increase in power produced by the amplifiers in the telescope receiver. Be careful not to confuse the gain of a radio telescope with the gain of the amplifiers; both gains serve to increase the power passed to the detector, but they function in very different ways. Moreover, in the case of the reflector, there are two different mathematical definitions to describe the gain. One definition of telescope gain is directly related to DPFU, introduced in Section 4.3, which is given by

$$G = \text{DPFU} = \frac{A_{\text{eff}}}{2k} \tag{4.31}$$

A larger effective area of the telescope allows it to collect more of the radiation from the source and hence yields a larger response from the telescope, which appears as a larger antenna temperature. A large gain means that even a weak source with a small flux density will produce a measurable antenna temperature. Note that A_{eff} is the only parameter that affects this gain, although there are many factors that influence the effective area, including the edge taper and the irregularities of the surface.

The other standard definition of the telescope gain is based on the directivity—how small a solid angle the telescope is sensitive to in comparison to the entire sky. This gain is defined as the ratio of the solid angle of the entire sky (4π sr) to the antenna solid angle, Ω_A, thus:

$$G = \frac{4\pi}{\Omega_A}$$

Unlike the telescope gain defined in Equation 4.31, which has units of K/Jy, gain with this definition is unitless. Using the antenna theorem, we can rewrite this gain as

$$G = \frac{4\pi A_{\text{eff}}}{\lambda^2} \tag{4.32}$$

This definition of gain not only depends on the effective area but also depends on the wavelength of the observation. Observations at shorter wavelengths (with the same dish) have smaller beams, and so yield larger values of G in Equation 4.32.

An accurate value for the telescope gain is critical to convert the measured antenna temperature to the flux density of the radiation. Be careful to know which gain is being used. Since there are so many sources of loss in a radio telescope (see Sections 3.1 and 3.2), it is hard to accurately predict the performance based on the telescope's design. The gain of a telescope, therefore, *must* be measured. In the following, we use the definition given in Equation 4.31.

As we previously discussed, the gain can be measured by observations of a point source of known flux density. Quite simply, the DPFU is the ratio of measured antenna temperature to the flux density of the source

$$DPFU = \frac{T_A}{F_\nu}$$

The effective area of the telescope can then be calculated with Equation 4.31. Since we also know the geometrical area of the primary reflector, this measurement will also give us the telescope's aperture efficiency, η_A, via $\eta_A = A_{eff}/A_{geom}$.

The telescope gain is actually not a fixed parameter. Because of deformations of the dish due to gravity, especially when the telescope is aimed close to the horizon or zenith, the telescope's effective aperture, and hence its gain, depends on the elevation of the pointing direction. The set of gain variations as a function of elevation is called the *gain curve* of a telescope, which is obtained by measuring the gain with known point sources at assorted positions around the sky, and fitting a function to the gain versus elevation. The gain of a telescope may also vary with time of day due to thermal deformation in the primary reflector from sunlight. Obtaining and maintaining a complete and accurate gain curve, although simple in concept, requires vigilance and is often an important task on a radio telescope operator's to-do list.

The gain of a telescope is, clearly, an important parameter when considering observations; it is a parameter with which you should be familiar and comfortable. One can use a telescope's reported gain, for example, to calculate the feasibility of a proposed observation, as outlined in Section 4.7.

Example 4.12:

We wish to observe a point source known to have a flux density of 7.0 Jy using our colleague's radio telescope. Our colleague provides us with Table 4.3, listing measurements of the telescope's gain. Our source has a declination of 20° and our colleague's observatory is at latitude 40°.

If we make our observation when the source is transiting the observatory, what antenna temperature can we expect to measure?

Answer:

Since our colleague gave us the gain in units of K/Jy, we can assume that it is the gain defined by Equation 4.31, in which case the antenna temperature will simply be the flux density times the gain. (Note if the source was resolved, we would have to determine what fraction of its flux density was detected.) We need, then, to calculate the source's elevation at the observatory when it transits. (This is a calculation discussed in Chapter 1.) At a latitude of 40°, the celestial equator will be 50° above the horizon,

TABLE 4.3   Gain versus Elevation

Elevation	G (K per Jy)
10°	0.0080
20°	0.0095
30°	0.0110
40°	0.0112
50°	0.0114
60°	0.0115
70°	0.0115
80°	0.0116
90°	0.0116

and this source is 20° further north and so will have an elevation angle of 50° + 20° = 70°. The gain of the telescope, then, will be 0.0115 K Jy^{-1}, and therefore the expected antenna temperature is

$$T_A = 0.0115 \text{ K Jy}^{-1} \times 7.0 \text{ Jy} = 0.0805 \text{ K}$$

4.6.3 Measuring the Beam Pattern and Main Beam Efficiency

In Example 4.9 (Section 4.5.1), we showed that when mapping a point source, the resulting image provides a direct measure of the telescope's beam pattern. This, therefore, provides a straightforward method for measuring both the entire beam function $P_{bm}(\theta, \phi)$ and also the FWHM of the main beam. Although one can derive the expected beam pattern based on the size of the reflector and the illumination of the reflector by the feed horn (see Chapter 3 and Appendix II), because the physical structure is never a perfect match to the design, it is best to directly measure the beam.

With a sufficiently strong point source, it is possible to measure the response of the sidelobes relative to the main beam. A strong point source is required, as the sidelobe response may be smaller than 1% of the main beam response. The error pattern is much more difficult to measure, because the response in any one direction is even smaller than the sidelobe response. Nevertheless, the total power in the error pattern can be significant because it has an angular size much larger than the main beam.

For sources that are extended, but not so large that they fill much of the error pattern, there is a simple relation between the antenna temperature and source brightness temperature that depends on the telescope's main beam efficiency at that wavelength (see Equation 4.26). Observations of such sources can be used to determine the telescope's main beam efficiency. This is accomplished by measuring the antenna temperature when observing a source of uniform intensity and of sufficient angular size to fill the main beam, and then applying Equation 4.26. At shorter radio wavelengths (less than about 1 cm) the larger planets, such as Jupiter, make ideal sources for such a measurement. At longer wavelengths, there are no good sources for this measurement, and so one must calculate the main beam efficiency from other measurements, as we discuss in Section 4.6.4.

4.6.4 Calculating Antenna Solid Angle and Main Beam Efficiency

Once we know the effective area of our telescope (measured by observations of a known point source, as described in Section 4.6.2) and the FWHM of the main beam (measured by mapping a point source, as described in Section 4.6.3) we can calculate both the antenna solid angle and the main beam efficiency. The antenna solid angle is given by the antenna theorem (Equation 4.11)

$$\Omega_A = \frac{\lambda^2}{A_{\text{eff}}}$$

and the solid angle of the main beam is given by (Equation 4.13)

$$\Omega_{\text{main}} = \frac{\pi}{4 \ln 2} \theta_{\text{FWHM}}^2$$

The main beam efficiency (introduced in Section 4.2) is simply the ratio of these two.

Example 4.13:

Let us review this calculation for our 7-m telescope.

1. How did we measure its effective area?
2. How did we determine the FWHM of the main beam?
3. What, then, are antenna solid angle and main beam efficiency at 22.2 GHz?

(Note: All these values depend on the observing wavelength.)

Answers:

1. In Example 4.4, we observed a point source of known flux density, measured the antenna temperature, and then used Equation 4.21 to calculate the effective area. We found $A_{\text{eff}} = 23.1$ m^2.
2. In Example 4.9, we made a map when observing a point source, which yielded a plot of the beam pattern. We then measured the FWHM. We found that this equaled what we expected by using Equation 4.14, which gives the main beam FWHM for a 10-dB edge taper, as a function of the diameter of the reflector. We found $\theta_{\text{FWHM}} = 0.13°$, which is $= 0.00227$ radians.
3. The antenna solid angle is then given by the antenna theorem (Equation 4.11 and shown above) which, in this case, is

$$\Omega_A = \frac{\left(3.00 \times 10^8 \text{ m s}^{-1} / 22.2 \times 10^9 \text{ Hz}\right)^2}{2.1 \text{ m}^2} = 7.91 \times 10^{-6} \text{ sr}$$

and the main beam solid angle, given by Equation 4.13 (and shown above) is

$$\Omega_{\text{main}} = \frac{\pi}{4\ln 2}(0.00227 \text{ radians})^2 = 5.83 \times 10^{-6} \text{ sr}$$

The main beam efficiency, then, is

$$\eta_{\text{mb}} = \frac{5.83 \times 10^{-6} \text{ sr}}{7.91 \times 10^{-6} \text{ sr}} = 0.737$$

4.7 TELESCOPE SENSITIVITY CONSIDERATIONS IN PLANNING AN OBSERVATION

Before you set out to use a particular telescope to make an observation, you should first calculate whether that telescope can succeed in making that observation. If you want to use a telescope at a national facility, for example, you will need to submit a proposal, in which you make the calculation that shows that the sensitivity of that telescope is sufficient for you to succeed and how much time you will need for the observation.

The capability of a particular telescope to detect a given signal is essentially a question of whether the signal is strong enough relative to the noise. We discussed the uncertainty in an antenna temperature due to noise in Section 4.1.4. We can now apply this calculation to the observation of a source of given flux density.

We will first review the main points of Section 4.1.4. Generally, one wants the expected signal to be *at least* three times (and often five or ten times) the rms of the noise. This is often discussed in terms of the SNR. Now, according to Equation 4.9, the 1σ noise level for a switched observation is given by

$$\sigma(T_{\text{A}}) = 2\frac{T_{\text{sys}}}{\sqrt{\Delta t_{\text{obs}}\Delta\nu}}$$

where:

Δt_{obs} is the total observation time

Consider an observation of a point source of flux density F_ν. We can convert the rms uncertainty in the antenna temperature into an rms uncertainty in the flux density using Equation 4.21.

$$\sigma(F_\nu) = \frac{2k}{A_{\text{eff}}}\left[2\frac{T_{\text{sys}}}{\sqrt{\Delta t_{\text{obs}}\Delta\nu}}\right] \tag{4.33}$$

In planning an observation, we often choose the minimum SNR that will satisfy the science goals. Therefore, the minimum flux density that can be detected for a given SNR is given by the following:

$$F_\nu(\min) = \text{SNR}\left(\frac{4k}{A_{\text{eff}}}\frac{T_{\text{sys}}}{\sqrt{\Delta t_{\text{obs}}\Delta\nu}}\right) \tag{4.34}$$

This is a very important equation. You will want to refer to this equation whenever you plan an observation. Choose your minimum SNR and plan your observation by adjusting Δt_{obs} and $\Delta\nu$ so that the right side of this equation is smaller than the flux density you expect from your source.

It is important to realize that our sensitivity is really in terms of flux density in each beam, and not total flux density. For example, if a source of uniform brightness has a flux density of 100 Jy, but its angular size is such that it takes 50 pointings of the telescope to cover it, then any one beam area will only see 2 Jy of flux density.

An alternative (but equivalent) way to describe the system performance is through the *system equivalent flux density* (SEFD). This is a quantity that can be thought of as the flux density that a point source must have to produce an antenna temperature equal to the system temperature. Consider Equation 4.21, which relates the flux density of a source to the antenna temperature it produces, and set the antenna temperature equal to the system temperature. The resultant flux density, then, is the SEFD, which we find is given by

$$\text{SEFD} = \frac{2k}{A_{\text{eff}}}T_{\text{sys}} \tag{4.35}$$

or, since DPFU $= A_{\text{eff}}/2k$,

$$\text{SEFD} = \frac{T_{\text{sys}}}{\text{DPFU}}$$

Note that SEFD has units of Jy.

The SEFD provides an alternative sensitivity measure for calculating the minimum detectable flux density. Using Equation 4.35 we can substitute in Equation 4.34 and rewrite this equation as

$$F_\nu(\min) = \text{SNR}\left[2\frac{\text{SEFD}}{\sqrt{\Delta t_{\text{obs}}\Delta\nu}}\right] \tag{4.36}$$

The SEFD is a quantity that combines the intrinsic sensitivity of the telescope, DPFU, with a measure of the noise in the observation, T_{sys}, which also depends on frequency, weather, and receiver temperature, among other factors. We stress that use of Equation 4.36 is really no different from using Equation 4.34. Equation 4.36 may be considered more convenient simply because the two important parameters for planning an observation, A_{eff} and T_{sys}, are combined into one, the SEFD.

Example 4.14:

Imagine you come across a science news article detailing the detection of a bizarre object at infrared wavelengths. You decide to use our 7-m radio telescope to see if you can detect it at 22.2 GHz. You check the list of sources from a 22-GHz survey done previously and learn that no radio source was detected at this location, with a minimum flux density limit of 0.2 Jy. In case you fail to detect the source, it would be good to at least decrease the upper limit of its flux density. Using five times the rms as a conservative upper limit in a null detection, how long should you plan to observe this source to ensure that your rms is less than one-fifth of the current detection limit? The maximum bandwidth at 22.2 GHz of our telescope is 500 kHz and we know, from prior examples, that $A_{eff} = 2.31 \times 10^5$ cm². Also, the expected $T_{sys} = 50$ K.

Answer:

Using Equation 4.34, with the known values of A_{eff} and T_{sys}, and setting SNR = 5, and $(F_\nu)_{expected} = 0.2$ Jy, we have,

$$0.2 \times 10^{-26} \text{ W m}^{-2} \text{ Hz}^{-1} = 5 \frac{4(1.38 \times 10^{-23} \text{ J K}^{-1})}{23.1 \text{ m}^2} \frac{50 \text{ K}}{\sqrt{\Delta t (500 \times 10^3 \text{ Hz})}}$$

and so

$$\Delta t = \left(\frac{20 \times 1.38 \times 10^{-23} \text{ J K}^{-1} \times 50 \text{ K}}{0.2 \times 10^{-26} \text{ W m}^{-2}\text{Hz}^{-1} \times 23.1 \text{m}^2} \right)^2 \frac{1}{500 \times 10^3 \text{ Hz}}$$

or

$$\Delta t = 1.78 \times 10^5 \text{ s} = 49.6 \text{ h.}$$

Considering that the source is above the horizon only about half the time, this observation would occur over about 5 days. This is the *minimum* time you would need, making a switched observation, just to match the current detection limit. So, you would have to observe even longer for the observation to accomplish anything. The conclusion is that this telescope, with a diameter of only 7 m and aperture efficiency of 60%, cannot do the job. Imagine, though, if you set out to make the observation without making the calculation first. What a waste of time that would be! The moral to this example is that one should always make a feasibility calculation, possibly using Equation 4.34 before embarking on any observation.

Example 4.15:

1. What is the SEFD for the observation in Example 4.14, in which $T_{sys} = 50$ K and $A_{eff} = 23.1$ m²?
2. What flux density point source could be detected at a 5σ level in a switched observation with a bandwidth of 2 MHz and a total observation time of 6 hr?

Answers:

1. We can calculate the SEFD using Equation 4.35,

$$\mathrm{SEFD} = \frac{2(1.38 \times 10^{-16}\,\mathrm{erg\,K^{-1}})}{2.31 \times 10^{5}\,\mathrm{cm^{2}}}\, 50\,\mathrm{K} = 5.97 \times 10^{-20}\,\mathrm{erg\,s^{-1}\,cm^{-2}\,Hz^{-1}} = 5970\,\mathrm{Jy}$$

Note how large this value is. Recall that the noise almost always dominates over the source signal and the SEFD is the flux density a point source must have for it to produce an antenna temperature equal to the system temperature. We see, in this case, that a source of nearly 6000 Jy would be needed to produce as much power in our receiver as the system noise.

2. Using Equation 4.36 we have

$$F_{v}(\min) = 5\left[2\frac{5970\ \mathrm{Jy}}{\sqrt{(6\ \mathrm{h})(3600\ \mathrm{s\ h^{-1}})(2 \times 10^{6}\ \mathrm{Hz})}}\right] = 0.287\,\mathrm{Jy}$$

It is interesting to note that this minimum flux density is almost as small as the 0.2 Jy upper limit we hoped to achieve in Example 4.14, and this example involved only 6 hr of observing time. The moral, here, is that using a larger bandwidth is of great value.

4.8 POLARIZATION CALIBRATION*

One last calibration process that we need to discuss is that involved in measuring the radiation's polarization. We introduced polarization of radiation in general terms in Section 2.7. Some astronomical radio sources, such as some active galaxies, are known to be linearly polarized at centimeter wavelengths; others, such as Jupiter's long-wavelength bursts, are circularly polarized, and some, for example, astrophysical masers, show both types of polarization. A general-purpose radio telescope, therefore, needs to be able to measure all types of polarization. A successful polarization measurement requires an observation that is simultaneously sensitive to two complementary polarizations (orthogonal linear polarizations or oppositely directed circular polarizations), which can then be combined to yield parameters, such as the Stokes parameters (see Section 2.7), that fully characterize the polarization. The two polarizations will pass through independent receiver paths. If these paths do not treat the signals in identical ways, then the instrument will introduce an apparent polarization into the observation. To avoid this we must go through a careful calibration process to compensate for the small differences that inevitably exist.

The polarization calibration process depends on the specifics of both the telescope (e.g., if it has an Alt–Az or an equatorial mount) and the observation (e.g., if it is minutes long or hours long). With this in mind, we will not give a rigorous and detailed discussion here. Rather, we will present the fundamental concepts to provide you with a basic understanding, and apply the principles to one particular setup as an example. Detailed information is provided in higher level texts and references should you need to perform a polarization calibration in some future radio observation. More thorough discussions are given in the chapter by Heiles in *Single-Dish Radio Astronomy: Techniques and Applications*[*] and in *Astronomical Polarimetry* by Tinbergen.[†]

We start with a discussion of the hardware required. A dual-polarization radio telescope has two separate receivers with feeds, which respond to complementary polarizations. Feeds that respond to linearly polarized waves are called *native linear* feeds and, of course, those that are sensitive to circularly polarized waves are *native circular* feeds. Typically, a single circular feed will pass both polarizations; they are immediately separated by an ortho-mode transducer, and the two resulting signals are then passed along two independent receiver paths. The signals are detected independently and recorded separately. Accurate processing of dual polarization observations, of course, requires that the polarization sensitivities of each feed be well known. We will first explain the ways in which the polarization sensitivities can be in error. Then we will briefly outline methods used to correct the errors, which is done by correctly calibrating the polarization measurements.

To make this discussion more concrete, we focus on a specific set-up, and we choose the simpler case to understand: two native linear feeds. For discussion sake, we let one feed be oriented vertically and the other horizontally. Imagine, now, that these feeds are not perfectly oriented. If both feeds are rotated by an angle φ relative to what is believed, as in Figure 4.8, then the inferred polarization angle will also be off by this angle. Even worse is if the orientations of the feeds are not exactly 90° apart, as depicted in Figure 4.9. In this case, a fraction of the radiation that is polarized in the direction of one feed will *leak* into the supposedly complementary feed. The detected signals, then, do not accurately represent the complementary polarization signals. This type of error can produce apparent net polarizations even when the radiation is completely unpolarized. This, therefore, is a critical source of error that must be eliminated through good calibration.

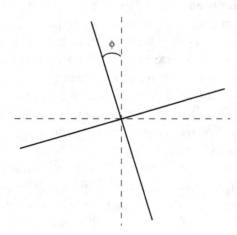

FIGURE 4.8   Native linear feeds tilted at an angle φ, relative to the vertical and horizontal.

[*] Heiles. C., A heuristic introduction to radioastronomical polarization. In *Single-Dish Radio Astronomy: Techniques and Applications. ASP Conference Proceedings*, Vol. 278, eds. Snezana Stanimirovic, Daniel Altschuler, Paul Goldsmith, and Chris Salter, Astronomical Society of the Pacific, San Francisco, CA, 2002.

[†] Tinbergen. J., *Astronomical Polarimetry*. Cambridge University Press, Cambridge, 1996. 158 pp.

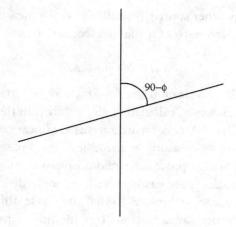

FIGURE 4.9   Native linear feeds with relative orientation of less than 90°.

An additional problem is that the signals in the two polarizations, due to instrumental factors, can differ in phase. The cables and electronics from the feeds to the computer are not identical along the two independent paths, and so the different polarizations will arrive at the computer with different phases. These phase differences introduce significant errors in the Stokes parameters calculated by cross multiplications, that is, U and V. This error will cause an unpolarized source to appear to have a net circular polarization (V) as well as a net linear polarization (due to the resulting nonzero U).

The basic approach for calibrating the polarization response of a single-dish radio telescope is to observe a source of known polarization, measure the Stokes parameters, and compare with the known values for the source. The coupling between all input and output Stokes parameters are then calculated and gathered together in a matrix called the Mueller matrix, represented by $\overline{M}$. The measured Stokes parameters can then be expressed in terms of the input (or true) Stokes parameters by a matrix equation, in which each set of Stokes parameters composes a 4-vector.

$$\vec{S}_{\text{measured}} = \overline{M}\vec{S}_{\text{input}}$$

or, in more explicit form,

$$\begin{pmatrix} I_{\text{measured}} \\ Q_{\text{measured}} \\ U_{\text{measured}} \\ V_{\text{measured}} \end{pmatrix} = \begin{bmatrix} M_{II} M_{IQ} M_{IU} M_{IV} \\ M_{QI} M_{QQ} M_{QU} M_{QV} \\ M_{UI} M_{UQ} M_{UU} M_{UV} \\ M_{VI} M_{VQ} M_{VU} M_{VV} \end{bmatrix} \begin{pmatrix} I_{\text{input}} \\ Q_{\text{input}} \\ U_{\text{input}} \\ V_{\text{input}} \end{pmatrix}$$

where:

M_{ij} terms couple the input Stokes parameter S_j to the measured Stokes parameter S_i
    Once the Mueller matrix terms are all determined, one can then calculate the true

polarization of any other source by multiplying the measured Stokes vector by the inverse of the Mueller matrix for that telescope, that is,

$$\vec{S}_{input} = \ddot{M}^{-1} \vec{S}_{measured}$$

The quantitative effects of the polarization errors we mentioned above are all contained within the Mueller matrix terms. However, calculating all the terms in the Mueller matrix directly is not so straightforward. The 16 Mueller matrix terms are linear combinations of 8 different parameters, but there are only 4 measurable quantities—the 4 measured Stokes parameters. The actual polarization calibration process, therefore, involves some additional steps.

The phase difference problems are easily solved, in single-dish observations, by inserting a correlated Cal source into both paths so that the phase difference can be measured directly. (For aperture synthesis observations this method is useless, and so the phase errors are solved by different means.)

Measurement of leakage between polarizations is accomplished by a trick making use of the Alt–Az mounts of radio telescopes. When an Alt–Az mount tracks a source, the entire telescope rotates relative to the sky. When one observes a polarized source at a number of different positions in the sky, the apparent polarization angle of the incoming radiation rotates relative to the antenna. Any leakage between the feeds will cause an additional apparent change in the polarization of the radiation. Since the way in which the telescope rotates as it moves to different positions is well known, it is easily removed. One assumes, reasonably, that the source's polarization relative to the sky is constant throughout the observation, so any remaining apparent change in the source's polarization angle at different times in the day is inferred to be due to leakage between the feeds. The leakage terms can then be solved for. If the feeds are native linear, for example, the relative orientation angle of the feeds can be treated as a single unknown to be fit to make the polarization of the calibrator source constant throughout the duration of the observation. An alternative approach is to observe a known *unpolarized* source, and ensure that the amplitudes of the orthogonal signals are the same.

Finally, the absolute polarization angle (which is affected by the absolute orientations of the feeds in the case of native linear feeds) is determined by comparing the inferred polarization angle of the known polarized source with the expected value, and applying a correction term to all other measured polarization angles.

It should be mentioned that even if you perform a perfect polarization calibration, at low frequencies, the inferred polarization of an astronomical source can be affected by the Earth's ionosphere, and also by the interstellar medium.

QUESTIONS AND PROBLEMS

1. Discuss the calibration procedure for a single-dish measurement of the flux density of a source. Start with the fact that the raw measurement made is of a voltage and explain:

    a. How the system temperature is determined

    b. How the antenna temperature is calculated

    c. How the antenna temperature is converted to a flux density

2. What effective area of a reflector is needed to provide a gain or DPFU of 1 K/Jy? Assuming an aperture efficiency of 0.600, and a parabolic reflector, what is the diameter of the dish?

3. The Green Bank Telescope has an effective area at 1400 MHz of approximately 6000 m². If one observed a relatively bright radio point source with a flux density of 0.100 Jy, how much power would be collected by the telescope in a bandwidth of 100 MHz? What would be the antenna temperature of this radio source?

4. A novice radio observer spends many hours making on-off position switching cycles, each one comprised of 5 min *on* source and 5 min *off* source. It occurs to the observer that so much time spent *off* source must somehow reduce the sensitivity of the observation, and decides to use 7.5 min *on* source and 2.5 min *off* source in subsequent switching cycles.

    a. Determine whether the uncertainty in the measured T_A increases or decreases. Use Equation 4.7 and remember that the rms from both the *on* and *off* scans contribute to the uncertainty in the final T_A measure and add in quadrature ($\sigma_{total}^2 = \sigma_{on}^2 + \sigma_{off}^2$). Consider that the total observing time, Δt_{obs}, is fixed and that some fraction of that time, Δt_{on}, is spent *on* source and the rest of the time is used for the *off* source scans. In the initial observing pattern half the time was spent *on* source and in the later observing pattern three-fourth of the observing time was on source. Which leads to a smaller uncertainty?

    b. Derive an expression for the uncertainty in T_A as a function of the fraction of the time spent *on* source and determine for what value of the fraction is the uncertainty the smallest.

5. What is beam dilution? Imagine you observed a source that has a brightness temperature of 300 K and is circular with an angular diameter of 4.00 arcsec with a 12.0-arcsec beam. What would the antenna temperature be?

6. a. Evaluate Equation 4.10 for a Gaussian main beam (Equation 4.12) to derive Equation 4.13.

    b. When observing a resolved source, what relative error is introduced into a brightness measurement (see Equation 4.24) if $\pi\,(\theta_{FWHM}/2)^2$ is used as the main beam solid angle instead of the correct form of Equation 4.13?

7. Describe how an image of a source is obtained with a single-dish telescope.

8. A two-component radio source with a one-dimensional brightness distribution as depicted in Figure 4.10a, below, is observed with a radio telescope with a beam pattern given by Figure 4.10b. The source consists of a point source on the left, separated by 0.5 angle units from an extended component that has a triangular intensity versus angle profile. The peak intensity of the extended component is B and has a flux density of F, equal to the flux density of the point source.

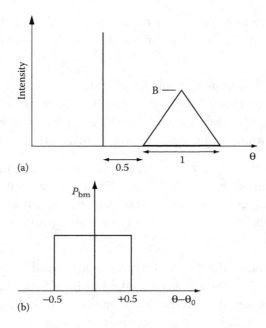

FIGURE 4.10 (a) Point source is separated by 0.5 angle units from an extended source with a triangular intensity versus angle profile. Both sources have flux density F. (b) Square beam of width 1 angle unit.

    a. Calculate the received flux density when the center of the beam is halfway between the edges of the two components, that is, 0.25 angle units to the right of the point source.

    b. Draw a plot of the observed flux density as a function of θ.

9. Discuss the net effect of the observing beam on the observed image size. For a Gaussian beam and a Gaussian source, what is the shape and full width at half maximum of the observed brightness distribution (in terms of the widths of the beam and source)?

10. An observation at a wavelength of 6.00 cm of a 10.0-Jy radio point source is made with a 36.0-m telescope when the position of the source corresponds to the maximum in the telescope's gain curve. The measured antenna temperature is 3.20 K.

    a. What is the DPFU of the telescope?

    b. What is the effective area of the telescope?

    c. What is the aperture efficiency?

11. A *drift scan* is made by pointing the telescope at a sky position where a short time later an astronomical object will pass by, owing to the rotation of the Earth. If the object is much smaller than the telescope beam size (i.e., a point source), the resulting trace on the computer (of T_{ant} vs. time) is a one-dimensional slice through the telescope beam pattern. As we discussed in Section 4.3, this pattern is well described by a Gaussian profile. If the source is only slightly smaller than the beam, then the observed profile will be somewhat

wider than the beam profile. Consider observations of a Gaussian source, with a range of source sizes. Apply Equation 4.29 ($\theta^2_{obs} = \theta^2_{beam} + \theta^2_{source}$) for varying source sizes by setting $\theta_{source} = f\theta_{beam}$ and letting f take on values from 0.1 to 1.0 in steps of 0.1. Calculate and tabulate the ratio of the observed FWHM to the beam FWHM as a function of f. For what value of f is the observed profile 10% wider than the beam profile?

As we mentioned in Section 4.5.2 (Equation 4.30), the angular precision obtained in an observation is inversely proportional to the square root of the SNR of the observation. If the precision with which we can determine the FWHM of a Gaussian source also has this dependency, what SNR is needed to measure the angular size of a Gaussian source of width $f\theta_{beam}$, where f is the answer to the question above (i.e., a source in which the beam width is increased by 10%)?

12. A telescope known to have a DPFU of 0.750 K/Jy at 2.00 cm is used to observe an unresolved source known be circular with a diameter of 2.00 arcsec. The observation yields an antenna temperature of 3.00 K. Assume that the source has uniform brightness across the 2-arcsec circle.

   a. What is the flux density of the source?

   b. What is the brightness temperature of the source?

   c. We observe the source again when it is at a lower elevation. The antenna temperature this time is 2.50 K, what is the DPFU of the telescope at this elevation?

   d. Do any of your answers to parts (a)–(c) depend on whether the source has uniform brightness? Would uniform brightness matter if the source were resolved?

13. A 32.0-m telescope with aperture efficiency of 0.620 is used to map a source at 6.00-cm wavelength, which is later learned to be a point source. The resulting image is of a circular feature with a Gaussian profile of FWHM of 7.40 arcmin.

   a. What is the solid angle of the main beam?

   b. What is the main beam efficiency?

14. A spherical, opaque, thermal radio source has a temperature $T = 2400$ K and an angular diameter of 18.0 arcsec, as shown in Figure 4.11. An observation is made at $\lambda = 15.0$ cm with a bandwidth of $\Delta v = 2.00$ MHz. The radio telescope used has a diameter of $D = 10.0$ m, and aperture efficiency $\eta_A = 0.62$. The system temperature is measured to be $T_{sys} = 150$ K.

   a. What is the brightness temperature of the radiation?

   b. What is the flux density of the radiation?

   c. What is the observed antenna temperature?

   d. What is the minimum integration time needed to detect this radiation in a position switched observation at a level five times the noise?

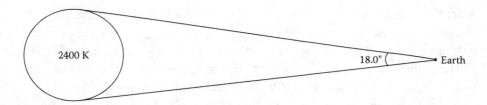

FIGURE 4.11   Spherical, 2400 K source with an angular diameter of 18.0 arcsec.

15. A telescope has effective area of 100 m², and a maximum bandwidth of 2.00 MHz. The system temperature is measured to be 70.0 K.

   a. What is the minimum detectable antenna temperature for a 2-min switched observation?

   b. What is the minimum detectable flux density for a point source in a 2-min switched observation?

   c. If you wish to observe a point source with a flux density of 1 Jy, how long must you integrate to produce a detection that is five times the noise level?

   d. By changing receivers, you can increase the bandwidth to 200 MHz, which will increase the sensitivity. By what factor would an increase in the telescope diameter produce the same increase in sensitivity offered by the wider bandwidth?

FIGURE 1.17   View through the Arecibo dish from below. The clouds and sky are clearly visible, as is the feed. (Courtesy of Clara Thomann.)

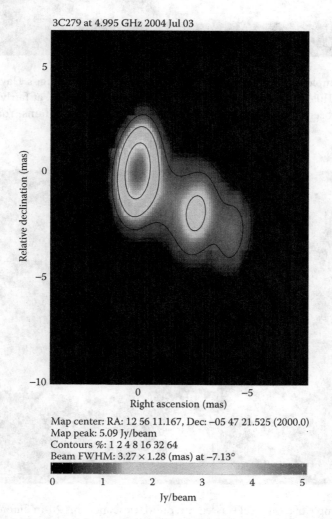

FIGURE 1.21   False color map of the 6.0 cm emission from 3C279. The color wedge at the bottom indicates the transfer function relating color and brightness.

FIGURE 3.3   Example of a prime focus radio telescope; Union College's Haystack Small Radio Telescope (SRT). This 2-m diameter radio telescope is made available, at fairly low cost, through MIT's Haystack Observatory radio astronomy education outreach programs. Your school may have a similar one.

FIGURE 3.4   Example of a classical Cassegrain radio telescope. This 20-m diameter radio telescope of the National Radio Astronomy Observatory is located in Green Bank, West Virginia. (Courtesy of B. Saxton, NRAO/AUI/NSF.)

FIGURE 3.23   LWA dipole antenna pair. This inverted dipole is actually two orthogonal dipole pairs. The construction is from aluminum tubing, and it is designed to work from 10 to 88 MHz. For comparison, a traditional parabolic reflector can be seen in the background.

FIGURE 5.1   The Atacama Large Millimeter/submillimeter Array (*ALMA*) located in the Atacama Desert in the Andes of northern Chile. (Courtesy of W. Garnier, ALMA [ESO/NAOJ/NRAO].)

FIGURE 5.2   The Jansky VLA on the Plains of San Agustin, New Mexico. (Courtesy of NRAO/AUI/NSF.)

CHAPTER 5

Aperture Synthesis Basics

Two-Element Interferometers

IN THIS CHAPTER AND in Chapter 6, we discuss the extremely important observing method known as *aperture synthesis*. This technique uses a large number of telescopes arranged in an array in order to produce high-resolution images. In this chapter, we consider a simplified situation to illustrate the basic idea of how the technique works and highlight its fundamental principles. We focus on the functioning of the fundamental unit of an aperture synthesis array: a pair of telescopes used as a two-element interferometer. We will also limit the mathematical complexity by considering only simple sources in one dimension and using only real numbers. We will extend the discussion in Chapter 6 to provide a fuller understanding of aperture synthesis, including the treatment of two-dimensional sources and the use of complex numbers. Although Chapter 6 is more mathematically sophisticated, we will still omit many details and focus our discussion on the understanding needed to pursue aperture synthesis techniques as an observer.

In the preface of this book, we mention a set of table-top interferometry labs, using the Haystack Very Small Radio Telescope (VSRT), and web pages that contain information about the equipment assembly and lab instructions. There are also computer-based simulations that can be performed in lieu of the labs, which lead to the same results (see Appendix VII). These labs and activities are intended to accompany the presentation in this chapter and provide intuitive experience without the mathematical underpinning of interferometry. We recommend the completion of these activities before reading Section 5.2, as the experience will make the subsequent explanations more concrete and easier to follow.

The importance of aperture synthesis cannot be overstated. It has had a profound impact on radio astronomy and the studies of the universe. In recognition of its importance, half of the 1974 Nobel Prize in Physics was awarded to Martin Ryle for his development of the method; the other half was awarded to Anthony Hewish for the discovery of pulsars. As evidence of its continuing importance, one of the largest recent international telescope projects is the *Atacama Large Millimeter/submillimeter Array* (*ALMA*), shown in Figure 5.1. Although only recently completed (in 2013) it has already yielded many significant results.

You have probably seen photos of the *Jansky Very Large Array* (or *VLA*, shown in Figure 5.2), which has appeared in several movies, including *2010* and *Contact*. Perhaps

FIGURE 5.1 **(See color insert.)** The Atacama Large Millimeter/submillimeter Array (*ALMA*) located in the Atacama Desert in the Andes of northern Chile. (Courtesy of W. Garnier, ALMA [ESO/ NAOJ/NRAO].)

FIGURE 5.2 **(See color insert.)** The Jansky VLA on the Plains of San Agustin, New Mexico. (Courtesy of NRAO/AUI/NSF.)

you wondered what is accomplished with a large number of telescopes as opposed to one large telescope. This is the subject of this chapter and Chapter 6.

5.1 WHY APERTURE SYNTHESIS?

The main advantage of aperture synthesis is angular resolution. The long wavelengths of radio astronomy limit single-dish observations to very poor resolution, as explained in Chapter 3 and revisited in Chapter 4. Because resolution depends inversely on the telescope diameter, you might imagine the construction of an extremely large radio telescope, so that its diameter could compensate for the long wavelengths. If we want to observe at a wavelength of 10 cm, for example, what size telescope is needed to obtain a resolution similar to that of visible-wavelength observations? Ground-based visible observations can obtain a resolution of order 1 arcsec, or 2×10^{-5} radians. So, the radio telescope diameter, D, needed to match this resolution, using Equation 3.2, is approximately

$$D \approx \frac{10 \text{ cm}}{2 \times 10^{-5}} \approx 5 \text{ km}$$

or about 3 miles! Can you imagine the engineering involved just trying to make a dish that large? And then consider the motors and mount needed to move such a dish to point in different directions! Obviously, this is not possible.

Fortunately, radio astronomers *can* achieve this resolution by combining a number of ordinary-sized radio telescopes. With the radio telescopes laid out in an array and treating each pair of telescopes as an *interferometer*, we can effectively *synthesize* a new telescope with a very large diameter. This is done by observing the same source with all telescopes in the array, combining the outputs *from each pair* of telescopes, and mathematically processing the data in a way that produces the equivalent resolution of a single large telescope. The result of this process is an effective resolution that is approximately equal to that of a telescope whose diameter equals the largest distance between antennas in the array. Considering our calculation above, we can achieve arcsecond resolution at a wavelength of 10 cm if we use an array of radio telescopes spread out over 5 km. The Jansky VLA, for example, has 27 telescopes, each with a diameter of 25 m, that can be distributed with the largest distance between telescopes as large as 36 km. Moreover, a particular method of aperture synthesis known as *very long baseline interferometry* (or VLBI) uses telescopes spread over Earth and can achieve the resolution of a telescope 12,000 km, in diameter! Space-based antennas can improve the resolution with VLBI by a factor of 20 or more. The resolution of VLBI observations, therefore, is typically thousands of times better than that of ground-based visible-wavelength observations.

All the individual radio telescopes in the array are combined to compose a single observing instrument, so the whole array can be called a *telescope*. Therefore, this word now is ambiguous. To avoid confusion, in the rest of this chapter and in Chapter 6, we will use the word *telescope* to mean the entire array, and use the words *antenna* or *element* to refer to the individual dishes in the array. We will refer to the instrument used in observations made with a single antenna (as in Chapters 3 and 4) as a *single-dish telescope*. Although the meaning is usually clear from the context, be careful to avoid ambiguity when using the words *telescope* and *antenna* in discussions of aperture synthesis.

The concept of aperture synthesis is to synthesize a large telescope with widely spaced, ordinary-sized antennas. In a large reflector telescope, light from different parts of the reflector surface is brought together at the focus where it is combined, including the effects of interference, resulting in the final beam pattern (see Chapter 3). In aperture synthesis, light is collected by each individual antenna and then combined with that collected by other antennas. The interference of the light occurs for each pair of antennas, and the net result of all the pairs determines the *synthesized beam* pattern. Therefore, in aperture synthesis, we view each pair of antennas in the array as a piece of the entire aperture. The resultant synthesized telescope, then, has a beam commensurate with a telescope of diameter equal to the size of the array.

Keep in mind that aperture synthesis produces only the *resolution* of a large telescope, not the sensitivity. The sensitivity depends on the total collecting area, which, for an array of antennas, is the sum of the effective areas of all the individual antennas. With the Jansky VLA, for example, that would be an increase of sensitivity of 27 times over that with just one VLA antenna, which is a much smaller increase than the resolution improvement—which is over a 1000-fold.

5.2 TWO-ELEMENT INTERFEROMETER

The basic unit of an aperture synthesis telescope is the two-element interferometer, composed of two antennas separated by a well-known distance with a well-known orientation. The vector describing the separation of the antennas is known as the *baseline*, which we will symbolize as $\vec{b}$. In an observation, both antennas are pointed at the astronomical source, which is located in some direction on the sky that can be defined relative to the zenith of the midpoint of the baseline. To simplify the discussion in this chapter, we will consider a special case in which the source, the antennas, and the zenith of the baseline midpoint are all contained in the same plane. This is the case, for example, if the interferometer is on the Earth's equator, has an east–west baseline, and the source is located on the celestial equator. In this case, when the source transits, it passes through the zenith of the interferometer; the source direction, then, can be described by a single angle, θ, relative to the zenith. Additionally, this constrains the baseline to only one possible orientation, and so the distance, b, between the antennas, will suffice to describe the baseline.

A depiction of an interferometric observation in this two-dimensional situation is shown in Figure 5.3. The radio signal from an astronomical source in direction θ relative to the zenith approaches both antennas along parallel paths, since the source is, effectively, infinitely far away. The antennas respond to the oscillating electric fields, $E_1(t)$ and $E_2(t)$, of the electromagnetic waves, which are filtered to pass a small bandwidth centered at some specific frequency, ν, and then sent along identical cables where they meet and are combined.

Our first goal is to answer the following questions:

1. What is the output from the interferometer?

2. What are the adjustable parameters?

3. What is the relation between the interferometer output, the source structure, and position?

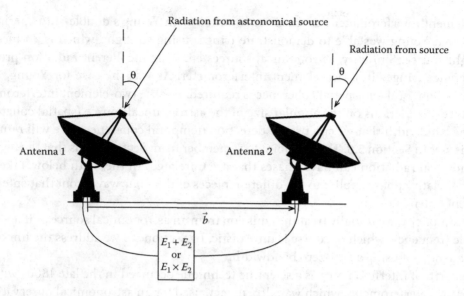

FIGURE 5.3   An interferometer, composed of two antennas separated by a distance b, receives signals from a radio source at an angle θ relative to the zenith.

Again, you may want to perform the labs and/or simulations before reading further.

We start by addressing the way that the signals are combined. Conceptually, the simplest way to combine signals is to *add* them. Thus, we first examine the case of the *additive interferometer* (Section 5.3.1), in which the *sum* of the two signals is averaged in time to produce the interferometer output. We then examine the case of the *multiplicative* or *cross-correlation interferometer* (Section 5.3.2), in which the *product* of the two signals, averaged in time, is the interferometer output. As we will demonstrate, the two methods lead to nearly the same result. Although the multiplicative (cross-correlation) interferometer is slightly more complicated in operation, the slightly different form of its output offers a significant advantage over that of the additive interferometer, as we show in Section 5.3.3. In fact, most modern interferometers are of the cross-correlation type, and in the remaining sections of the chapter we will focus only on the multiplicative case.

In an additive interferometer, the sum of the electric fields received by the two antennas, $E_1(t) + E_2(t)$ is sent to a square-law detector (introduced in Section 3.2.3), which averages the summed signal in time and converts it to a power. In a cross-correlation interferometer, the signals are sent to a *correlator*, where the electric fields are multiplied and averaged over time. In both cases, the average occurs over an *integration time* that must be much longer than the period of the waves. Both types of interferometers output a detected power; the additive interferometer output comes from the square law detector, and the cross-correlation interferometer involves the product of the electric fields (recall that the power of an electromagnetic wave is proportional to E^2). The detected power is calibrated and converted to a flux density in both cases.

An issue that deserves mention before we begin our detailed explanations is that of *coherence*. If you recall the discussion of this topic in Chapter 2, you may realize that the two-element interferometer has much in common with Young's double-slit experiment. Recall that Young was able to demonstrate fringes using sunlight, which is an incoherent light source. Similarly, astronomical sources emitting incoherent radiation produce interference fringes in a two-element interferometer. As was the case for Young, some degree of temporal and spatial coherence is required. For the two-element interferometer, this places restrictions on the angular size of the astronomical source (spatial coherence) and the bandwidth that we can use for detection (temporal coherence). We will return to these issues in Section 5.8. The key point to remember from Section 2.6 is that the incident astronomical radiation can have phases that are unrelated. In the math below, take note that the relative phase applies to the different pieces of the *same* wave front that enters the different antennas.

For simplicity, we initially treat the emission from an astronomical source as if it were at a single frequency, which of course is unrealistic. In Section 5.8 we address the important consequences of using a non-zero bandwidth.

The birth of interferometry as a scientific technique occurred in the late 1800s with the Michelson interferometer, which was also, in fact, used for an astronomical observation at visible wavelengths. There is much interesting history in the development and use of the Michelson interferometer, including the discovery of the constancy of the speed of light by Michelson and Morley in 1887, but little of that information is relevant to understanding modern aperture synthesis, and so we will not discuss those details here.

And that is it for our preamble. We now jump to a detailed, mathematical description of interferometry, starting with the simple situation of a point source.

5.3 OBSERVATIONS OF A SINGLE-POINT SOURCE

Consider two identical antennas, separated by a distance b, and a single astronomical radio source so small in angular extent that we can model it as a point. Let this point source have flux density F_ν, be located in the plane defined by the two antennas and the baseline zenith, and have an angle relative to the zenith, θ, as depicted in Figure 5.3.

Consider what happens to a single electromagnetic wave, of frequency ν, as it reaches and enters the two antennas. Keep in mind that the electromagnetic wave is not simply a ray of light, but is three dimensional with the wave front extending in both directions perpendicular to the direction of travel, so different parts of the same wave front will enter the two antennas.

As illustrated in Figure 5.4, the radiation entering antenna 1 will need to travel an extra distance, Δs, compared to the radiation entering antenna 2. So the arrival of the wave front at antenna 1 will involve a time *delay*, which we denote as τ. The delay is given, simply, by

$$\tau = \Delta s / c$$

where:
  c is the speed of light

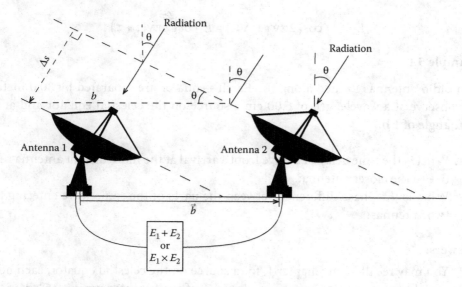

FIGURE 5.4 Observations of a single-point source in a direction at angle θ with respect to the zenith of the antennas. The antennas are separated by a *baseline* distance b. The extra path length that the wave must travel to reach antenna 1, on the left, is the length Δs, which is the side opposite angle θ in a right triangle with hypotenuse equal to b.

Using the right triangle shown in Figure 5.4, we can express the delay in terms of the direction toward the source. The extra distance Δs is one side of the right triangle in which b is the hypotenuse. So by simple trigonometry, $\Delta s = b \sin \theta$, or

$$\tau = \frac{b \sin \theta}{c} \tag{5.1}$$

This time delay in the arrival of the wave front means there will be a phase difference, ΔΦ, between the two signals, given by

$$\Delta \Phi = 2\pi \Delta s / \lambda$$

If Δs equals one wavelength, then the wave front entering antenna 2 will be 2π radians ahead of the same wave front entering antenna 1. We can also express the phase difference in terms of the time delay by

$$\Delta \Phi = 2\pi \nu \tau \tag{5.2}$$

If we choose the time when the wave enters antenna 1 as defining zero phase, then we must add the phase shift to the wave entering antenna 2 (to make it further along in its cycle before combination). Therefore, we have that the electric fields arriving at the combination point are given by

$$E_1 = E_0 \cos\left(2\pi \nu\, t\right) \tag{5.3}$$

and

$$E_2 = E_0 \cos(2\pi\nu\, t + \Delta\Phi) = E_0 \cos\left[2\pi\nu(t+\tau)\right] \tag{5.4}$$

Example 5.1:

Two radio antennas located along the Earth's equator are separated by 30.0 meters and observe at a wavelength of 6.00 cm a source on the celestial equator and at an hour angle of 1 h.

1. What is the time delay of a wave front's arrival at the more distant antenna relative to the closer antenna?
2. What is the phase difference between the parts of the same wave entering the two antennas?

Answers:

1. You may recall from Chapter 1, for a source on the celestial equator, each hour of right ascension (RA) corresponds to 15° of arc. So, this source is 15° from the zenith of the baseline. The time delay is given by Equation 5.1, so

$$\tau = \frac{30.0\,\text{m}}{3.00\times10^8\,\text{m s}^{-1}}\sin 15° = 2.59\times10^{-8}\,\text{s}$$

2. We can calculate the phase difference using Equations 5.1 and 5.2, or

$$\Delta\Phi = 2\pi\frac{b}{\lambda}\sin\theta = 2\pi\frac{30.0\,\text{m}}{0.0600\,\text{m}}\sin 15° = 258.82\pi \text{ radians}$$

We have kept five digits because both the full value of the phase and the fractional remainder after removing multiples of 2π are important. The full phase value indicates how far from the zenith the source is. The fractional remainder provides the phase of the cosine in Equation 5.4; this angle is equivalent to 0.82π radians.

We note that due to the Earth's rotation, the angle θ will slowly change with time, and so will the phase difference or time delay between the signals from the two antennas.

Now we need to add or multiply these two signals and then average. We consider the additive interferometer first.

5.3.1 Response of the Additive Interferometer

The sum of the electric fields given by Equations 5.3 and 5.4 is

$$E_1 + E_2 = E_0 \cos(2\pi\nu\, t) + E_0 \cos\left[2\pi\nu(t+\tau)\right]$$

This sum is then squared in the detector, yielding

$$(E_1 + E_2)^2 = \left\{E_0 \cos(2\pi\nu\, t) + E_0 \cos\left[2\pi\nu(t+\tau)\right]\right\}^2$$

Performing the square on the right hand side, this becomes

$$(E_1 + E_2)^2 = E_0^2 \cos^2(2\pi v\, t) + 2E_0^2 \cos(2\pi v\, t)\cos\left[2\pi v(t+\tau)\right] + E_0^2 \cos^2\left[2\pi v(t+\tau)\right]$$

We can factor out E_0^2 from each term. The remaining expression can be expanded using the trigonometric identities:

$$\cos^2 A = \frac{1}{2}\left[\cos(2A)+1\right]$$

$$\cos(A+B) = \cos A\cos B - \sin A\sin B$$

and

$$\cos A\sin A = \frac{1}{2}\sin(2A)$$

to become

$$\frac{1}{2}\cos(4\pi vt)+\frac{1}{2}+\cos(4\pi vt)\cos(2\pi v\tau)+\cos(2\pi v\tau)-\sin(4\pi vt)\sin(2\pi v\tau)$$

$$+\frac{1}{2}\cos(4\pi vt)\cos(4\pi v\tau)-\frac{1}{2}\sin(4\pi vt)\sin(4\pi v\tau)+\frac{1}{2}$$

This signal is then averaged over time. Even after being mixed down to the intermediate frequencies (see Section 3.2.2), all the terms containing $\cos(4\pi vt)$ or $\sin(4\pi vt)$ oscillate about zero with frequencies of order 10^6 Hz or higher, and so when averaged over even a small fraction of a second, these terms average to zero. This leaves only $1 + \cos(2\pi v\tau)$. (*Note*: Be careful to distinguish τ from t. The added signals are averaged over time t, while τ, the delay for the wave front to reach antenna 1, is relatively constant over the integration time (although it changes slowly as Earth rotates). The term $\cos(2\pi v\tau)$, therefore, does *not* go to zero when time averaged.) The time-averaged response for an additive interferometer, then, is

$$\left\langle (E_1 + E_2)^2 \right\rangle = E_0^2\left[1+\cos(2\pi v\,\tau)\right]$$

where we use the angle brackets $\langle\ \rangle$ to indicate a time average.

We can also express this response in terms of the direction toward the source by using the form of τ in Equation 5.1. It is also convenient to use $c = \lambda v$ to eliminate the observing frequency, and, in effect, express the baseline length in units of the observing wavelength. In this form, the additive interferometer's response is

$$\left\langle (E_1 + E_2)^2 \right\rangle = E_0^2\left[1+\cos\left(2\pi\frac{b}{\lambda}\sin\theta\right)\right] \tag{5.5}$$

Recall, for a moment, our prior discussion about the radiation not all being in phase. Take note that the interferometer response given in Equation 5.5 contains no information about the phase of the waves coming from either antenna. It does involve the phase *difference* between the two paths, but *all* waves of the same frequency will have this same phase difference. Therefore, each individual wave will elicit an interferometer response described by Equation 5.5, so the total response is of the same form as Equation 5.5.

5.3.2 Response of the Multiplicative Interferometer

We again start with Equations 5.3 and 5.4 to represent the electric fields entering the two antennas. This time the electric fields are multiplied, so we have

$$E_1 \cdot E_2 = E_0{}^2 \cos 2\pi\nu\, t \cos\left[2\pi\nu\left(t+\tau\right)\right]$$

We can use the trigonometric identity $\cos(A) \cdot \cos(B) = 1/2[\cos(A + B) + \cos(A - B)]$ to write this as

$$E_1 \cdot E_2 = \frac{E_0{}^2}{2}\cos\left(4\pi\nu\, t + 2\pi\nu\tau\right) + \frac{E_0{}^2}{2}\cos\left(2\pi\nu\tau\right)$$

(Again, be careful not to confuse the ts and τs.) The first cosine term oscillates between positive and negative values with a frequency of 2ν, while the second cosine term varies slowly in time due to the Earth's rotation.

As with the additive interferometer, the product of the electric fields is averaged over a period of time, called the *integration time*, which must be long compared to $1/\nu$. The first term averages to zero, and so only the second term remains. We have then that the detected power of the multiplicative interferometer is

$$\left\langle E_1 \cdot E_2 \right\rangle = \frac{E_0{}^2}{2}\cos\left(2\pi\nu\tau\right) \tag{5.6}$$

As before, the delay is related to the extra path length by $\tau = \Delta s/c$, where $\Delta s = b\sin\theta$, allowing us to write

$$\left\langle E_1 \cdot E_2 \right\rangle = \frac{E_0{}^2}{2}\cos\left(2\pi\frac{b}{\lambda}\sin\theta\right) \tag{5.7}$$

Comparing this equation with the result for the additive interferometer (Equation 5.5), we note that (neglecting the factor of 2 in the cosine amplitude, which we can account for in the calibration) both forms have the same dependence on θ. The only substantive difference between them is the presence of the constant offset term $E_0{}^2$ in the additive form. Since θ slowly varies with time due to the Earth's rotation, the response of an interferometer is quasi-sinusoidal, and this oscillating response is often referred to as *fringes*, which are discussed in detail in Section 5.4.

5.3.3 Effect of Noise

The lack of the offset term in the multiplicative interferometer is a substantial advantage over the additive interferometer, as it allows the multiplicative interferometer to detect fainter astronomical sources than is possible with a similar additive system. To understand this, we need to discuss noise. In real interferometer observations, the signals from the two antennas, before being added or multiplied, are processed by the receivers in the antennas (as discussed in Chapter 3) and therefore now contain contributions due to the receiver noise as well as the astronomical source. The offset term in the additive interferometer is related to the total power, which is dominated by noise. Even though we state that this offset is a constant, it varies in time because the noise power from each of the receivers varies in time, thus making it difficult or impossible to detect the much weaker slowly oscillating term due to the astronomical source.

This is not to say that the multiplicative interferometer is not affected by noise. We stated above that some of the response terms of the interferometer (both additive and multiplicative) average to zero and thus we eliminated these terms. However, because of the noise contribution, these terms average to zero only for infinitely long averaging time. In practice, our observations cannot be infinitely long, so there may not be sufficient time for the averaged value to approach zero. In addition, even the slowly varying cosine term has inherent fluctuations whose average over a finite time is not zero. Therefore, noise is always a concern, and it is often the limiting factor in determining how faint an astronomical source can be detected with an interferometer. Hence, there is as much attention paid to constructing low noise receivers for interferometers as there is for the single-dish telescopes, whose noise is discussed in Section 3.3. A more detailed description of noise in interferometry is discussed in Chapter 6.

Since the multiplicative interferometer does not have the extra noise term (the so-called constant in the additive interferometer), it is more sensitive to the signal from astronomical sources. For this reason, almost all modern interferometers are multiplicative (or cross-correlation) interferometers. Additive interferometers are still sometimes used by amateur radio astronomers because of their extreme simplicity of construction. We will only discuss the multiplicative or cross-correlation interferometer in the rest of this chapter.

5.4 FRINGE FUNCTION

Examination of Equation 5.7 reveals that the interferometer output is dependent on both θ, the location of the source, and b, the spatial separation of the antennas, divided by the observing wavelength, λ. Because of the Earth's rotation, the source position changes with time. If the interferometer is located at the equator and has an east–west baseline, and if the source is on the celestial equator, then the source position relative to the baseline zenith is given simply by

$$\theta = \omega_E t \tag{5.8}$$

where:

$\omega_E = 7.29 \times 10^{-5}$ radians s^{-1} is the angular rotation rate of Earth, and we have chosen $t = 0$ to represent the moment when the source transits the midpoint of the baseline.

Thus, the multiplicative interferometer response is a function of time that can be expressed by

$$\langle E_1 \cdot E_2 \rangle = \frac{E_0^2}{2} \cos\left(2\pi \frac{b}{\lambda} \sin \omega_E t \right) \tag{5.9}$$

For a given b and λ, then, the interferometer's response to a single-point source oscillates with time. Note that the oscillation has a very slow variation with time, unlike the incident electromagnetic waves, which oscillate with periods of 10^{-8} to 10^{-10} s. At small angles, $\sin\theta \sim \theta$, so the interferometer response is sinusoidal, but the frequency changes as the source moves away from the $\theta = 0$ position. An example of this function is plotted in Figure 5.5.

The interferometer's oscillatory response can also be described in terms of the source position. The time on the horizontal axis in Figure 5.5, for example, can be converted to position angle by use of Equation 5.8. This is a simple linear relation and so the interferometer response has the same shape when plotted with respect to source position, which is demonstrated in Figure 5.6.

The pattern displayed in Figure 5.6 is reminiscent of the double-slit interference pattern; waves from the same source are passed through two apertures and recombined and yields an oscillating pattern that depends on angle. These oscillations, therefore, are commonly called *fringes*, and the time-dependent function

$$\cos\left(2\pi\frac{b}{\lambda}\sin\omega_E t\right) \qquad (5.10)$$

is called the *fringe function*. The angular size of a fringe (shown as $\Delta\theta_{\text{fringe}}$ in Figure 5.6), which is known as the *fringe spacing*, is defined as the angular separation between adjacent zeros. The condition for zeros is that the argument of the cosine in Equation 5.10 is $\pm\pi$, which occurs when

$$2\pi\frac{b}{\lambda}\sin\theta = \pm\pi$$

Because the primary beam of most parabolic dishes is rather small, in practice we are always observing at small angles from the telescope axis, so the small angle approximation,

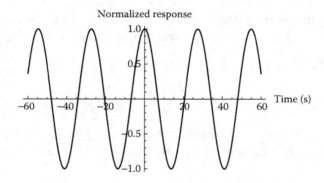

FIGURE 5.5   Response of a calibrated multiplicative interferometer to a point source, as given by Equation 5.9, divided by the source's flux density, plotted versus time, where $t = 0$ is the time of transit of the source. The displayed fringes correspond to that detected with a baseline that is 500 times the wavelength, with the antennas, zenith, and source all located in the same plane.

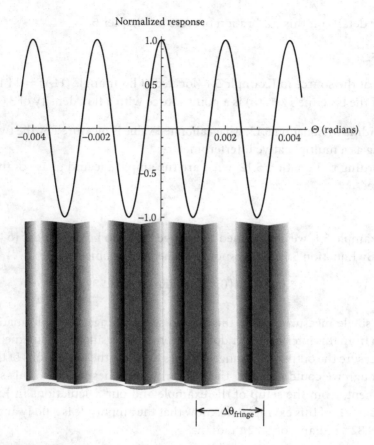

FIGURE 5.6 Oscillating pattern of the interferometer response as a function of the position of the point source. This is known as the *fringe pattern*. The fringe spacing, $\Delta\theta_{\text{fringe}}$, is indicated.

$\sin\theta \sim \theta$, may be applied. We see, then, that the angular separation between zeros of the fringe pattern is

$$\Delta\theta_{\text{fringe}} = \frac{\lambda}{b} \tag{5.11}$$

The fringe spacing is an important concept for understanding the functioning of an interferometer; we will refer to it as needed.

We can calibrate a multiplicative interferometer by observing a point source of known flux density and fitting the fringe oscillations to the expected time dependence, given by Equation 5.9. This calibration then provides a conversion factor to relate the fringe amplitudes to a flux density. The calibrated response, R, of a multiplicative interferometer is then

$$R = F_v \cos\left[2\pi\frac{b}{\lambda}\sin\left(\omega_{\text{E}}t\right)\right] \tag{5.12}$$

We give more details on this calibration process in Chapter 6.

Example 5.2:

Imagine that the source in Example 5.1 (located at hour angle (HA) = +1 h, or 15° to the west of the baseline's zenith) is a point source with a flux density of 3.00 Jy.

1. According to Equation 5.12, what calibrated value is measured by the two antennas acting as a multiplicative interferometer?
2. According to Equation 5.12, what are the amplitude and phase of the detected fringe?

Answer:

1. In Example 5.1, we determined the fringe phase to be equivalent to 0.82π radians. By Equation 5.12, the response of the interferometer is

$$3.00\,\text{Jy}\,\cos(0.82\pi\ \text{radians}) = -2.53\,\text{Jy}$$

   As a single measured value, this should seem strange. The information we can glean from this interferometer does not come with a single measurement. We need to measure the output for a range of times to detect the oscillations of the fringes.
2. Although we could not infer the answer to this question from this single measurement, from the setup of the example and our calculations in Example 5.1 and part (1) of this example, we know that the amplitude is 3.00 Jy and the phase is 258.82π radians, or 0.82π radians.

Example 5.3:

For the multiplicative interferometric observation in Example 5.2, how long must we wait to measure one full oscillation of the fringes?

Answer:

We need to determine the value of t when the phase in Equation 5.12 changes by 2π. We determined in Example 5.1 that the direct calculation of the fringe phase yielded a value of 258.82π radians. This occurred when the source was at HA = 1 h, or when $t = 3600$ s. Also in Example 5.1, we were given the values for b (30.0 m) and λ (0.06 m). Therefore, the completion of one full cycle will occur when

$$\Delta\Phi = 2\pi = 2\pi\frac{30.0}{0.06}\sin\left(7.29 \times 10^{-5}\ \text{radians s}^{-1}t\right) - 258.82\pi\ \text{radians}$$

which gives us

$$\sin\left(7.29 \times 10^{-5}\ \text{radians s}^{-1}t\right) = 0.261$$

We find, then, that a full fringe oscillation is completed at $t = 3622$ s, or 22 s later.

5.5 VISIBILITY FUNCTION

Let us examine Equation 5.12 in more detail, for it contains much useful information. First, consider keeping b/λ fixed, as when using a single antenna pair to observe a point source. We have, then, an oscillating function of time that describes the fringes. When the interferometer is calibrated, the amplitude of the fringes equals the flux density of the point source. The argument of the cosine, that is, the *fringe phase*, depends on the position of the source relative to the transit position. So, if we observed the source for sufficient time to see a full oscillation, we can infer the flux density of the source from the fringe amplitude, and, by fitting the oscillations to Equation 5.12, we can infer the exact time of transit of the source and hence the source position. Note that the argument of the cosine contains $\sin(\omega_E t)$ and therefore the frequency of the oscillations is time-dependent; hence the time of transit has a single solution.

Imagine, now, that we observe a point source thinking it is located at some sky position—let us call it θ_0—and thus we can predict the fringe phase. However, if the source is located a small angle away, at $\theta_0 + \Delta\theta$, then when we try to fit the data to Equation 5.12, we find that the phase is incorrect and must be adjusted. In other words, we do not actually need to know, precisely, the location of the source in advance. We pick a location in the sky as our reference position, and if the source is not at that location, we will discover this by the needed phase shift. At times near transit, when θ is very small, and using the small angle approximation ($\sin\theta \approx \theta$), the phase shift of the observed oscillations compared to the expected oscillations would be

$$\Delta\Phi = 2\pi \frac{b}{\lambda} \Delta\theta$$

Hence, we can think of the argument of the cosine as containing both a time dependence due to the Earth's rotation and a second term, which gives the dependence on position relative to the reference point. Therefore, in this case, we can write the response of the multiplicative interferometer as

$$F_\nu \cos\left[2\pi \frac{b}{\lambda} \sin\left(\omega_E t\right) + 2\pi \frac{b}{\lambda} \Delta\theta \right] \tag{5.13}$$

The fringe function corresponding to the reference position can be fitted to the response of the interferometer. In this example, because the source is not located at the reference position, there will be a phase offset from what we would expect. This phase difference tells us the position of the point source. Remember that the amplitude is equal to the flux density of the point source. After the fitting process and solving for amplitude and phase, the interferometer response can be rewritten as the following:

$$R = F_\nu \cos\left(2\pi \frac{b}{\lambda} \Delta\theta \right) \tag{5.14}$$

We will refer to this process as *fringe function removal*. This has the same effect as if the reference position were moved to the transit position ($t = 0$) and Earth was stopped from rotating. Equation 5.14 contains the information that we have been after. In this example, in which we observe a single-point source, we now have an equation that contains only b/λ and information about the point source—its flux density and its position relative to the reference position. Equation 5.14, therefore, is fundamentally important when using interferometers to learn about astronomical sources.

This function is called the *visibility function*. We can think of Equation 5.14 as being a function of the independent variable b/λ that contains information about the source. The need to measure the visibility over a range of baselines, to yield a measure of the visibility *function* is the reason that interferometers, such as the Jansky VLA, involve a number of antennas. In the set of antennas, which is referred to as an *array*, each pair makes a different baseline, and so visibility data is obtained for many baselines at once.

This interpretation of Equation 5.14 deserves extra emphasis. The interferometer response yields a function of b/λ that can be used to infer information about the position and intensity of the source. If the source is more complicated than a point source, as we will discuss in Sections 5.4 and 5.5, then the visibility function will also give information about the source structure. The amplitude of the oscillations (the factor in front of the cosine) is the *visibility amplitude*, and the argument of the cosine is the *visibility phase*. The independent variable of this function, b/λ, is called the *spacing*. We will write the visibility amplitude as V_A and the visibility phase as Φ_V, and the entire function as $V(b/\lambda)$, which will help us remember that it is a function of the spacing b/λ. Equation 5.15 expresses this mathematically.

$$V\left(b/\lambda\right) = V_A \cos\Phi_V \qquad (5.15)$$

And, so far, we know that for a single, unresolved source, the visibility amplitude and phase are given by

$$V_A = F_v \qquad (5.16)$$

and

$$\Phi_V = 2\pi\frac{b}{\lambda}\Delta\theta \qquad (5.17)$$

If you completed the VSRT labs or the VSRTI_Plotter activities (described in Appendix VII), you will discover that the visibilities hold information about the source structure.

It is important to recognize that the observer *chooses* the reference position on the sky, and that this is the point at which the fringe function is defined. The phase offset in Equation 5.13 reveals where the point source is located *relative* to the reference position. The reference position, therefore, becomes the *map center*. Notice, also, that the choice of map center dictates the position that has zero visibility phase, so this position is also called the *phase center*. The fringe function is, in essence, a reflection of the baseline, observing wavelength, and choice of phase center, and *not of the source*.

It is also important to note that the phase center can be changed *even after the observations have occurred*. The choice of phase center is a mathematical selection we make when processing the data. If we later wish to use a different point as the reference position, we can do so by introducing a phase shift into the data. Provided that the new reference position is not too far from the center of the primary beam of the individual antennas, we can thus *re-point* our telescope after the observations have occurred.

Since the discussion above is so important, it is worth a quick summary. The interferometer response is a function of time, because of the Earth's rotation. The amplitude of the oscillations equals the visibility amplitude. A reference position, called the *phase center*, is chosen and the fringe function is the time dependent response of the interferometer for that reference position. The time-dependent part of the phase is fitted for a chosen reference position, and the extra phase needed to fit the data equals the visibility phase. The visibility function contains information about the source structure; it is a function only of the spacing b/λ. Remember that the fringe function is defined *relative to a chosen map center*. (If you have some experience with aperture synthesis, you may have seen visibilities plotted vs. time in contradiction to our statement that the time dependence is removed. This time dependence is likely due to another effect of the Earth's rotation, which we will discuss in Chapter 6. This additional time dependence occurs because the Earth's rotation causes a change in the orientation and the effective, that is, *projected*, baseline length.)

Example 5.4:

In the previous examples, we have discussed an observation at 6.00 cm wavelength of a 3.00-Jy point source located 15° to the west of the zenith of our baseline of 30.0 m. Imagine that the phase center was chosen to be the sky position currently located 14.98° to the west of the zenith.

1. What is the phase of the fringe function in this measurement? What are the visibility amplitude and phase?
2. The phase center is adjusted to be the location of the point source. How does the visibility change from what it was in (1)? If the visibility contains information about the source, why can our visibilities in (1) and (2) differ when observing the same source?

Answer:

1. The fringe function is defined relative to the phase center, so its phase in this measurement is

$$\Phi_{\text{fringe}} = 2\pi\frac{b}{\lambda}\sin(\theta) = 2\pi\frac{30.0\text{ m}}{0.0600\text{ m}}\sin(14.98°) = 258.48\pi\text{ radians}$$

The visibility phase is the measured phase minus the fringe function phase, or

$$258.82\pi - 258.48\pi = 0.34\pi\text{ radians}$$

The visibility amplitude, meanwhile, equals the flux density of the source, or 3.00 Jy.
2. According to Equation 5.15, the visibility amplitude will not change and will still be 3.00 Jy. The phase of the fringe function, now, will be equal to observed

phase, so the inferred visibility phase will be zero. We get a different visibility phase from (1) only because of the choice of the phase center. The visibility phase depends not only on the source but also on the choice of the phase center.

Example 5.5:

The same observation as in Example 5.4, part 2 is made on two more baselines, of lengths 10.0 m and 40.0 m.

1. What are the visibility amplitudes and phases in each of these two measurements?
2. Describe the observed visibility function.

Answer:

1. The visibility amplitude of a point source, as demonstrated in Equation 5.15, depends only on the point source's flux density, so both these baselines have the same amplitude as the first baseline, that is, 3.00 Jy. Since the visibility phase depends on the angular distance between the phase center and the point source, and the phase center is coincident with the point source, the visibility phase on both of these baselines is zero, the same as on the first baseline.

2. This is a simple constant function. The phase on all baselines equals zero and amplitude on all baselines equals the flux density of the point source. As we explain next, this is the nature of the visibility function of a point source. More complex sources produce more complicated visibility functions.

5.5.1 Analysis of Visibilities for a Single-Point Source

Let us examine the equations for the visibility of a point source to see precisely what form the visibility function takes. Remember that the independent variable of the visibility function is the spacing, b/λ. Therefore, we consider Equations 5.16 and 5.17 for a range of baseline lengths, or spacings.

We start with the amplitudes. As seen in Equation 5.16, the amplitude depends only on the flux density of the source; it is independent of spacing. All pairs of antennas in the array will measure the same amplitude. Therefore, *the visibility amplitude of a point source is constant, independent of* b/λ, *and equal to the flux density.*

If you completed the Simulation Activities discussed in Appendix VII, note the appearance of the plot in Step 4 of Activity 1. You should have found that the visibility amplitude was constant for all values of b/λ.

Now consider the visibility phase, as given in Equation 5.17. The phase depends on both the source position, $\Delta\theta$, and on the spacing, b/λ. For a point source at position $\Delta\theta$ relative to the map center, the phase varies linearly with b/λ. Therefore, the rate of change of Φ_V, with respect to a change in b/λ, is constant, and is given by

$$\frac{d\Phi_V}{d(b/\lambda)} = 2\pi\Delta\theta \qquad (5.18)$$

Recall that the small angle approximation was used to obtain Equation 5.17; as a consequence, Equation 5.18 is valid only for small values of $\Delta\theta$. Note that, if the point source is at the phase center of the map, that is, $\Delta\theta = 0$, then the visibility phase is also zero. Therefore, *the visibility phase of a point source located at the phase center is zero.*

The behavior of V_A and Φ_V for a point source at the phase center is important for the calibration of interferometric data, as we will see in Chapter 6.

If we used an assortment of interferometer pairs, with differing baselines, to observe a source, and we found that all pairs yielded the same visibility amplitude, and that the phases varied, but had a constant rate of change of phase with respect to spacing, $d\Phi_V/d(b/\lambda)$, then we could infer that we have observed a point source, the flux density of the point source, and its location in the sky. Figure 5.7 displays the plot of visibility phase versus b/λ for a situation like this.

Example 5.6:

The observation discussed in Example 5.4, part 1 ($\lambda = 6.00$ cm, $b = 30.0$ m, a point source with $F_v = 3.00$ Jy, and 1°W of the phase center; therefore, $V_A = 3.00$ Jy and $\Phi_V = 16.90\pi$) is also made on the two additional baselines of lengths 10.0 m and 40.0 m. If the observers suspect that the source is indeed a point source located 0.02° from the phase center, how should they expect the visibility phase to change with baseline?

Answer:

Applying Equation 5.18, we should expect the visibility phase to change with baseline at a rate of

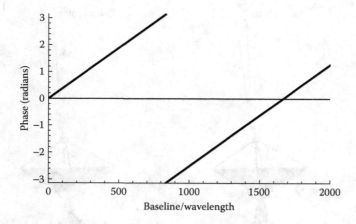

FIGURE 5.7   Visibility phase plotted versus baseline/wavelength for a point source located 0.0006 radians (2 arcmin) from the phase center. Because of the redundancy in angles differing by 2π, the phases are seen to *wrap* around when reaching $+\pi$ radians and reappearing at $-\pi$ radians.

$$\frac{d\Phi_V}{d(b/\lambda)} = 2\pi \times 0.02° \times \frac{\pi}{180°} = 2.19 \times 10^{-3} \text{ radians per spacing}$$

With more complicated sources, the visibility phase is not simply the location of the source relative to the phase center; therefore, the observers could not conclude this information from the single-baseline measurement. However, with the observation on additional baselines, if they also find that the measured visibility phases change at a rate consistent with Equation 5.18, then they have confirmation that the phase is due solely to the source location.

Before ending the discussion of observations of a point source, we emphasize that the equations we have given so far apply only to the special situation that we have described. In particular, the equations apply only to the case where the two antennas and the one-dimensional astronomical source are coplanar. The conceptual conclusions are generally correct; the more general equations are given in Chapter 6.

We now consider a slightly more complicated source, to show how the visibility amplitudes and phases are affected by source structure.

5.6 OBSERVATIONS OF A PAIR OF UNRESOLVED SOURCES

Consider, a pair of point sources, A and B, of flux densities F_A and F_B and located at angles $\Delta\theta_A$ and $\Delta\theta_B$ from the phase center, separated by a small angle $\Delta\Theta$ ($\Delta\Theta = \Delta\theta_A - \Delta\theta_B$). The configuration for this situation is sketched in Figure 5.8.

We wish to calculate the visibility function when each antenna is receiving waves from *both* sources. It is fairly easy to show that the additive and multiplicative interferometers again yield the same visibility function. But, as in Section 5.5, we will derive the visibilities only for the multiplicative interferometer.

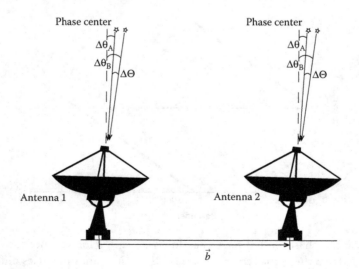

FIGURE 5.8 A pair of antennas is used as an interferometer to detect a pair of point sources, located at angles $\Delta\theta_A$ and $\Delta\theta_B$ from the phase center, and separated by $\Delta\Theta$ from each other.

So, consider now the signal received by antennas 1 and 2, composed of the electromagnetic waves coming from sources A and B. When the signals from the two antennas are multiplied, there will be contributions from both sources in each antenna's signal. Since sources A and B are independent, the phase variations of the two signals are uncorrelated; therefore, when the waves from source A received by antenna 1 are multiplied by the waves from source B received by antenna 2, the product will average to zero. Likewise, for the product of the waves from source A received by antenna 2 with waves from source B received by antenna 1. Hence, in the multiplication of the signals from the two antennas, only the products of waves from the same source will survive, that is,

$$E_A{}^2 \cos(2\pi v\, t)\cos\big[2\pi v(t+\tau_A)\big] + E_B{}^2 \cos(2\pi v\, t)\cos\big[2\pi v(t+\tau_B)\big]$$

The total output, then, is just the sum of the interferometer responses obtained from sources A and B individually.

We can use Equation 5.14 for each source, so the total visibility is

$$V_{AB}\cos\Phi_{AB} = F_A \cos\left(2\pi\frac{b}{\lambda}\Delta\theta_A\right) + F_B \cos\left(2\pi\frac{b}{\lambda}\Delta\theta_B\right) \tag{5.19}$$

We can manipulate Equation 5.19 to obtain the total visibility amplitude, V_{AB}, and phase, Φ_{AB}. A mathematical trick we can use here is to recognize that since each term on both sides of Equation 5.19 involves a number multiplied by a cosine, each term is like the x-component of a vector, where the two vectors on the right sum to the vector on the left, that is, if

$$\vec{C} = \vec{A} + \vec{B}$$

then

$$C\cos\Phi_C = A\cos\Phi_A + B\cos\Phi_B$$

This is relatively easy to solve, but the math is significantly easier, and just as enlightening, if we let vector $\vec{A}$ lie along the x-axis (so that $\Phi_A = 0$), as depicted in Figure 5.9. (Note that we can rotate this triangle, so that $\Phi_A \neq 0$, yielding a general solution, but the triangle stays in the same shape, so we will obtain the same result.) Therefore, the contribution to the inferred visibility amplitude and phase arising from each source is

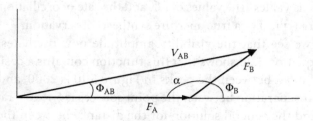

FIGURE 5.9   The relation between the visibilities of two sources, V_A and V_B, and the total visibility, V_{AB}, treated as a vector addition.

$$V_{AB} \cos \Phi_{AB} = F_A + F_B \cos \Phi_B$$

This mathematical relation is the same as the vector addition shown in Figure 5.9.

With the three sides of the triangle in Figure 5.9 having lengths equal to F_A, F_B, and V_{AB}, we can use the law of cosines to determine the value of V_{AB}; therefore,

$$V_{AB}^2 = F_A^2 + F_B^2 - 2 F_A F_B \cos \alpha$$

where:

α is the angle opposite V_{AB}.

As shown in Figure 5.9, α must equal $\pi - \Phi_B$. We have, then,

$$V_{AB}^2 = F_A^2 + F_B^2 - 2 F_A F_B \cos(\pi - \Phi_B) = F_A^2 + F_B^2 + 2 F_A F_B \cos \Phi_B$$

By taking the square root, we get an expression for the total visibility amplitude:

$$V_{AB} = \sqrt{F_A^2 + F_B^2 + 2 F_A F_B \cos \Phi_B}$$

More generally, we must allow for a non-zero Φ_A, in which case the triangle in Figure 5.9 will rotate with the left corner fixed at the origin. But, since the separation angle between the sources is a given, the difference in the phases is fixed

$$\Phi_B - \Phi_A = 2\pi \frac{b}{\lambda} \Delta\Theta$$

and so the triangle retains its shape. The angle represented by Φ_B in Figure 5.9, therefore, is actually the phase difference, $\Phi_B - \Phi_A$. Therefore, we substitute in for Φ_B and get

$$V_{AB} = \sqrt{F_A^2 + F_B^2 + 2 F_A F_B \cos\left(2\pi \frac{b}{\lambda} \Delta\Theta\right)} \qquad (5.20)$$

Figure 5.10 shows a plot of the visibility amplitude, V_{AB}, versus b/λ with $\Delta\Theta = 5 \times 10^{-4}$ radians, as an example.

If you completed the Simulation Activities discussed in Appendix VII, compare your plot from Activity 2 to that shown in Figure 5.10. These plots should be identical, except for the different x-axis values. The values of b/λ and the rate of oscillation in the visibility in Figure 5.10 are more typical of a true aperture synthesis observation.

In Figure 5.10, we see that the visibility amplitude now oscillates, though not as a simple sinusoid. Equation 5.20 shows that this function contains a cosine inside a square root. The distance in b/λ between the peaks in Figure 5.10 is 2000, and this corresponds to when the angular separation between the point sources is 5×10^{-4} radians. We can use Equation 5.20 to find the general solution for the distance between the peaks of the visibility amplitude when observing a pair of point sources. The peaks occur when the cosine equals +1, which occurs when the argument of the cosine equals an even number times π.

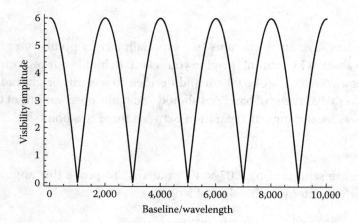

FIGURE 5.10 A plot of the visibility amplitude vs. spacing (b/λ) when observing two point sources, each with flux density = 3 Jy and separated by 5×10^{-4} radians (1.7 arcmin).

Therefore, if we call $P_{b/\lambda}$ the period in b/λ-distance of the visibility amplitude oscillation, we have

$$2\pi P_{b/\lambda} \Delta\Theta = 2\pi$$

or

$$P_{b/\lambda} = \frac{1}{\Delta\Theta} \qquad (5.21)$$

We find, therefore, that the inverse of the angular separation of the two sources defines the scale of b/λ needed for inferring this two-source structure. To obtain the information that there is a pair of sources we need to observe the oscillation of the visibilities over a range of baselines of this scale. Recall from Section 5.4 (see Equation 5.11) that λ/b is also the fringe spacing. The fringe spacing, therefore, can be viewed as the angular size scale that a given interferometer probes. If b_{min} and b_{max} are the smallest and largest baselines in our data, then we can only infer information about structure of angular sizes smaller than λ/b_{min} and larger than λ/b_{max}.

The key points here are that for a pair of equal point sources with an angular separation of $\Delta\Theta$:

1. The visibility amplitude as a function of the antenna spacing, b/λ, is a periodic function.

2. That periodic function has a spacing period, $P_{b/\lambda}$ given by Equation 5.21, that is,

$$P_{b/\lambda} = \frac{1}{\Delta\Theta}$$

   where:
   $\Delta\Theta$ is in radians

3. The first zero in the visibility occurs at $b/\lambda = 1/(2\Delta\Theta)$.

Example 5.7:

The radio galaxy Cygnus A contains two especially bright points of radio emission separated by about 0.71 arcmin. Imagine you set out to build your own interferometer to observe at a wavelength of 6.00 cm and decided to use this bright radio galaxy to test it out. Over what range of baselines should you make observations of Cygnus A to be able to measure the angular separation between these two points?

Answer:

These points are separated by 2.07×10^{-4} radians. To probe this angular scale, we need an interferometer with baseline of length

$$b = \frac{0.0600 \text{ m}}{2.07 \times 10^{-4}} = 290 \text{ m}$$

This represents the baseline distance for one oscillation period. We could obtain a good determination of the oscillation of the visibility amplitude due to the pair of sources, therefore, by using baselines ranging from 150 to 450 m, or from the smallest baseline we can arrange (equal to the diameter of our antennas) up to 300 m.

Example 5.8:

In an observation of a pair of point sources, using an array of antennas, the visibility amplitudes are seen to oscillate with a spacing period equal to that expected (i.e., equal to one over the believed separation angle of the two sources). However, the amplitude never goes to zero, oscillating between 2 and 4 Jy. Why? What does this mean in terms of the sources?

Answer:

With careful examination of Equation 5.20, we see that the visibility amplitude only goes to zero if the two point sources have equal flux densities. Knowing that the maximum and minimum amplitudes are 2 and 4 Jy, we can infer what the individual flux densities are. The maximum occurs when the cosine equals 1; therefore,

$$4 \text{ Jy} = \sqrt{F_A^2 + F_B^2 + 2F_A F_B} = \sqrt{(F_A + F_B)^2} = (F_A + F_B)$$

And the minimum occurs when the cosine equals -1, and so

$$2 \text{ Jy} = \sqrt{F_A^2 + F_B^2 - 2F_A F_B} = \sqrt{(F_A - F_B)^2} = (F_A - F_B)$$

We find, then, that $F_A = 3$ Jy and $F_B = 1$ Jy.

We have, so far, shown that for a single-point source, the visibility amplitude is *constant for all baselines* and is equal to the flux density of the source; explained that two separate

point sources must be incoherent; and stated that the interferometer response to two incoherent sources is *the sum* of the individual response from each source. Nevertheless, in
Figure 5.10, we see that the visibility amplitude is, for most spacings, *less than* the sum of
the flux densities of the two sources. Is this a contradiction? No, it is not. It may be more
enlightening if we add the fringes graphically, which we show in Figure 5.11. Because of
their different positions, the sources transit at different times, which causes a slight shift in
the plot of one fringe pattern relative to the other. Because of this shift, the maxima of the
two patterns do not align, so the maxima of the total interferometer response are less than
the sum of the maxima of the two individual fringes. The amplitude of the total oscillations, therefore, is less than the sum of the flux densities.

Also according to Figure 5.10, for sources separated by 5×10^{-4} radians, when $b/\lambda = 1000$,
the visibility amplitude equals zero. This occurs because, at this spacing, the delays are
such that the fringes of the two sources are exactly out of phase.

This brings us to another very important principle to keep in mind. Consider using a
number of different baseline lengths to observe a double radio source. If all the baselines are
much shorter than $\lambda/\Delta\Theta$, so that for all baselines.

$$\Delta\Theta < \frac{\lambda}{b}$$

then the periodicity in the visibility amplitudes will not be detected. This set of baselines,
in short, will *not* have the *resolution* needed to resolve the source into its two constituents.
We see, therefore, that with an array of telescopes,

$$b_{max}/\lambda$$

is a measure of the resolution of the array, where b_{max} is the longest baseline in the array.
Note that this is also the fringe spacing associated with the longest baseline.

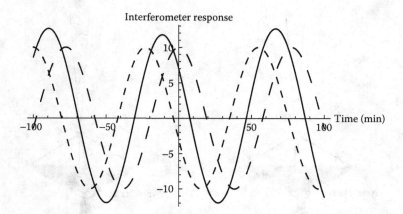

FIGURE 5.11   Interferometer response, plotted versus time (where $t = 0$ corresponds to the time of
transit), for a pair of 10-Jy point sources separated by 0.1 radians and $b/\lambda = 3$. The short and long
dashed lines represent the visibility amplitude for two individual sources at slightly different locations and the solid line shows the total visibility amplitude when observing both sources.

Remember, though, that we are still only considering sources and baselines along the equator. In general, our statement of the resolution of an array only applies along the direction of the baseline. For example, if all our baselines are in the east–west direction, then the resolution in the north–south direction is only that given by the main beam of an individual antenna.

5.7 OBSERVATIONS OF A SINGLE EXTENDED SOURCE

We now imagine observing a single source that extends over an angle $\Delta\Theta_{size}$. To simplify the details, we assume the source has uniform intensity. Electromagnetic waves now enter both antennas, but coming from a small range of directions. This situation is depicted in Figure 5.12.

Activity 1 in Appendix VII provides insight into the visibility amplitude of a single source with uniform intensity that has a finite angular size. For an extended, uniform source with an angular width of $\Delta\Theta_{size} = 2\times10^{-4}$ radians, the visibility amplitude as a function of b/λ is plotted in Figure 5.13. The shape of the visibility function is a sinc function (see Section 3.1.5) and the first zero in the visibility amplitude occurs at $b/\lambda = 1/\Delta\Theta_{size}$. We will demonstrate in Chapter 6 that there is a Fourier transform relationship between the intensity distribution of the source on the sky and the visibility as a function of baseline length.

The shape of the visibility amplitude versus baseline shown in Figure 5.13 resembles the beam pattern derived in Section 3.1.5 for a uniformly illuminated one-dimensional aperture (which is the diffraction pattern of a single slit). This should not be a surprise, as all of these involve a Fourier transform relation. The electric field as a function of angle on the sky is related to the Fourier transform of the electric field distribution in the aperture plane (for the case worked out in Section 3.1.5 the electric field was uniform), while the visibility amplitude versus baseline is related to the inverse Fourier transform of the source intensity

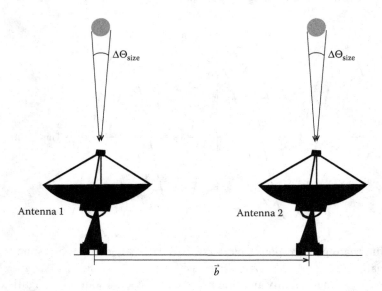

FIGURE 5.12 Interferometer detects an extended source with angular size $\Delta\Theta_{size}$.

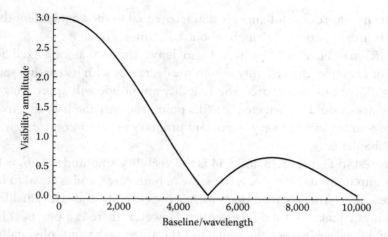

FIGURE 5.13 Visibility amplitude from an observation of a single, resolved (one-dimensional) source of angular width 2×10^{-4} radians (40 arcsec) and total flux density of 3 Jy is plotted versus spacing.

distribution (which in the case shown in Figure 5.13 is also uniform). The first zero of the beam pattern of an aperture of width a occurs at $\theta = \lambda/a$, which you will note has a similar relationship to the first zero of the visibility function.

For a smaller source, then, we should expect the visibility amplitude to decrease more slowly and reach zero at a larger spacing. We show this to be the case in Figure 5.14, which displays the visibility amplitude for a source half as large as that in Figure 5.13.

Notice that in the caption of Figure 5.14 we qualified the source as *one dimensional*. As with an antenna beam, for which the diffraction pattern of a circular aperture differs slightly from that of a one-dimensional aperture, a circular source has a slightly different visibility function than a one-dimensional source. The b/λ of the first zero in the visibility

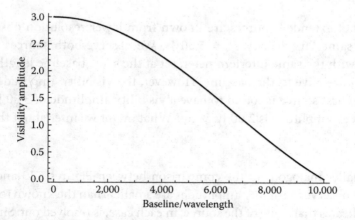

FIGURE 5.14 Visibility amplitude in an observation of a single, resolved (one-dimensional) source of angular width 1×10^{-4} radians (20 arcsec) and total flux density of 3 Jy is plotted versus spacing.

for a uniform circular source (of angular diameter equal to the angular length of the one-dimensional source) is, not surprisingly, about 1.22 times larger.

The comparison of Figures 5.13 and 5.14 also demonstrates that for a fixed b/λ (consider $b/\lambda = 4000$, for example) the visibility amplitude decreases with increasing source size. In fact, for a very large uniform source, the visibility amplitude will approach zero—meaning that no fringes would be detected. At this point, the baseline has, effectively, *resolved out* the entire source. This is a very important property of interferometers, which we will return to in Chapter 6.

Also of interest in Figures 5.13 and 5.14 is the visibility amplitude at $b/\lambda = 0$. Note that the visibility amplitude at this point is the same in both cases, and is equal to the total flux density of the source. This is called the *zero-spacing* visibility and is a valuable data point, because it always equals the total flux density. However, there is a practical limit to how small a spacing is possible since the centers of the antennas cannot physically be located zero meters apart. For the Jansky VLA, the antenna diameters are 25 m and therefore the centers cannot be placed closer than 25 m. For a source at the zenith, it is therefore impossible to have a baseline of less than 25 m. If observing a source overhead with the Jansky VLA at a wavelength of 20 cm, the shortest b/λ possible would be 125 and therefore sources with an angular size greater than about $1/125 = 0.01$ radians cannot be adequately measured. We will return to this topic in Chapter 6.

For extended sources, the reduction in the fringe amplitude or even the absence of fringes is similar to what happens in Young's double-slit experiment. Remember in Chapter 2 that we mentioned that some *spatial coherence* was needed to produce fringes in the double-slit experiment and Young passed the light through a pinhole to produce the needed spatial coherence. Without the pinhole, the fringe amplitude in the double slit would be greatly reduced. Likewise, when an astronomical source has an angular extent similar to or larger than λ/b, the fringe amplitude will be diminished or eliminated. Interferometers are not good instruments for observing sources of large angular extent.

Example 5.9:

Two single but extended sources are known from lower resolution observations to contain the same flux density, $F_v = 5.00$ Jy. We observe both sources at the same wavelength with the same interferometer, set at the same baseline length, and at the same position relative to the baseline. However, the visibility amplitudes are found to differ. The first source is found to have a visibility amplitude of 3.00 Jy, while the second source's amplitude is 2.00 Jy. Why? What might we infer about the sources?

Answer:

This is, actually, analogous to the comparison between Figures 5.13 and 5.14. First, note that the observed visibility amplitudes are smaller than the known total flux density. This indicates that some of the source, in each case, is resolved out. Since we know that the sources have the same total flux density, the smaller visibility amplitude of the second source indicates that more of it is resolved out than with the first source.

Therefore, we could infer that the first source contains more of its flux density within the observed fringe angle. This would result if the second source is larger, for example.

5.8 COHERENCE AND THE EFFECTS OF FINITE BANDWIDTH AND INTEGRATION TIME

In our discussion of coherence in Chapter 2, we mentioned the issue of *spatial* coherence; we also discussed the need for *temporal* coherence and explained that the degree of coherence decreases when the radiation extends over a range of frequencies. We have, so far, demonstrated that when the radiation extends over a larger angle the visibility amplitude is decreased, and we have commented that this can be viewed as a decrease in the spatial coherence of the source. In this section, we consider the impact that a decrease in temporal coherence has on interferometric observations.

All of our prior discussion in this chapter assumed a monochromatic receiver system. But, in reality, we need a reasonably large bandwidth in order to detect astronomical sources (remember that the amount of detected power, as given in Equation 3.1, is proportional to the bandwidth). A large bandwidth, though, causes a decrease in the temporal coherence. While making observations, we need to be wary of the amount of temporal coherence lost, since it can have a negative impact on our ability to measure visibilities. This effect in interferometry is known as *bandwidth smearing*. A common practice to reduce the incoming data rate is to increase the integration time per visibility. This also can have a negative impact on our ability to measure visibilities and is referred to as *time smearing*. In this section, we will explain the origin of both of these effects, and provide guidelines for avoiding their deleterious effects.

5.8.1 Bandwidth Smearing

To obtain an intuitive feel for bandwidth smearing, consider the following argument. In Section 5.4, we defined the fringe function, given by Equation 5.10, as

$$\cos\left(2\pi\frac{b}{\lambda}\sin\omega_E t\right)$$

and we noted the similarity of this function to a two-slit interference pattern. In other words, the response of a two-element interferometer is a fringe pattern on the sky, as shown in Figure 5.6, where the light areas indicate maximal response and the shaded areas indicate minimal response. The separation between adjacent minima (shown by $\Delta\theta_{fringe}$ in the figure) is just the fringe spacing, $\Delta\theta_{fringe} = \lambda/b$, defined in Equation 5.11. Bandwidth smearing occurs because of the dependence of $\Delta\theta_{fringe}$ on the observing wavelength, λ—or, equivalently, on the observing frequency, ν. At different frequencies, the fringes in Figure 5.6 will have different widths and when added will spread out and flatten the fringes at sky positions far from $\theta = 0$. For example, if the bandwidth is sufficiently wide, so that the wavelength at the band edge is 1.5 times the wavelength at the band center, then maxima from the band center will fall on minima from the band edge, and the fringes will cancel.

To make the argument more quantitative, consider the effect that different frequencies will have on the phase difference between the signals arriving at the two antennas. At the center of the bandpass, the phase shift between the two antennas of a two-element interferometer depends on frequency as given by Equation 5.2

$$\Delta\Phi = 2\pi\nu\tau$$

Therefore, $\Delta\Phi$ will vary across the bandpass for non-zero delays. Remember that delay is related to sky position. If the variation in $\Delta\Phi$ becomes large, then the fringes produced by the source at different frequencies within the bandpass will be out of step and will start to cancel. In this case, the fringe amplitude will be reduced, eventually becoming undetectable. Figure 5.15 displays the fringe oscillations of a point source versus sky position at two different frequencies, along with their sum. We see in the figure that away from the transit position, indicated by $\theta = 0$, the two fringe oscillations get out of sync, so their sum has a decreasing amplitude for positions away from the zenith. In Figure 5.16, we show the total fringe oscillation plotted versus time and delay of a point source observed with five different frequencies. In this figure, it is clear that no fringes will be detected at large delays.

When the phase difference between the center and the edge of the bandpass is π radians, then the edges will be exactly out of phase with the center of the band. This occurs when

$$\Delta\Phi(\nu_0) - \Delta\Phi\left(\nu_0 - \frac{1}{2}\Delta\nu\right) = 2\pi\nu_0\tau - 2\pi\left(\nu_0 - \frac{1}{2}\Delta\nu\right)\tau = \pi$$

and thus when $\tau = 1/\Delta\nu$. Therefore, when the time delay is as large as $1/\Delta\nu$, the fringes will be completely washed out. The time $1/\Delta\nu$ is called the *coherence time*. For example,

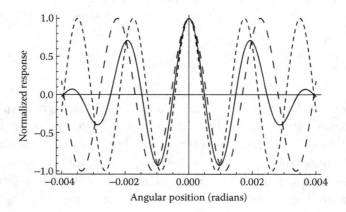

FIGURE 5.15 Fringe oscillation of a point source at two frequencies (short and long dashed lines) and the average (solid line) of the two. (To demonstrate the effect clearly, we show fringe functions with $\Delta\nu/\nu \sim 0.25$.)

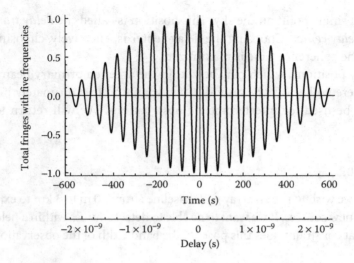

FIGURE 5.16   Normalized interferometer response due to a point source averaged over five close frequencies (spread over a total $\Delta v \sim 0.067\ v$).

a bandwidth of 100 MHz has a coherence time of 10^{-8} s, which corresponds to a path length of about 3 m. The path length difference, Δs, is equal to the baseline length, b, times the sine of the zenith angle, θ, of the source (see Figure 5.4). Even for relatively short baselines, angles of more than a few degrees will have path length differences greater than 3 m. Does this mean that we must always work with very small bandwidths to avoid suppression of the fringe amplitude? By this argument, the bandwidth must be smaller (preferably *much* smaller) than the inverse of the largest delay, that is,

$$\Delta v < 1/\tau \qquad (5.22)$$

What bandwidth limit is implied by this argument? The Jansky VLA has a maximum baseline of 36 km, so when observing a source at a zenith angle of 45°, the delay would be

$$\tau = b \sin \theta / c = 8.5 \times 10^{-5}\ s$$

suggesting a maximum bandwidth of about 12 kHz, which is well below the bandwidth needed for successful detection of most radio sources.

We solve this problem by using a technique called *delay tracking*. If we add some extra path length (an *instrumental delay*) to the signal from antenna 2 (see Figure 5.4), then the time delay between the two antennas is significantly shortened—so that it is much less than the coherence time. Modern interferometers insert these delays digitally, adjusting the instrumental delay continually to account for the changing zenith angle of the source. The net delay ($\tau_{geometrical} - \tau_{instrumental}$) of the signals arriving at the correlator then stays small. Larger bandwidths can then be used without loss of visibility amplitude due to the decrease in temporal coherence. Note, though, that the delay can only be perfectly eliminated for a single point on the sky. This position is called the *delay tracking center*, or often just the *delay center*. Choosing the delay center is, effectively, choosing the place on the sky where the coherence time is maximized.

For other sky positions, which may be within the antenna primary beam but not at the delay center, there will always be a small time delay, and this ultimately limits the bandwidth that can be used for interferometric observations. We will return to this topic in Chapter 6.

Example 5.10:

Imagine that we wish to use an array with baselines from 10 m to 3 km to explore an area of the sky for new sources. If we want to be able to detect sources within a field of view 10′ on a side, what constraint does this pose on the bandwidth of the observation?

Answer:

Our bandwidth constraint is that the largest delay we want to be sensitive to must be smaller than the inverse of the bandwidth. With delay tracking, the center of the field of view will always be at zero delay. The largest delay, then, is for a point 5.0′, or 1.45×10^{-3} radians, from the center and on the largest baseline. This delay, by Equation 5.1, is

$$\tau = \frac{3.00 \times 10^3 \text{ m} \times \sin(5.82 \times 10^{-4} \text{ radians})}{3.00 \times 10^8 \text{ m s}^{-1}} = 1.45 \times 10^{-8} \text{ s}$$

By Equation 5.22, we need to use a bandwidth smaller than the inverse of this, so we set an upper limit to our bandwidth as 68 MHz.

5.8.2 Time Smearing

A problem similar to bandwidth smearing arises if we use large integration times when taking the data. We can reduce the data volume generated by an interferometer array by integrating the data over a period of time before making the correlation. But, if this integration time is too long, there will be a negative effect on the fringe amplitude similar to the effect of the finite bandwidth. In particular, because Earth continues to rotate during the integration time, the fringe phase (see Equation 5.9),

$$2\pi \frac{b}{\lambda} \sin(\omega_E t)$$

continually shifts. Intuitively, one can image the pattern of Figure 5.6 shifting with time.

Let us first consider the effect of integration time on the fringe phase without delay tracking. Using the same limiting criterion as with bandwidth smearing, we require that the fringe phase at the center of the integration time and the fringe phase at the beginning or end of the integration time differ by less than π, or, equivalently, that the phases at the beginning and end of the integration period be within 2π of one another—and preferably much less than 2π. This requires that the integration time be short enough that

$$2\pi \frac{b}{\lambda} \sin\left(\omega_E t_{int}\right) < 2\pi$$

or

$$\sin\left(\omega_E t_{int}\right) < \frac{\lambda}{b}$$

For any reasonable integration time, we can use $\sin(\omega_E t_{int}) \sim \omega_E t_{int}$. Thus, the integration time constraint becomes

$$t_{int} < \frac{\lambda}{b} \frac{1}{\omega_E}$$

This equation shows that observations with longer baselines and at shorter wavelengths have tighter constraints on the integration time. The rotation rate of Earth is 7.3×10^{-5} radians per second, so if observing at a wavelength of 0.01 m and with a baseline of 1 km, for example, we get that the integration time in the cross-correlation must be shorter than 0.1 s. Such a short integration time would pose problems for successfully detecting fringes to faint sources.

As with bandwidth smearing, delay tracking solves this problem for sources at the delay center. However, there is still a shift of the fringe phase at positions away from the delay center. Recall that in Section 5.4, we saw that the fringe function is a quasi-sinusoid because its frequency changes with time (see Equation 5.9), or with the position of the source. Therefore, when the fringe phase to the delay center is kept constant, the fringe phase continues to shift at other delays, albeit at a smaller rate than otherwise. The phase shift is proportional to the angular distance from the delay center, and so the integration time limit is given by

$$t_{int} < \frac{\lambda}{b} \frac{1}{\omega_E} \frac{1}{\Delta\theta} \tag{5.23}$$

where:
  $\Delta\theta$ is the largest angular offset from the delay center that we wish to be sensitive to

Now, if observing at a wavelength of 0.01 m and with a baseline of 1 km, and we wanted to be sensitive to a source offset by 0.001 radians, we need the integration time in the cross-correlation to be shorter than 4 min.

Example 5.11:

In Example 5.10, we calculated the maximum bandwidth to be sensitive to sources out to 1.45×10^{-3} radians from the phase center with baselines as long as 3 km. If we make this observation at 6.00 cm, what is the maximum integration time for this observation?

Answer:

Using Equation 5.23 we have

$$t_{\text{int}} < \frac{0.06\ 00\text{m}}{3.00 \times 10^3\ \text{m}} \frac{1}{7.3 \times 10^{-5}\ \text{radians s}^{-1}} \frac{1}{1.45 \times 10^{-3}\ \text{radians}}$$

This observation requires, therefore, that our integration time be shorter than 3 min.

5.9 BASIC PRINCIPLES OF INTERFEROMETRY

In this chapter, we have presented the basic principles of interferometry. Here we summarize the key points, for an east–west interferometer observing a source on the celestial equator, at wavelength λ.

1. The vector describing the separation distance of the antennas is called the *baseline* and is denoted as $\vec{b}$. When measured in units of the observing wavelength, it is the independent variable of the data in interferometry, and is called the *spacing*: b/λ.

2. Depending on the source position, θ, the radiation experiences an extra time *delay*, τ, to reach one of the antennas, as given by Equation 5.1

$$\tau = \frac{b \sin \theta}{c}$$

3. While observing a source as it moves across the sky, the interferometer response is an oscillating function of time due to the rotation of Earth. The time dependence of the interferometer response due to the Earth's rotation is known as the *fringe function*.

4. A reference position on the sky, called the *map center* or *phase center*, is chosen and the fringe function for that position is fitted to and removed from the data.

5. To ensure that the fringe is detected considering the effects of observing with a finite bandwidth, *delay tracking* is used, which keeps all net delays small.

6. The amplitude of the fringe oscillations is known as the *visibility amplitude*, and the phase of the cosine that remains after fitting (and removing) the fringe function is the *visibility phase*.

7. For a single-point source, the visibility amplitude is constant and equals the flux density of the point source, while the visibility phase depends on the position of the point source and is given by

$$\Phi_V = 2\pi \frac{b}{\lambda} \Delta\theta$$

where:

$\Delta\theta$, in radians, is the (small) angular distance of the source from the map center

8. For more complex sources, the visibility amplitude depends on both the source structure and the spacing.

9. For a pair of point sources, the visibility amplitude is a periodic function of spacing, with a spacing period

$$P_{b/\lambda} = \frac{1}{\Delta\Theta}$$

where:

$\Delta\Theta$ is the angular separation of the point sources

10. For extended sources, the visibility amplitude decreases at longer baselines, as the source is *resolved out* and the larger the source is, the faster the amplitudes decrease with spacing.

11. For a single, extended source with uniform intensity over an angle $\Delta\Theta_{size}$, the visibility amplitude reaches zero at the following spacings:

   a. First null at $b/\lambda = (1/\Delta\Theta_{size})$ for a one-dimensional source.

   b. First null at $b/\lambda = 1.22 \times (1/\Delta\Theta_{size})$ for a circular source.

The signals from the two antennas are multiplied and averaged over an *integration time*, which must be much longer than the period of the waves. The result of the multiplication and average is affected by the relative time *delay*, τ. To obtain the visibility function, the fitting of the fringe function in item #4, above, requires correct knowledge of the delay to the phase center. Fitting and removing the fringe function, mathematically, is identical to subtracting the delay from the signal coming from the further antenna. If no delay is subtracted, the interferometer output contains the fringe function; if the correct time delay is removed, we get the visibility function. The implication of all this is that the multiplication and average of the signals is, actually, a function of the delay. We have two functions of time, one of which we allow to be delayed by a variable τ, and these two functions are multiplied, integrated for a period of time, and then divided by the integration time. This describes a mathematical operation known as a *cross-correlation*, and so modern interferometers are commonly called *cross-correlation interferometers*. We will discuss cross-correlation interferometers in Chapter 6.

QUESTIONS AND PROBLEMS

1. What is the *visibility function*?

2. What is meant by the *phase center*?

3. An east–west interferometer on the Earth's equator with a baseline of 12.0 m observes a point source on the celestial equator with flux density 5.00 Jy at a wavelength of 3.00 cm.

   a. When the source is 15° to the east of the transit position, what is the delay between the signals coming from the antennas?

   b. What is the phase difference?

   c. What is the calibrated output value of the multiplicative interferometer?

   d. With the phase center set to be at the position of the point source, what are the visibility amplitude and phase of this observation at this time?

   e. Why is the answer to (d) not equal to that answer to (c)?

4. Consider an interferometric observation of two point sources at a wavelength of 5.00 cm. One has a flux density of 2.00 Jy and is located at RA = 04 h 00 m and Dec = 0°, and the other has a flux density of 2.50 Jy and is located 2.00×10^{-3} radians (6.88 arc-min) to the east of the first. The interferometer is composed of two antennas on the Earth's equator, separated by 20.0 m.

   a. If the phase center is set to be at RA = 04 h 00 m and Dec = 0°, what are the visibility amplitude and phase of each source?

   b. Solve for the total visibility phase and amplitude.

   c. With the phase center moved to be exactly halfway between the two sources, recalculate the answers to (a) and (b).

5. Describe what a one-dimensional visibility function looks like when observing

   a. A single, unresolved source.

   b. A single, resolved source. How does the shape of the visibility function change as the angular size of the source increases?

   c. A pair of unresolved sources. How does the shape of the visibility function change as the separation of the two sources increases?

6. Imagine we obtain extensive aperture synthesis observations of two different sources.

   a. When we examine the visibilities of the first source, we find that they all have roughly the same amplitude and all the phases are, essentially, zero, regardless of the baseline distance or orientation. Figure 5.17 shows our plot of visibilities versus b/λ. What can we infer about the structure of this source?

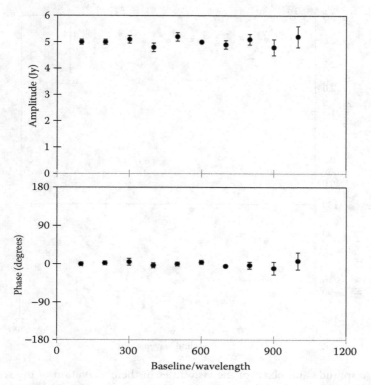

FIGURE 5.17 Amplitude and phase of the visibilities in the first observation of Question 6 are plotted versus b/λ distance. The amplitude is roughly constant, at 5 Jy, and the phases are all close to zero.

b. Our observation of the second source produces almost identical visibilities as in (a) except that the phases increase at a roughly constant rate, suddenly appear at the bottom of the plot and then continue increasing. The visibilities for this observation are shown in Figure 5.18. What can we infer about this source?

7. In the visibility data of an unknown source, the amplitudes are periodic, as shown in Figure 5.19. What can we infer about the source in this observation?

8. The visibility data of a source shows amplitudes that decrease with b/λ as shown in Figure 5.20. What can we say, qualitatively, about the structure of this source?

9. Draw and explain a plot of the visibility amplitude function for a source that contains a pair of sources separated by 5×10^{-4} radians (as in Figure 5.10) and in which each source has an angular size of 1×10^{-4} radians (as in Figure 5.14). Recall that a two-slit interference pattern is enveloped by the diffraction pattern of the individual slits.

10. Choose (and argue) the range (give the minimum and maximum) of baselines to use at a wavelength of 6 cm to infer the structure of the source shown in Figure 5.21. The source contains a large, relatively faint component, with an angular size of 90″, and inside this component is a smaller, brighter feature with width 30″.

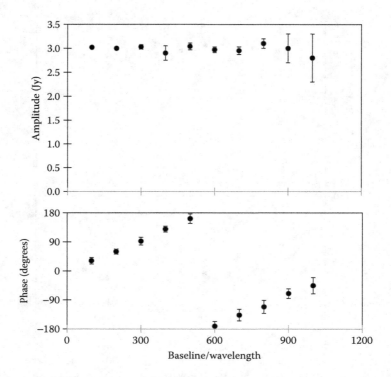

FIGURE 5.18 Amplitude and phase of the visibilities in the observation of the second source of Question 6. The amplitude, again, is roughly constant, this time around 3 Jy, but the phases vary, increasing monotonically, and then wrapping to the bottom and then increasing again.

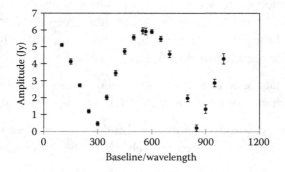

FIGURE 5.19 Visibility amplitudes in the observation in Question 7 oscillate up and down, with a period of $b/\lambda \sim 550$.

11. Consider the visibility amplitude function shown in Figure 5.22. What can you infer about the structure of this source?

12. Suppose antenna 1 detects an astronomical signal a while contributing a receiver noise A, while antenna 2 detects a signal b from the same astronomical source while

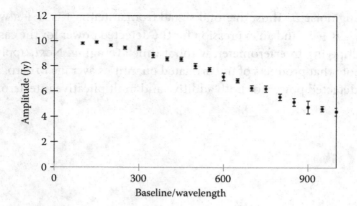

FIGURE 5.20   Visibility amplitudes in the observation in Question 8.

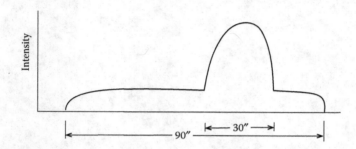

FIGURE 5.21   Profile of a source, discussed in Question 10, containing two components: one low-intensity, broad (90 arcsec) component and a brighter, narrower (30 arcsec) component.

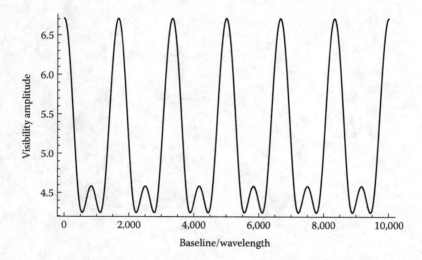

FIGURE 5.22   Visibility amplitude function of an unknown source, mentioned in Question 11.

contributing noise B. Thus, the total signal from antenna 1 is $A + a$ while that from antenna 2 is $B + b$. Find an expression for the detected power for the case of an adding and a multiplying interferometer, without time averaging. Next, apply time averaging, and note that products of uncorrelated quantities average to zero. Find the time-averaged detected power for both additive and multiplicative interferometers.

CHAPTER 6

Aperture Synthesis

Advanced Discussion

I N CHAPTER 5, WE introduced the fundamental principles of aperture synthesis by considering a two-element interferometer and restricting our discussion to a simplified situation in which the radio sources, any source structure, and the antennas are all contained in the equatorial plane. This simplified the mathematics, but of course does not describe the most general situation. In this chapter, we expand our presentation of radio interferometry with the goal of providing a deeper—but still basic—understanding that will serve for embarking on aperture synthesis observations. More advanced and in-depth discussions are provided in *Interferometry and Synthesis in Radio Astronomy* by Thompson et al. (2001)[*] and *Synthesis Imaging in Radio Astronomy II Volume CS-180* edited by Taylor et al. (1999).[†]

Let us first review some important definitions. The configuration we set up in Chapter 5 involves two antennas separated by a distance b, called the *baseline*, and they receive radiation from a source at an angle θ relative to the zenith. Because of the extra path length distance to one of the antennas, the arrival of the wave fronts to one antenna involves a time *delay*, τ, given by Equation 5.1 as $\tau = b \sin \theta / c$, which causes a phase difference between the two signals given by $\Delta \Phi = 2 \pi \nu \tau$. The signals from the antennas are multiplied and averaged over an *integration time*, which is much longer than the period of the waves, but shorter than the time for any significant change in the delay, and typically of order a few seconds. The phase difference between the two signals changes as the source moves across the sky (due to the Earth's rotation), and so the product of the two signals results in an oscillation in time called *fringes*. The amplitude of the fringes is called the *visibility amplitude*. For a point source, the *visibility amplitude*, after calibration, equals the source flux density. A reference position, called the *phase center*, is chosen by the observer, and the fringe function for that position is removed from the interferometer output versus time (see Section 5.5).

---

[*] A. R. Thompson, J. M. Moran, and G. W. Swenson. 2001. *Interferometry and Synthesis in Radio Astronomy*. New York: John Wiley & Sons.

[†] G. B. Taylor, C. L. Carilli, and R. A. Perley (eds.). 1999. *Synthesis Imaging in Radio Astronomy II Volume CS-180*.

The remaining phase is the *visibility phase*. In an observation of a point source, the visibility phase is determined by the point source's position relative to the phase center.

For aperture synthesis observations, a number of observing parameters, such as resolution, field-of-view, and sensitivity, depend on the telescope in a different way than in single-dish observations. For example, the resolution of a single-dish telescope is given by the width of the telescope's main beam, which is proportional to λ/D, where D is the telescope diameter. In this chapter, we will see that an interferometric array produces a *synthesized beam* that determines the resolution of the observation. The size of this beam is proportional to λ/b, where b is the baseline length. Because $b \gg D$, the angular resolution of an array can be much greater than the angular resolution of a single antenna.

The field-of-view of a telescope—whether single-dish or interferometer—depends on the details of the optics and the receivers. The term *field-of-view* is taken to mean the solid angle on the sky from which a single pointing of the telescope can capture information. If the telescope has a single feed and receiver, then the field-of-view is generally taken to be the size of the main beam of the individual antenna(s), measured at the half-power points. For the case of a 10-dB edge taper, this is given by Equation 3.2 as $\theta_{FWHM} = 1.15\lambda/D$, where D is the diameter of the reflector. Astronomical objects can be detected outside of this angular region, but the telescope will be less sensitive to them because of the lower main beam response. If the main beam power pattern is well known, then a correction factor can be applied to sources outside the half-power points. The relative uncertainty in the correction increases outside the half-power points, so θ_{FWHM} is usually adopted as the standard size for the field-of-view. For an interferometer, this solid angle is called the *primary beam*, to distinguish it from the synthesized beam. An interferometer array can image objects within the primary beam at an angular resolution corresponding to the synthesized beam. A single-dish telescope can detect objects within the main beam but cannot image them at higher angular resolution.

To explain how aperture synthesis works in three dimensions, and to provide a deeper understanding of some important details, we need to expand our discussion in several ways. We need to include vectors in our equations, with some vectors defined relative to the sky and others relative to the Earth's surface. The Earth's rotation will affect the relative orientations of these vectors. Furthermore, as we explain below, a complex correlator is used to produce visibilities that contain both real and imaginary parts. We will revisit the issue that the antennas receive radiation over a range of frequencies and this will affect the interferometer output. To facilitate the comprehension of all these complexities, we will extend our presentation one detail at a time. We start by discussing the concept of the fundamental operation of an interferometer, which is the cross-correlation of the signal from the two antennas, a mathematical concept that we only briefly mentioned near the end of Chapter 5.

6.1 CROSS-CORRELATION OF RECEIVED SIGNALS

In Chapter 5, we showed that fringes occur because the delay in the arrival of waves from the phase center to the more distant antenna changes as the Earth rotates. We also showed that the visibility phase is the remaining phase in the oscillations after fitting the fringe function for a source at the phase center. The removal of the fringe function is mathematically identical to compensating for the delay at the position of the phase center. Consider, again, the situation in Figure 5.3, in which the wave fronts reach antenna 2 first. The delay in the arrival of the wave fronts to antenna 1 is determined solely by the geometry and is well known. We can remove the fringe function, then, by introducing an instrumental time delay in the signal path of antenna 2 equal to the geometric delay between antennas 1 and 2. The instrumental delay is adjusted as the Earth rotates to stay equal to the geometric delay, and hence the response of the interferometer does not oscillate.

Let us now represent this mathematically. We let the instrumental delay be applied to the path from antenna 2 and denote it by τ. When the waves are combined at time t, the waves from antenna 2 will have a phase corresponding to the earlier time, $t-\tau$. The mathematical statement of this multiplication and average is

$$\frac{1}{t_{int}} \int_0^{t_{int}} E_1(t) E_2(t-\tau) dt \tag{6.1}$$

where:

t_{int} is the integration time

The delay, τ, is determined for each integration time, and changes as the Earth rotates. The user chooses a phase center (the reference position in the sky for defining the fringe function) and the delay is calculated for that sky position, baseline, and wavelength using Equation 5.1.

Consider the form of Equation 6.1. This equation takes two functions of time, one of which is delayed by a variable, τ, and the two functions are multiplied, integrated for a period of time, and averaged over that time period. This is the mathematical definition of a *cross-correlation*, which we will denote by Γ. Because the cross-correlation is a function of τ, we write Equation 6.1 as

$$\Gamma(\tau) = \frac{1}{t_{int}} \int_0^{t_{int}} E_1(t) E_2(t-\tau) dt$$

The concept of a cross-correlation is important for the understanding of many details of aperture synthesis, so it is worth some extra discussion. A cross-correlation function, basically, is a measure of how similar the two input functions are to one another *as a function of the delay* (i.e., the shift in time between the two functions). We briefly discussed an *autocorrelation* (or *self*-correlation) in Chapter 3; a *cross*-correlation is similar, but measures the similarity between two *different* functions. Imagine, for example, detecting two different sound signals at two different times and wondering if the second sound was the result of the first sound echoing off a wall. By performing a cross-correlation of the signals, we could determine if the functional shape of the two sound signals vs. time are identical. If the second was a perfect echo of the first, then there would be 100% correlation for some particular time shift, and that time shift is the extra travel time required for the reflection. A cross-correlation, therefore, yields two pieces of information: how similar the two functions are, and by how much one is shifted relative to the other.

Consider a calibrated interferometric observation of a point source. The visibility amplitude equals the source flux density when the correct delay (given by Equation 5.1) is applied. By finding the delay value that yields the maximum output, we can determine both the flux density of the point source and the geometric delay, which is related to the source position.

In Section 5.6, in discussing the multiplicative interferometer's response to a pair of unresolved point sources, we made the point that the two sources are incoherent with respect to each other, and because of their relative incoherence, the product of their waves necessarily averages to zero. We see now that this multiplication and average *are* the cross-correlation, so we can state the following important principle:

The cross-correlation of mutually incoherent sources must be zero.

There is a potential source of confusion here. Keep in mind the difference between the cross-correlation of the signals from *different sources* and the cross-correlation of signals from the *same source* but detected by *different antennas*. It is the former that is always equal to zero and the latter correlation that is the key to aperture synthesis.

The zero cross-correlation of incoherent sources provides an alternative conceptual view of many of the results we presented in Chapter 5. The multiplicative interferometer's response can be interpreted in terms of coherence. For example, consider interferometric observations of sources with extended structure. In this case, there is a range of geometric delays and no single delay value can correct for all of them. In Section 5.7, we discussed how the decrease in the visibility amplitude for a resolved source can be viewed as a consequence of a decrease in the spatial coherence of the source. Now, we can also say that a lower level of spatial coherence causes a smaller cross-correlation. In other words, the electromagnetic waves from one part of the source do not cross-correlate with those from other parts because of their relative incoherence, in the same way that the radiation from two distinct point sources does not cross-correlate.

The maximum amount of cross-correlation, therefore, contains information about the source structure. We will show later in this chapter how the maximum cross-correlation relates to the source structure, and how to extract information about the source structure from the cross-correlations.

6.2 COMPLEX-VALUED CROSS-CORRELATION

We have, now, that the cross-correlation gives us the same visibility function as we found for the multiplicative interferometer that we discussed in Chapter 5. We will now show that the visibility function that we presented in Chapter 5, in fact, is insufficient for inferring some source structures and that we obtain a better visibility function by performing a complex-valued cross-correlation.

For simplicity, we assume waves of a single frequency, ν. In Chapter 5, we found the visibility of a point source of flux density F_ν that is offset from the phase center by an angle $\Delta\theta$ to be given by Equations 5.15 through 5.17 as

$$V(b/\lambda) = F_\nu \cos\left(2\pi \frac{b}{\lambda}\Delta\theta\right)$$

It is a significant limitation that the source position dependence is contained solely within the argument of a *cosine*. The cosine is an even function, meaning that it has the same value for positive and negative arguments, that is, $\cos\theta = \cos(-\theta)$. The cross-correlation defined by Equation 6.1, then, gives the same result for sources on opposite sides of the phase center, where $\Delta\theta = 0$. In an observation of a point source that is displaced from the phase center, then, the visibility cosine function is consistent with the source being located at either $+\Delta\theta$ or $-\Delta\theta$. One might conclude that there are two point sources, one at $+\Delta\theta$ and the other at $-\Delta\theta$. This ambiguity can be removed by placing the phase center at the position of the point source so that $\Delta\theta = 0$. The cosine is then constant and the visibility function is independent of baseline. This reveals that there is a single point source and the location of the source is at the position of the phase center. With a proper choice of phase center, then, the cosine function can accurately reveal a *single* point source. However, as we will now show, with more complicated sources the ambiguity of the cosine function leads to unresolvable ambiguities about the sources.

Consider the simple case of two point sources. The visibility in this case, as given by Equation 5.19, is

$$V_{AB}\cos(\Phi_{AB}) = F_A\cos\left(2\pi\frac{b}{\lambda}\Delta\theta_A\right) + F_B\cos\left(2\pi\frac{b}{\lambda}\Delta\theta_B\right)$$

where:

F_A and F_B are the flux densities

$\Delta\theta_A$ and $\Delta\theta_B$ are the positions of the two point sources relative to the phase center

The visibility, therefore, appears as the sum of two cosine functions. Again because of the ambiguity with the cosine, this is the same visibility function as if there were sources located at $-\theta_A$ and $-\theta_B$, or if there were four sources, located at $+\theta_A$, $+\theta_B$, $-\theta_A$, and $-\theta_B$. We can eliminate the confusion about number of sources by placing the phase center exactly halfway between the two sources, so that points A and B are located such that $\Delta\theta_A = -\Delta\theta_B$. The visibility function is then

$$V_{AB}\cos(\Phi_{AB}) = F_A\cos\left(2\pi\frac{b}{\lambda}(\Delta\theta_A)\right) + F_B\cos\left(2\pi\frac{b}{\lambda}(-\Delta\theta_A)\right)$$

$$= (F_A + F_B)\cos\left(2\pi\frac{b}{\lambda}(\Delta\theta_A)\right)$$

The visibility data now contain only one cosine function, and so we could infer that there are just two sources. There is no phase center for which the visibility is constant, so we know that there cannot be a single point source. However, we still have a significant problem. Since we only have a single cosine function with amplitude of $F_A + F_B$, it would be impossible to determine how much of the flux density comes from either side of the phase center.

With even more complicated sources, the limitation of the cosine function can cause more drastic problems. If the object consists of three point sources, for example, with unequal separation angles, there is no phase center location that would yield information to unambiguously infer that there were three sources. We see, therefore, that we cannot use Equation 5.14 alone as our visibility function and obtain reliable maps.

The solution to this problem is to *also* calculate a cross-correlation using an *odd* function, for which $f(-x) = -f(x)$. This is accomplished through a fairly simple trick of electronics. The signals from each antenna are split into two equal parts, and to one signal a delay of a quarter cycle, or 90°, is implemented. This turns the cosine into a sine since $\sin(2\pi\nu t) = \cos(2\pi\nu t - \pi/2)$. Finally, cross-correlations are performed between the original two cosine signals and also between the cosine signal and its corresponding sine signal. The product of two cosines is even, but the product of a cosine with a sine is an odd function, which is sensitive to asymmetric structures. In terms of complex numbers, the 90° phase shift of a cosine to a sine is equivalent to the multiplication by i, and so the cross-correlation of the cosine and sine terms is treated like an imaginary term, while the correlation of the cosines is real. Figure 6.1 shows the schematic of a complex correlator detailing the outputs of the real and imaginary parts.

We have, then, a complex-valued cross-correlation and, thus, it is convenient to use the complex exponential form to describe the electromagnetic waves and the complex arithmetic to describe the cross-correlation. For readers who are unfamiliar with the complex description of waves, we provide a primer in Appendix IV. For the rest of this chapter, we will express the EM waves in the complex exponential form, given by

$$E(t) = E_0 e^{i\,2\pi\nu t}$$

So, another important point of aperture synthesis is as follows:

The complex cross-correlation involves both cosine and sine functions.

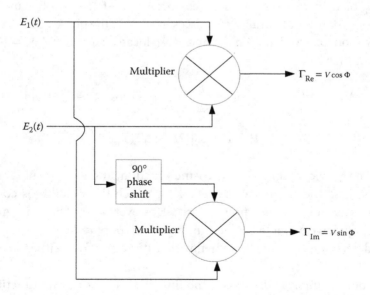

FIGURE 6.1 A schematic of a complex multiplier. E_1 and E_2 are the inputs from the two antennas and the outputs from the multipliers give the real and imaginary cross-correlation terms.

6.3 COMPLEX CORRELATION OF A POINT SOURCE AT A SINGLE FREQUENCY

Using the complex exponential form, let us return to the observation of a single point source. For the moment, we continue to limit the antenna and source positions to the equatorial plane; hence the geometry of Figure 5.4 still applies. We now express the electric fields at the two antennas in the complex exponential form,

$$E_1(t) = E_0 e^{i\,2\pi\nu t}$$

$$E_2(t) = E_0 e^{i\,2\pi\nu(t-\tau)}$$

When calculating the cross-correlation of these signals, Γ, we must remember to use the complex conjugate of the second function. Appendix IV explains why a complex conjugate must be used when multiplying two complex numbers. So the cross-correlation becomes

$$\Gamma = \frac{1}{t_{int}} \int_0^{t_{int}} E_0\,e^{i\,2\pi\nu t}\,E_0\,e^{-i\,2\pi\nu(t-\tau)}\,dt$$

In this form, we get the answer quite easily—we just add the exponents, yielding

$$\Gamma = \frac{1}{t_{int}} \int_0^{t_{int}} E_0^2\,e^{i\,(2\pi\nu t - 2\pi\nu t + 2\pi\nu\tau)}\,dt = \frac{1}{t_{int}} \int_0^{t_{int}} E_0^2\,e^{i\,2\pi\nu\tau}\,dt$$

Recall that τ is *delay* not *time* (even though its units are seconds). So there is no time-dependence in the integrand; we can remove it from the integral, leaving just the integration of dt, which is then divided by the integration time and so cancels out. We have, then,

$$\Gamma = E_0^2 e^{i\,2\pi\nu\tau} \tag{6.2}$$

In terms of the source position, and using Equation 5.1 for the delay, we have

$$\Gamma = E_0^2 e^{i\,2\pi b \sin\theta/\lambda} \tag{6.3}$$

Examine Equation 6.3 closely and compare it to Equation 5.7. Note that the phase of the complex exponential in Equation 6.3 is identical to the phase of the cosine term in Equation 5.7. In fact, the real part of Equation 6.3 is nearly identical to Equation 5.7 except for the factor of 2 in the amplitude. Recall, though, that the amplitudes are converted to flux density via calibration, so this factor of 2 is irrelevant. After calibration, then, we have that the complex cross-correlation of a point source is given by

$$\Gamma = F_\nu e^{i\,2\pi\nu\tau} \tag{6.4}$$

Let us now examine the visibility function. Following the same steps as in Chapter 5—using the small-angle approximation and subtracting the angular position of the phase center— the visibility function in this situation is

$$V(b/\lambda) = F_\nu e^{i\,2\pi b \Delta\theta/\lambda} \tag{6.5}$$

where $\Delta\theta$ is the source position relative to the phase center. Now compare the amplitude and phase in Equation 6.5 to the visibility amplitude and phase given in Equation 5.14. Writing the visibility function in the complex exponential form, therefore, yields the same visibilities we inferred using real numbers. Keep in mind that V in Equation 6.5 is not a single real-valued number; it contains two values—the real and imaginary parts. We can obtain the visibility amplitude and phase by combining the real and imaginary parts appropriately

$$V_{amp} = \sqrt{\Gamma_{Re}^2 + \Gamma_{Im}^2} \text{ and } \Phi_V = \tan^{-1}\left(\frac{\Gamma_{Im}}{\Gamma_{Re}}\right) \tag{6.6}$$

Note that the real and imaginary parts of the visibility are sufficient information to solve for the two unknowns of amplitude and phase, without the need of explicitly measuring the fringe function.

6.4 EXTENDED SOURCES AND THE FOURIER TRANSFORM

Now consider the visibility function of an extended source. The visibility from Equation 6.5 will be the sum of the visibilities from all the infinitesimal points of emission that make up the extended source. To represent the intensity in two dimensions, we will need to describe it as a function of two angular positions on the sky, defined relative to the phase center. To simplify the notation we will represent the two angles relative to the phase center as x and y so that the two-dimensional intensity distribution is then represented by $I_\nu(x,y)$. For the present discussion of a one-dimensional source, however, we define a similar quantity, which we denote as $i_\nu(x)$ and define by

$$i_\nu(x) = \frac{dF_\nu(x)}{dx}$$

where:
   $dF_\nu(x)/dx$ is the flux density per unit angle on the sky at position x relative to the phase center

Remember that we are restricting our discussion to the equatorial plane so that $i_\nu(x)$, conceptually, is the same as intensity, $I_\nu(x,y)$, but in one dimension. Integrating over the whole source, then, the visibility function due to the source distribution is

$$V(b/\lambda) = \int i_\nu(x) e^{i\,2\pi(b/\lambda)x} dx \tag{6.7}$$

We have replaced the sin x in the exponential of Equation 6.5 with x since the entire source will be contained within a small enough angle that the small angle approximation applies. Note that the dependence on x goes away with the integration, while the dependence on b/λ remains. Therefore, V must be a function of the baseline, b/λ. For any given value of b/λ, the integral in Equation 6.7 yields a single complex number. But for different values of b/λ we may get different values of V, depending on the structure of the source as given by $i_\nu(x)$. Imagine that we observe with a number of different baselines. Equation 6.7 tells us what the complex visibility function for each baseline will be. In other words, the integral in Equation 6.7 transforms a function of x, $i_v(x)$, into a function of b/λ, $V(b/\lambda)$.

In fact, Equation 6.7 is a very special transformation. This kind of integral is called a *Fourier transform*. To be precise, because of the positive argument in the exponential, Equation 6.7 shows an *inverse Fourier transform*. Thus, $V(b/\lambda)$ is the inverse Fourier transform of $i_v(x)$. Conversely, $i_v(x)$ is the Fourier transform of $V(b/\lambda)$. Note that the two independent variables, x and b/λ, are multiplied by each other in the exponential. The functions $V(b/\lambda)$ and $i_v(x)$ are a Fourier transform pair, and there are well-developed mathematical relations that permit transforming back and forth.

In Equation 6.7, we take the inverse Fourier transform of a *one*-dimensional source, and restrict the baseline to be in the east–west direction. In the general case, the astronomical source will be *two*-dimensional, and the baseline will be a vector, $\vec{b}$, that can point in any direction. Hence, in the general case, the integral is over two angles, x and y, and the $i_v(x)dx$ is replaced by $I_v(x,y)d\Omega$, where Ω is solid angle. Equation 6.7 then becomes a *two*-dimensional inverse Fourier transform, from intensity as a function of sky position to the visibility as a function of $\vec{b}/\lambda$. In addition, because of the Earth's rotation, $\vec{b}/\lambda$ is a function of time, the effect of which we see as the fringe function. After removal of the fringe functions from the cross-correlations (and calibration of the antennas), we get the two-dimensional complex visibility function, $V(\vec{b}/\lambda)$.

In summary, by measuring the cross-correlation for a range of baselines, and removing the fringe function from each, yielding the visibility function $V(\vec{b}/\lambda)$, we can then take the Fourier transform of the visibilities to obtain the sky intensity $I_v(x,y)$. Thus, we obtain an image of the source! This sky intensity image is the goal of aperture synthesis and so we now see the procedure that we must develop:

> *We measure the visibilities for a range of baselines and take their 2-D Fourier transform to yield an image of the source.*

Clearly, the Fourier transform is fundamentally important to the rest of this chapter, and so we encourage all readers unfamiliar with this mathematical tool to read the short primer we provide in Appendix V. We also encourage using the Tool for Interactive Fourier Transforms (TIFT; see the online materials) and playing with some of its accompanying exercises. In Section 6.5, we summarize some Fourier transform relations that are especially useful to know when doing aperture synthesis. Fourier transforms are easier to appreciate in one dimension than in two, so we will delay the use of two-dimensional transforms until Section 6.6.

In practice, a radio astronomer rarely needs to perform Fourier transforms; the data processing packages do it for us. Nevertheless, an understanding of the relationship between the visibility data and the resulting image is both important and useful.

6.5 FOURIER TRANSFORMS FOR SOME COMMON SOURCE SHAPES

In Chapter 5, we described the visibilities for a number of simple source configurations. Here, we revisit these configurations, and express the relation between the source structure and the visibility function in terms of the Fourier transform. Because these were one-dimensional sources, for now we restrict our discussion to one-dimensional Fourier transforms of the source intensity $i_v(x)$, and its inverse Fourier transform, the one-dimensional visibility function $V(b/\lambda)$. The latter is given by

$$V(b/\lambda) = \int i_v(x) e^{i\, 2\pi(b/\lambda)x}\, dx$$

where:

$i_v(x)$ is the Fourier transform of $V(b/\lambda)$

$$i_v(x) = \int V(b/\lambda) e^{-i\, 2\pi(b/\lambda)x}\, d(b/\lambda)$$

(We omit the limits of integration here in order to focus on the relation between the intensity and the visibility.)

6.5.1 Visibility Function of a Point Source

Consider a point source located an angular distance x_0 from the phase center. We represent the point source by a delta function (see Appendix V), with amplitude equal to the flux density, F_v. Therefore, the visibility function is given by

$$V(b/\lambda) = \int F_v \delta(x - x_0) e^{i2\pi(b/\lambda)x}\, dx$$

Following the example in Appendix V, we evaluate this integral, giving

$$V(b/\lambda) = F_v e^{i\, 2\pi(b/\lambda)x_0}$$

To interpret this visibility, consider a point source located at the phase center, so that $x_0 = 0$. In this case, the real part of the visibility is a constant, equal to F_v, and the imaginary part is zero. Thus, as we saw in Chapter 5, for a point source at the phase center, the visibility amplitude is constant and the visibility phase is zero for all baselines. If the point source is *not* at the phase center, the amplitude is still constant, but the phase will depend on the source position and the baseline, as given by

$$\Phi = 2\pi\left(\frac{b}{\lambda}\right)x_0$$

6.5.2 Visibility Function of Two Point Sources

Now consider the visibility function of two point sources. Suppose that one source is located at the phase center, with $x = 0$ and flux density F_1, while the other source is located at position x_0 with flux density F_2. The visibility, then, is

$$V(b/\lambda) = \int \left[F_1 \delta(x - 0) + F_2 \delta(x - x_0) \right] e^{i\, 2\pi(b/\lambda)x}\, dx$$

$$= \int F_1 \delta(x - 0) e^{i\, 2\pi(b/\lambda)x}\, dx + \int F_2 \delta(x - x_0) e^{i\, 2\pi(b/\lambda)x}\, dx$$

As demonstrated in Appendix V.5 (see Equation V.8) the visibility amplitude is given by

$$V_A = \sqrt{F_1^2 + F_2^2 + 2F_1 F_2 \cos\left[2\pi\left(\frac{b}{\lambda}\right)x_0\right]}$$

which is an oscillating function of b/λ, with minima separated by $1/x_0$; c.f. Figure 5.10 and Equation 5.20. Therefore,

> *The visibility function of two point sources has an amplitude that oscillates with a period equal to the inverse of the separation of the two point sources.*

A related and useful rule (which results from the linearity of the Fourier transform) is that

> *The visibility function of a sum of point sources equals the sum of the visibility functions of the individual point sources.*

6.5.3 Visibility Function of a Gaussian Profile

In Appendix V.6, we show that the Fourier transform of a Gaussian is *also* a Gaussian, and that its width is inversely related to the width of the initial function. Hence, the width of the visibility function will be inversely related to the source size. If we compare the visibilities for both a large and a small Gaussian source, we will find that both sets of visibilities are described by a Gaussian pattern. But the larger *source* will have a *narrower* visibility pattern. That is, the visibility amplitudes of the larger source will decrease more rapidly than those of the smaller source, reaching zero at smaller values of b/λ (i.e., on short baselines). With decreasing source size, the Gaussian visibility amplitude pattern will be broader, decreasing more slowly with b/λ. When the source becomes small enough (a point source), the visibility amplitude will be infinitely wide; that is, it will be constant. In this case, the visibility amplitude is the same value for all baselines we can infer nothing about the source structure; this is another way of saying that a point source is unresolved.

This inverse relation between the widths of the two Fourier domains is a general principle; it applies to most functional shapes, not just to the Gaussian profile. It is also the best way to understand a special property of interferometers: they are blind to a constant background radiation level. A uniform sky intensity can be thought of as an infinitely wide function, whose Fourier transform will be a delta function, or an infinitely narrow visibility, which we will not be able to measure.

The important point to understand here is as follows:

> *The Fourier transform of a Gaussian function is another Gaussian function, whose width is inversely proportional to the width of the original Gaussian.*

6.6 THREE DIMENSIONS, THE EARTH'S ROTATION, AND THE COMPLEX FRINGE FUNCTION

We are now ready to consider the general case of the cross-correlation of a source in three dimensions (with antennas anywhere on Earth). This will naturally involve vectors. We now let the astronomical source be two-dimensional, given by $I_v(x,y)$, and the baseline can have any spatial orientation with respect to the source. To begin, we use the unit vector $\hat{n}$ to denote the direction of the source relative to the center of the baseline and the vector $\vec{b}$ to represent the *baseline vector* (distance and direction from antenna 1 to antenna 2). These vectors are shown in Figure 6.2. The *path-length difference*, $b\sin\theta$, in vector form, is now given by $\vec{b} \cdot \hat{n}$, that is, the component of $\vec{b}$ along the direction of $\hat{n}$. And so the *delay*, τ, is now given by

$$\tau = \frac{\vec{b} \cdot \hat{n}}{c} \tag{6.8}$$

Substituting Equation 6.8 into Equation 6.4 (and using $c = \nu\lambda$), the cross-correlation of a point source becomes

$$\Gamma = F_\nu e^{i2\pi\, \vec{b} \cdot \hat{n}/\lambda} \tag{6.9}$$

Note that both the real and the imaginary parts of the cross-correlation given by Equation 6.9 have an oscillatory dependence on the source position. The amplitude of this cross-correlation is constant, while its phase depends on the source direction—as expected for a point source.

Now let us include the rotation of the Earth. This means that the source position, with respect to each baseline, will continually change with time. Thus, $\vec{b} \cdot \hat{n}$, and hence Γ, will both be functions of time. The cross-correlation phase will then have one contribution due to the source position and another contribution due to its motion across the sky. We can distinguish between these two contributions by defining a reference position on the sky and expressing the position of the point source relative to this reference position. We define $\hat{n}_0$ as the direction of the phase center and so the direction of the point source relative to the phase center is the vector $\hat{n} - \hat{n}_0$. Note that the point source can be *at* the phase center, in which case $\hat{n} - \hat{n}_0$ is 0. By employing a reference position, we can maintain the source position as a variable in our general equation. In these terms, the cross-correlation of the emission from a single point source is

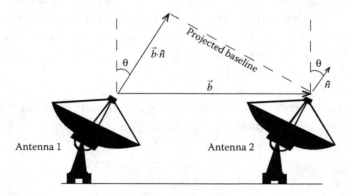

FIGURE 6.2 Baseline vector, $\vec{b}$, gives the distance and direction between antennas and the unit vector, $\hat{n}$, indicates the direction of the source. The dot product $\vec{b} \cdot \hat{n}$ is the extra path length that the signal must travel to reach antenna 1. The dashed line represents the projected baseline length.

$$\Gamma = F_v \exp\left[i2\pi \frac{\vec{b} \cdot \hat{n}}{\lambda} \right] = F_v \exp\left[i2\pi \frac{\vec{b} \cdot \left(\hat{n} - \hat{n}_0 + \hat{n}_0 \right)}{\lambda} \right]$$

$$= F_v \exp\left[i2\pi \frac{\vec{b} \cdot \left(\hat{n} - \hat{n}_0 \right)}{\lambda} \right] \exp\left[i2\pi \frac{\vec{b} \cdot \hat{n}_0}{\lambda} \right]$$

Here, the time dependent factor due to the Earth's rotation is completely contained in the last exponential. This factor is the complex-valued *fringe function*, given by

$$\Gamma_{\text{fringe}} = e^{i2\pi\left(\vec{b} \cdot \hat{n}_0 / \lambda\right)} \tag{6.10}$$

which can easily be divided out. The cross-correlation of the signal from a point source after dividing by the fringe function is

$$\Gamma / \Gamma_{\text{fringe}} = F_v e^{i2\pi\left[\vec{b} \cdot \left(\hat{n} - \hat{n}_0\right)/\lambda\right]} \tag{6.11}$$

Equation 6.11 contains *all* the information about the point source—its flux density and its position. This is the *visibility function*. We again find that the flux density of the point source is given by the *visibility amplitude*, and the position of the point source, relative to the reference position, is contained in the *visibility phase*. For a source located at the reference position, $\hat{n} = \hat{n}_0$, the visibility phase is zero, and so we see, again, why it is called the *phase center*.

In practice, the antennas track the source so that it remains at the center of the primary beam pattern of the individual antennas. This tracking position is called the *pointing center* and is conceptually very different from the phase center. The two positions are independent of one another, although in practice they are usually chosen to be the same. An important distinction between them is that the phase center can be changed after the observations, by introducing phase shifts into the data. The pointing center, on the other hand, cannot be changed. If the source was outside the primary beam during the observations, no amount of mathematical manipulation can repoint the antennas after the fact.

There is a subtle but important detail of our visibility function, now given by Equation 6.11. In particular, note that the baseline vector is now dotted by $\hat{n} - \hat{n}_0$. This is a vector in the sky plane and so, by definition, it is perpendicular to $\hat{n}_0$. Therefore, the *effective* baseline for our visibility function is the projection of $\vec{b}$ onto the sky plane, given by the dot product. This is known as the *projected baseline* and is depicted in Figure 6.2. As the Earth rotates, the direction of $\hat{n}_0$ changes, so the projected baseline will change as well. When an astronomical source is transiting, its direction $\hat{n}$ is normal to the baseline vector, and the baseline presents its full physical length to the source. At all other times, the baseline will have a component in the direction of the source, and so its projected length will be shorter than its physical length. The physical separation of the antennas may be many kilometers, but near rising and setting times, the projected baseline length may be substantially less. The salient point here is that the *projected* baseline length continually changes with time, and so the visibility function measured by an antenna pair is continually probing different values of $\vec{b}/\lambda$. As we saw in Section 6.5, the way in which the visibility amplitudes vary with $\vec{b}/\lambda$, is related to the source structure. Hence, during the course of a day, a single antenna pair will sample the visibilities for a range of baselines.

The conversion of the visibility function to the image of the source will be the main focus of the remainder of this chapter. First, let us examine the fringe function in greater detail. The fringe function is a complex exponential with unity amplitude and phase given by $2\pi\left(\vec{b}\cdot\hat{n}_0\right)/\lambda$, where $\vec{b}\cdot\hat{n}_0$ continuously changes because of the Earth's rotation. Therefore, both the real and imaginary parts of the fringe function vary with time. It is fairly clear why $\vec{b}\cdot\hat{n}_0$ must vary as the Earth rotates, but the exact nature of its time dependence can be difficult to recognize because it is buried in the vector dot product. The time variation of the fringe function was more apparent in our discussion in Chapter 5, but that was only the real part. We can now examine the imaginary part of the fringe function as well. To aid our discussion, we temporarily return to the simplified situation of a point source on the celestial equator with a purely east–west baseline on the Earth's equator. In this case, $\vec{b}\cdot\hat{n} = b\sin\theta$, where θ is the direction angle of the source relative to the meridian of the midpoint of the baseline.

In Figure 6.3, we plot the complex fringe function for this one-dimensional situation. The real and imaginary parts and the amplitude of the fringe function are shown in Figure 6.3a, while the phase is shown in Figure 6.3b. Figure 6.3 shows that the imaginary fringe term oscillates with the same amplitude and frequency as the real term, but is shifted by one-fourth of a period.

Note the slope of the curve in Figure 6.3b. The phase versus time slope is not constant and therefore the *rate* at which the phase changes depends on the sky position throughout the course of the day. In particular, when the source is close to transit the slope of phase vs. time, or *phase rate*, is greatest. The phase rate, therefore, is an indication of the position of the source in the sky.

The rate at which the fringe phases change is another useful measure, for which we will want a general, three-dimensional expression. The derivation involves spherical geometry, which is rather long and not particularly enlightening, so we have chosen to omit it, and merely state the result. The rate of change of the fringe phase angle is commonly called the *fringe frequency*, and is defined (in radians per second) by $d\phi/dt = \omega_F$. It is given by

$$\omega_F = \pm 2\pi\omega_E\frac{b}{\lambda}\left(\cos\gamma\cos\delta\cos\omega_E t\right) \tag{6.12}$$

where:

ω_E is the angular rotation frequency of the Earth (7.3×10^{-5} radians/s)

γ is the north–south tilt of the baseline ($\gamma = 0$ for a purely east–west baseline)

δ is the declination of the source

t is the hour angle in seconds ($t = 0$ corresponds to the moment of transit; $+$ is used before transit and $-$ after transit)

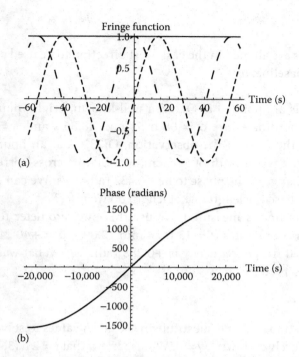

FIGURE 6.3 (a) Real (long dash) and imaginary (short dash) parts, the amplitude (solid line) and (b) phase of the fringe function for a position on the celestial equator with a baseline that is purely east–west. The time is defined so that $t = 0$ corresponds to the moment of transit. The plot in (a) shows the fringe function for a 2-min period centered on transit while (b) shows a 12-h period.

We assume that there is no elevation difference between the antennas. Any difference in elevation would warrant an extra term in Equation 6.12, but this component is likely to be much smaller than the other baseline components and so we do not concern ourselves with it here. (The complete expression, including elevation differences, is found in *Interferometry and Synthesis in Radio Astronomy* by Thompson, Moran, and Swenson.[*])

Note the dependence of ω_F on δ and t. If we know the baseline orientation, then a single measurement of the fringe frequency, as a function of time, will yield both the right ascension and declination of the source.

The fringe function is also extremely useful for measuring the baseline, since it depends on the baseline length and orientation. In fact, the baseline length and orientation can be determined more accurately by the fringe function of a point source than by conventional surveying or Global Positioning Systems. Such *geodetic interferometry* is used by geophysicists to monitor the movement of the Earth's tectonic plates and to determine the orientation of the planet. The technique involves measuring the positions of radio antennas that form intercontinental baselines. This technique belongs to the realm of very long baseline interferometry (VLBI), which we discuss briefly in Section 6.14.

[*] A. R. Thompson, J. M. Moran, and G. W. Swenson. 2001. *Interferometry and Synthesis in Radio Astronomy.* New York: John Wiley & Sons.

Example 6.1:

A pair of antennas are aligned in the east–west direction to be used as an interferometer with a fixed baseline, b.

1. The antennas are pointed at OJ287, a well-known, bright, point source whose equatorial coordinates are α = 08 h 54 min 48.86 s and δ = +20° 06′ 30″.6 (J2000).* At the time of the observation, OJ287 is at an hour angle of +2 h. Observing at a wavelength of 1.35 cm, we obtain a cross-correlation, and find the rate of change of the phase to be −0.432 radians/s. We can use these data to measure the baseline length; what length do we infer?
2. After the baseline is measured, we use this interferometer to observe a new source, with coordinates α = 13 h 24 min 12.09 s, δ = +40° 48′ 11″.8 (J2000). The measured fringe frequency is +0.288 radians/s. What was the hour angle when we observed this new source?

Answers:

1. We use Equation 6.12 and substitute in the given values to solve for b. Since the baseline is purely east–west, $\gamma = 0$. We also know that $t = 2$ h (3600 s/h) = 7200 s. For the declination, we convert the arcseconds and arcminutes to fractions of a degree:

$$\delta = (+20 + 6/60 + 30.6/3600)° = +20.1085°$$

The product of $\omega_E t$ gives 0.5256, but this is in radians, which we can express as 30.1°. Substituting these into Equation 6.12 and rearranging we have

$$b = \frac{0.432 \text{ radians sec}^{-1} \times 0.0135 \text{ m}}{2\pi\left(7.3 \times 10^{-5} \text{ radians sec}^{-1}\right)\cos 20.1085° \cos 30.1°} = 15.65 \text{ m}$$

Note: We did not need the source's right ascension (RA). The source RA is implicit in the t, the source's hour angle. The actual value of RA would be relevant if you also knew the LST; the hour angle would be the difference between RA and LST.

2. Again we use Equation 6.12, but with the new value of δ, and this time we provide a value for b and solve for t. In terms of degrees, δ = +(40 + 48/60 + 11.8/3600)° = 40.803°, and since the fringe frequency is positive, the source is still approaching transit. We get, then,

$$\cos(-(7.3 \times 10^{-5} \text{ radians/s}) \, t) = 0.7155$$

---

* Due to the precession of the Earth's rotational axis, the orientation of the equatorial coordinate system shifts relative to the celestial sphere, so any specific coordinates must be associated with a specific epoch, which is often given as the year and shown in parentheses after the coordinates.

$$\omega_F = \pm 2\pi \omega_E \frac{b}{\lambda} \left[\cos\gamma \cos\delta \cos\omega_E t \right]$$

and so

$$t = 10{,}300 \text{ s} = 2.87 \text{ h before transit}$$

6.7 NONZERO BANDWIDTH AND FINITE INTEGRATION TIME

We now account for the fact that observations do not occur at a single wavelength. In fact, because there is no energy contained in zero bandwidth, there *must* be a nonzero Δv. We discussed this issue briefly in Section 5.8.1, where we showed that a nonzero bandwidth has an important effect on the fringe amplitude for nonzero delays. We revisit this issue here, in the context of the complex correlation. As before, we would like to observe with large bandwidths to achieve better continuum sensitivity, but bandwidths that are too wide will lead to loss of coherence.

We can get a sense of the effect of a range of frequencies by examining the wavelength dependence of the fringe function, given in Equation 6.10. At slightly different wavelengths, the maxima of the real and imaginary parts of the fringe function occur at slightly shifted θs, corresponding to slightly shifted times relative to transit. The real parts at all frequencies have maxima at the transit position, when $t = 0$, but the maxima away from $t = 0$ occur at slightly different times, and so they do not coincide (see Figure 6.4a).

The imaginary parts at all frequencies *cross* zero at $t = 0$ (Figure 6.4(b)). But, similar to the real parts, the maxima of the imaginary parts occur at slightly different times, and do not coincide.

So, we see that the amplitude of the sum of the fringes decreases with increasing angular position, relative to the transit position. With regard to the delay, the fringe amplitude for a range of frequencies decreases as the delay moves away from $\tau = 0$. At delays far from zero, the peaks are sufficiently shifted that the fringes are undetectable. The fundamental effect of a finite bandwidth, then, is to cause the amplitude of the cross-correlation to decrease at large delays.

We can demonstrate this mathematically by adding up all the fringe patterns at the different frequencies. Because signals at different frequencies are incoherent with respect to one other, they cross-correlate to zero and do not contribute to the sum. We perform the sum by integrating in frequency. To obtain a generally correct expression, we allow for a nonuniform sensitivity with frequency by defining a *bandpass function*, $h(v)$, which we include in the integrand. The cross-correlation of a point source, then, becomes

$$\Gamma = \frac{1}{\Delta v} \int_{-\infty}^{+\infty} h(v) F_v e^{i2\pi\left(\bar{b}\cdot\hat{n}/\lambda\right)} dv \qquad (6.13)$$

The limits of the integral can be set to $\pm\infty$ because the function $h(v)$ will be zero outside of the bandpass. The λ in the exponent, of course, is related to v. The effect of this integral becomes clearer if we define a central frequency (just as we defined a central *position* in the previous section). We convert the λ in the exponent to frequency and express it relative to the central frequency, v_0, so that Equation 6.13 becomes

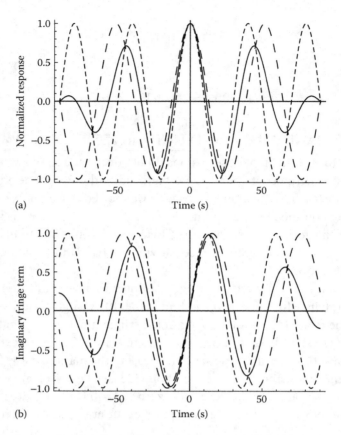

(a)

(b)

FIGURE 6.4 (a) Real part of two different fringe functions of higher (short dash) and lower (long dash) frequencies and their average (solid line) are displayed at times near transit. All real parts have a central maximum at $t = 0$ but the subsequent maxima do not coincide, and so the amplitude of the average decreases with increasing times away from $t = 0$. (b) The imaginary terms for the same fringe functions as in Figure 6.4(a) are shown. All imaginary parts cross zero at $t = 0$, as befitting the sine function, while the maxima are increasingly displaced from one another with increasing times away from $t = 0$, causing a decrease in the average.

$$\Gamma = \frac{1}{\Delta v} \int_{-\infty}^{+\infty} h(v) F_v e^{i2\pi(v_0 + v)(\vec{b}\cdot\hat{n}/c)} dv$$

We can separate the exponentials to give

$$\Gamma = \frac{1}{\Delta v} \int_{-\infty}^{+\infty} h(v) e^{i2\pi v(\vec{b}\cdot\hat{n}/c)} F_v e^{i2\pi v_0(\vec{b}\cdot\hat{n}/c)} dv$$

The second exponential is independent of the integration variable, v, and if we can approximate F_v as constant over the bandpass, it can be removed from the integrand along with the second exponential, giving us

$$\Gamma = F_v e^{i2\pi v_0\left(\vec{b}\cdot\hat{n}/c\right)} \frac{1}{\Delta v} \int\limits_{-\infty}^{+\infty} h(v) e^{i2\pi v\left(\vec{b}\cdot\hat{n}/c\right)} dv$$

The term outside the integral is just the single-frequency cross-correlation of a point source, while all the effects of the finite bandwidth are contained within the integral. We recognize this integral as a Fourier transform of the bandpass function. The transform yields a function whose independent variable is $\vec{b}\cdot\hat{n}/c$, which, by Equation 6.8, is the delay, τ. If we write the Fourier transform of $h(v)$ as $H(\tau)$, then the cross-correlation is

$$\Gamma = F_v e^{i2\pi v_0\tau} H(\tau)$$

For most telescopes, the bandpass sensitivity function, $h(v)$, is approximately described by a normalized top-hat function of unit amplitude, as depicted in Figure 6.5.

The Fourier transform, $H(\tau)$, of this square pulse is the bandwidth multiplied by a sinc function,

$$H(\tau) = \Delta v \frac{\sin(\pi\Delta v\tau)}{\pi\Delta v\tau}$$

which is plotted in Figure 6.6. We introduced the sinc function in Section 3.1.5, where we used a Fourier transform to calculate the beam pattern of a square antenna aperture. Because sinc 0 = 1, $H(\tau)$ will equal Δv when the delay is zero, and so it will cancel with the Δv in the denominator of Equation 6.13. Thus, at zero delay the fringe function is independent of the bandpass shape. But at nonzero delays, the amplitude of the cross-correlation is reduced by a factor of the sinc function as shown in Figure 6.6. This, of course, is an undesirable effect that needs further discussion.

With the Fourier transform relation between H and h, the width, $\Delta\tau$, of $H(\tau)$ is inversely proportional to the bandwidth, Δv. In particular, as with the single-dish beamwidth (see Section 3.1.5), the FWHM is

$$\Delta\tau_{\text{FWHM}} = 0.89/\Delta v$$

Because of the nonzero bandwidth, the cross-correlation becomes undetectable for sources whose delays are substantially larger than $1/\Delta v$. This is the same conclusion we reached in

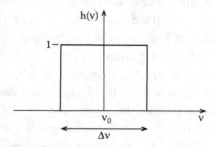

FIGURE 6.5   Top-hat bandpass function, $h(v)$.

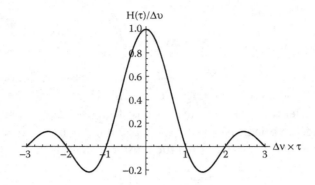

FIGURE 6.6 $H(\tau)$ when $h(\nu)$ is a square pulse, as shown in Figure 6.5.

Section 5.8.1, using a heuristic argument. As we discussed in Chapter 5, this is problematic because the bandwidths needed for reasonable sensitivity prevent us from detecting sources far from the phase center (where the delay is far from zero). The electromagnetic waves integrated over a bandwidth $\Delta\nu$ have a *coherence time* given by Equation 5.20, which we can rewrite as

$$t_{\text{coherence}} = \frac{1}{\Delta\nu} \tag{6.14}$$

Cross-correlating signals involving a delay greater than this will fail because of the loss of coherence. A bandwidth of 10 MHz, for example, requires delays less than 0.1 μs. In Section 5.8.1, we showed that a 36-km baseline of the VLA, observing at a zenith angle of 45°, corresponds to a geometric delay of 85 μs.

As we discussed in Section 5.8.1, this problem is largely resolved by a technique called *delay tracking*. For each baseline, the signal arriving to the antenna closer to the source is digitally delayed so that the signals from both antennas arrive at the correlator within a time range much smaller than the coherence time. As the source moves across the sky, these instrumental delays are adjusted to keep the delays on all baselines as small as possible.

But the delay tracking center is a single point on the sky; objects within the field-of-view, but away from the delay tracking center, will have relative delays that are not removed by this instrumental delay. If we wished to observe with a 10-MHz bandwidth, for example, the delays to all the sources in our observation should be kept to less than 0.1 μs. Thus our map should not extend too far from the delay tracking center. If this observation uses the VLA A-array, in which the longest baseline is 36 km, solving $\tau = b \sin\theta/c$ for θ, we obtain a maximum angle of 8×10^{-4} radians or 3 arcmin.

The integration time, as we mentioned in Chapter 5, also can cause a degradation of the visibility amplitudes for sources away from the delay center. As we can see in Figure 6.3b the fringe frequency depends on position in the sky and so the fringes due to objects away from the delay center will shift in phase while the fringes at the delay center are held constant. As we explain in Section 5.8.2, the rate that a fringe phase changes is proportional to the distance from the delay center and related to the fringe rate at the delay center without delay tracking. Without derivation, we found that the integration time limit is given by Equation 5.23, which we repeat here

$$t_{\text{int}} < \frac{\lambda}{b} \frac{1}{\omega_E} \frac{1}{\Delta\theta} \tag{6.15}$$

Let us apply this to our desired observation with the 36-km baseline. Since our bandwidth constrains the field of view to extend to less than 3 arcmin from the phase center, we can use Equation 6.15 to determine the maximum integration time. Let us consider observing at a wavelength of 1.4 cm. We get, then,

$$t_{\text{int}} < \frac{0.014 \text{ m}}{36,000 \text{ m}} \frac{1}{7.3 \times 10^{-5} \text{ radians sec}^{-1}} \frac{1}{8 \times 10^{-4} \text{ radians}}$$

or that the correlation integration time must be less than 6.7 sec.

6.8 SOURCE STRUCTURE AND THE VISIBILITY FUNCTION

6.8.1 Sky Coordinates and the Visibility Function

We are now ready to show how a source with extended structure can be imaged. Our ultimate goal is to produce a map of intensity versus position on the sky. Although the sky is curved, we can treat our image as occurring on a plane since the field of view is relatively small; the image will be the *projection* of the source onto the plane tangent to the sky at the point of the phase center.

We will assign map coordinates (x,y), where x and y are angles in radians, that specify a pixel position corresponding to a sky position relative to the phase center and map center. Each pixel will represent a small angular region of the plane tangent to the sky. In this coordinate system, $+x$ is a step to the right in the image. This direction is to the West, in the direction of smaller RA (recall from Chapter 1 that in a map of the sky, west is to the right). A step toward $+y$ is upward in the map, toward the north and larger declination.

Each pixel will have a solid angle $\Delta x \Delta y$ and so will contain a flux density given by $I_\nu(x,y) \Delta x \Delta y$, where $I_\nu(x,y)$ is the source intensity as a function of position. We can treat the flux density in each pixel as a point source and the cross-correlation of the extended source, then, is the sum of the cross-correlations of all these point sources. We can write this sum as the integration of the cross-correlation of a point source with flux density $I_\nu(x,y)dxdy$ over the area of the source. Thus, we start by integrating Equation 6.13 over the two spatial dimensions, giving us

$$\Gamma = \frac{1}{\Delta\nu} \iiint h(\nu) I_\nu(x,y) e^{i2\pi(\vec{b}\cdot\hat{n}/\lambda)} \, dx dy d\nu$$

where:

$I_\nu(x,y)\,dxdy$ assumes the role of F_ν from Equation 6.13.

The unit vector $\hat{n}$ indicates the direction of a position (x,y) while $\hat{n}_0$ indicates the direction of the map center (or phase center). To specify the different positions in the map, we define the vector $\vec{r}$ (lying in the plane of the sky) to point from the map center to the position of each part of the source. So the direction of each piece of the source, relative to the center of the array, is given by the vector sum

$$\hat{n} = \hat{n}_0 + \vec{r}$$

as depicted in Figure 6.7. Note that $\hat{n}$ and $\hat{n}_0$ are unit vectors, but $\vec{r}$ is not.

Now, $\vec{r}$ is just the vector sum of $\vec{x}$ and $\vec{y}$, the coordinate vectors in the map, that is,

$$\vec{r} = \vec{x} + \vec{y} = x\hat{x} + y\hat{y}$$

Substituting $\vec{x}$ and $\vec{y}$ into the correlation function, we have

$$\Gamma = \frac{1}{\Delta\nu}\iiint h(\nu)I_\nu(\vec{x},\vec{y})e^{i2\pi\vec{b}\cdot(\hat{n}_0+\vec{x}+\vec{y})/\lambda}\,d\vec{x}d\vec{y}d\nu$$

We can now expand the integrand by separating the exponentials into two terms to produce

$$\Gamma = \frac{1}{\Delta\nu}\iiint h(\nu)\,I_\nu(\vec{x},\vec{y})e^{i2\pi\vec{b}\cdot\hat{n}_0/\lambda}e^{i2\pi\vec{b}\cdot(\vec{x}+\vec{y})/\lambda}\,d\vec{x}d\vec{y}d\nu \qquad (6.16)$$

For simplicity, we neglect the frequency dependence; it is peripheral to the present discussion, and as we explain elsewhere, various methods can be used to mitigate the bandwidth effects. So the integration over frequency in the numerator cancels with $\Delta\nu$ in the denominator and the cross-correlation becomes

$$\Gamma = \iint I_\nu(\vec{x},\vec{y})e^{i2\pi\vec{b}\cdot\hat{n}_0/\lambda}e^{i2\pi\vec{b}\cdot(\vec{x}+\vec{y})/\lambda}\,d\vec{x}d\vec{y} \qquad (6.17)$$

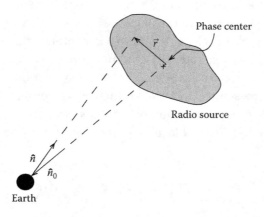

FIGURE 6.7 Unit vector $\hat{n}$, giving the direction of each piece of the source, is the vector sum of the unit vector $\hat{n}_0$, the direction of the phase center, and $\vec{r}$, the sky position vector of that piece of the source relative to the phase center.

The first exponential is independent of x and y, so we can take it outside the integral to give

$$\Gamma = e^{i2\pi\vec{b}\cdot\hat{n}_0/\lambda} \iint I_v\left(\vec{x},\vec{y}\right)e^{i2\pi\vec{b}\cdot(\vec{x}+\vec{y})/\lambda}\,d\vec{x}d\vec{y}$$

The exponential in front of the integral is the fringe function of a point source at the map center (see Equation 6.10). In practice, it is removed in real time while the signals are being collected and correlated. Each baseline vector is well known and the position of the phase center is specified by the observer, so the geometrical delay to each antenna due to the extra path length ($\tau = (\vec{b}\cdot\hat{n}_0)/c$) can be compensated for.

Now, with the fringe function removed, we get the general expression for the *visibility function*, given by

$$V = \Gamma e^{-i2\pi\vec{b}\cdot\hat{n}_0/\lambda} = \iint I_v\left(\vec{x},\vec{y}\right)e^{i2\pi\vec{b}\cdot(\vec{x}+\vec{y})/\lambda}\,d\vec{x}d\vec{y} \tag{6.18}$$

For future reference, remember that the more general definition includes the integration over frequency, that is,

$$V = \Gamma e^{-i2\pi\vec{b}\cdot\hat{n}_0/\lambda} = \frac{1}{\Delta v}\iiint h(v)I_v\left(\vec{x},\vec{y}\right)e^{i2\pi\vec{b}\cdot(\vec{x}+\vec{y})/\lambda}\,d\vec{x}d\vec{y}dv \tag{6.19}$$

We are almost to the main point now. We can simplify the vector math in the exponent by breaking the dot product into components as

$$\vec{b}\cdot\left(\vec{x}+\vec{y}\right) = b_x x + b_y y$$

where:
   b_x and b_y are the projections of the vector $\vec{b}$ onto the x and y axes. (Recall that x and y are the *sky* coordinate axes while $\vec{b}$ is a vector on Earth.)

This allows us to rewrite Equation 6.18 as the visibility function (ignoring the frequency dependence) as

$$V = \iint I_v(x,y)e^{i2\pi(b_x x + b_y y)/\lambda}\,dxdy$$

or

$$V = \iint I_v(x,y)e^{i2\pi(b_x/\lambda)x}e^{i2\pi(b_y/\lambda)y}\,dxdy \tag{6.20}$$

Consider what we have now. Equation 6.20 is a two-dimensional inverse Fourier transform that transforms $I_v(x,y)$, the intensity distribution on the sky, into $V(b_x/\lambda,b_y/\lambda)$, the two-dimensional visibility function. Equation 6.20 means that *the Fourier transform of the measured visibilities, as a function of the baseline vectors measured in wavelengths yields the sky intensity distribution*, that is, the image of the source.

The sky brightness distribution is inherently a purely real number. The visibility function, on the other hand, is a complex number with both amplitude and phase. The total visibility amplitude is obtained from the many individual visibilities by the addition of these complex numbers. This is completely analogous to the magnitude of a vector being determined by the sum of many vectors with different directions. We discuss the visibilities in greater detail below. For now, the salient point is that the visibilities have amplitudes and phases, and these are determined by the Fourier transform of the sky brightness distribution.

6.8.2 *uv*-Plane

The parameters b_x/λ and b_y/λ that appear in the Fourier transform integrals of Equation 6.20 are fundamental quantities in aperture synthesis and so they deserve simple notation. In radio interferometry they are assigned the variables u and v where

$$u = b_x/\lambda \text{ and } v = b_y/\lambda \tag{6.21}$$

We say that the visibility function is measured in the *uv*-plane, while intensity is a quantity in the xy-*plane* (or *image plane*), and we say that we *transform from the* uv-*plane to the image plane*.

The units of u and v are commonly expressed as *wavelengths* or *kilo-wavelengths* and denoted as $k\lambda$, as a reflection of the fact that they represent the components of the baseline in units of the wavelength. However, they can be thought of as inverse radians (although a radian, which is the ratio of two distances, is dimensionless). Consider that the resolution of a telescope, which is proportional to λ/D, gives an angular size on the sky in radians. The *uv* distances, meanwhile, are the inverse of this; they correspond to a telescope size in units of λ (Figure 6.8). This is as it should be, as we discussed above in the introduction of Fourier transforms—u and v must have inverse units of x and y, which are angular positions on the sky.

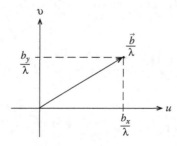

FIGURE 6.8   Orientation and length of the baseline vector divided by the observing wavelength is plotted on the *uv*-plane, where u is the component in the east–west direction and v is the north–south component.

The dimensions of the visibilities, from Equation 6.18, are intensity times $dxdy$, the latter being a solid angle. Visibilities, then, have dimensions of flux density, usually given as Jy. In the inverse Fourier transform, the visibilities (in Jy) are multiplied by $dudv$, which are inverse angles, which is the same (dimensionally) as dividing by two angles, and so we get units of Jy per solid angle, which is intensity.

The baselines are actually three-dimensional (i.e., an array in general is *not* strictly coplanar), and so they cannot be completely described by just u and v. A third dimension, perpendicular to the sky plane, is what causes the delay, and hence the fringe function. This dimension is assigned the variable w, defined by

$$w = \frac{\vec{b} \cdot \hat{n}_o}{\lambda}$$

With delay tracking, this so-called w-term is usually quite small. In some cases, particularly when observing large sky areas, it can be important and must be accounted for in the analysis.

In summary, we have learned the following:

1. The signals detected by each antenna pair during an integration time interval are cross-correlated with one another.

2. The cross-correlations after being corrected for the fringe function at the phase center gives the visibility function, $V(u,v)$.

3. A two-dimensional Fourier transform is performed on the visibility function to yield the sky brightness distribution, $I_v(x,y)$, resulting in an image of the source.

4. This is represented mathematically by

$$V(u,v) = \Gamma e^{-i2\pi \vec{b} \cdot \hat{n}_0/\lambda} = \iint I_v(\bar{x}, \bar{y}) e^{i2\pi(ux+vy)} dxdy \qquad (6.22)$$

and

$$I(x,y) = \iint V(u,v) e^{-i2\pi(ux+vy)} dudv \qquad (6.23)$$

As a final note, and to illustrate the role of different terms, we rewrite Equation 6.16 in the following way:

$$\Gamma = \left[e^{i2\pi v_0(\vec{b} \cdot \hat{n}_0/c)} \right] \left[\int \frac{h(v)}{\Delta v} e^{i2\pi v_0(\vec{b} \cdot \hat{n}_0/c)} dv \right] \left[\iint I_v(x,y) e^{i2\pi(ux+vy)} dxdy \right]$$

The square brackets delineate the three factors that contribute to the cross-correlation. The first factor is the fringe function, which is defined for a given phase center and central frequency and so is independent of all the integration variables and can be removed from the integral. This factor is removed in real time by adding a geometric delay to one of the two signals to keep the relative delay equal to zero. The second factor accounts for finite bandwidth, through the function $H(\tau)$ as shown in Figure 6.6. The third quantity is the visibility function, $V(u,v)$.

6.8.3 Visibility Functions of Simple Structures

With the derivation of Equations 6.22 and 6.23, we have shown the Fourier transform relation between the visibilities and the source brightness distribution. In a way, we already derived and analyzed these relations in Sections 5.5 through 5.7, although our discussion there was constrained to the equatorial plane; now we have more general equations that include all three dimensions. We can apply the relations we found in Chapter 5 to the general case by considering slices through the uv- and xy-planes. For example, if we observed a pair of point sources separated by an angle $\Delta\theta$ in the east–west direction, then the visibility amplitudes would have oscillations in the u direction. Stripes of constant amplitude would appear parallel to the v-axis.

The important relations between visibilities and simple source structure were summarized in Section 5.9. In Example 6.2, we show how several simple source structures can be inferred from the corresponding visibilities.

Example 6.2:

1. We obtain visibility data for a particular astronomical object. A 3D plot of visibility amplitudes is shown in Figure 6.9. The amplitudes are shown in gray scale; u is plotted on the horizontal axis and v on the vertical axis. Stripes are seen at a 45° angle, running from northeast to southwest. A slice through this 3D plot, going perpendicular to the stripes is plotted as V_{amp} versus uv distance in Figure 6.10. (Remember that the uv-plane represents the spacings on the Earth's surface, and so west is to the left in Figure 6.9.) What can we infer about this source?

2. A second astronomical object is observed. Its 3D visibility plot shows an amplitude decreasing with uv distance in all directions, as shown in Figure 6.11. A 2D plot of the amplitude versus uv distance is shown in Figure 6.12. What can we infer about the structure of this source?

Answers:

1. In this data set, the visibility amplitudes oscillate with a period of 550 inverse radians. This must correspond to a pair of sources separated by 1/550 radians (see Section 5.9). The peak amplitude is about 6 Jy and the minimum amplitudes are about 0. This region must have two point sources, each with a flux density of about 3 Jy and separated by 1/550 radians = 6.3 arcmin.

   The direction in which the amplitude varies indicates the direction that the sources are separated. The oscillations occur in the NW–SE direction while the

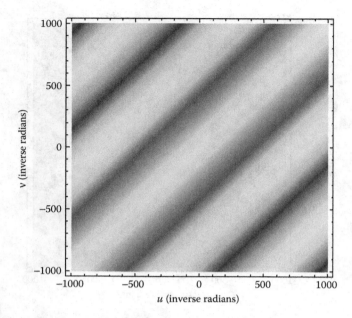

FIGURE 6.9   Gray scale plot of the visibility amplitude of the first source, with u on the horizontal axis, v on the vertical, and the visibility amplitude represented by the level of gray (where black = 0).

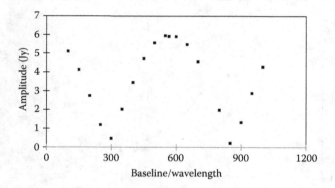

FIGURE 6.10   Northwest–southeast slice through Figure 6.9, showing visibility amplitude versus u. The amplitude is a periodic function with a period of about 550 inverse radians.

stripes, or maxima, run along the northeast–southwest direction. Thus, the sources are separated by 6.3 arcmin in the northwest–southeast direction.

2. The fact that the visibility amplitude continually decreases with total uv distance means that as the resolution increases less flux density is detected. Therefore, this source is resolved. Additionally, since the visibility amplitudes decrease symmetrically in both u and v, the source is resolved in all directions equally, and we can infer that the source is circular.

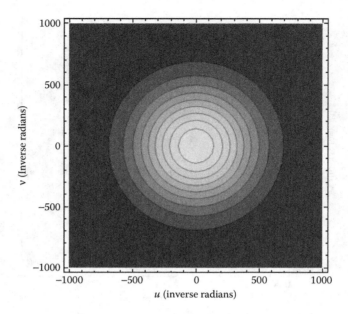

FIGURE 6.11   Gray scale plot of the visibility amplitude of the second source, showing a circular shaped $V(u,v)$ that decreases with uv distance.

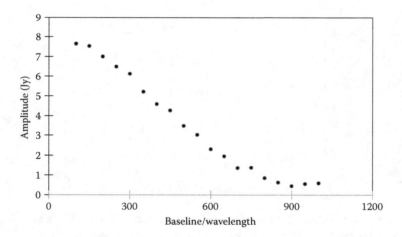

FIGURE 6.12   Visibility amplitude versus total uv distance of the second source.

Without performing a fit to the curve in Figure 6.12, we do not know the exact brightness distribution of this source. But, we can get an estimate of its diameter by noting that the visibility is roughly Gaussian in shape, with a full width at half maximum of about 450 wavelengths. So the source size is roughly Gaussian with a FWHM of about 1/450 = 0.0022 radians or about 7.6 arcmin.

Lastly, we know that at zero uv distance, the visibilities are sensitive to the entire source, and so we can also infer the total flux density of the source from the visibility amplitude at the smallest uv spacing. Therefore, the flux density of the source is about 8 Jy.

6.9 THE EARTH'S ROTATION AND *UV* TRACKS

We form an image using many individual Fourier components, each one of which is a visibility corresponding to a particular *uv* point. We can improve the quality of the image by including greater numbers of Fourier components, coming from different points in the *uv*-plane. We achieve this by observing with as many different projected baselines as possible. In fact, mathematicians will tell you that unless you know *V* for *all uv* points, covering $(-\infty, -\infty)$ to $(+\infty, +\infty)$, the Fourier transform will necessarily produce a flawed image. We cannot measure *V* at every single *uv* position, but as we will see, this is not a serious problem for producing useful images.

There are several ways to sample a large number of *uv* points. One way is the use of a large number of antennas, arranged in a well-designed array so that each pair of antennas creates a distinct baseline vector, of unique length and orientation. A second way is to make use of the Earth's rotation, which changes the orientation and projected length of each baseline relative to the sky during the course of a day. A third method, called multifrequency synthesis (which we will only mention and not discuss further), is to observe over a wide bandwidth. From Equation 6.21, we see that the different values of λ will give different values of *u* and *v*.

The second method is often referred to as *Earth rotation synthesis*. Recall that *u* (given by b_x/λ) and *v* (given by b_y/λ) are projections of the baselines onto the sky plane. If the source is at a celestial pole and the baseline has an east–west orientation, seen from the source, this baseline vector will trace out a circle as the Earth rotates. Now consider the situation of Figure 6.13, in which the source is not at a celestial pole, and the baseline is randomly oriented. As the Earth rotates, there will be times when $\vec{b}$ has a component directed toward the source and so the projection of the baseline onto the sky will be shortened. The baseline, then, will trace out an ellipse. Also, there will be times when the baseline and the source are blocked from view by the Earth, so only a portion of the ellipse is obtained. Additionally, since it does not matter which antenna is placed at each end of the baseline, the visibility data must be symmetric about the origin, that is, $V(u,v) = V^*(-u, -v)$. (The complex conjugate is needed on the right side because of the change of sign of *u* and *v*.) In 12 h (if the source is above the horizon at least half the day), each baseline yields *uv* data over two half-ellipses.

The paths in the *uv*-plane traced by the baselines are called *uv tracks*. Figure 6.14 depicts a pair of *uv* tracks traced by a single baseline in a 12-h observation. Note that the two half ellipses in Figure 6.14 do not connect. This occurs because the geometric center of the ellipse is not at the origin. Origin-centered ellipses occur only if the baseline is purely east–west. Visibility data are recorded at every integration time, typically every 1 to 10 seconds, and so each baseline can measure a visibility at about 4300 to 43,000 points in the *uv*-plane during a 12-h observation.

Equations for *u* and *v*, as functions of time, antenna positions, and source position, can be obtained by transforming vectors from the Earth's spherical coordinate system to the sky spherical coordinate system. The process is rather tedious, though, and unnecessary for the purposes of this text. A complete discussion is provided in *Interferometry and Synthesis in Radio Astronomy* by Thompson, Moran, and Swenson.[*]

---

[*] A. R. Thompson, J. M. Moran, and G. W. Swenson. 2001. *Interferometry and Synthesis in Radio Astronomy.* New York: John Wiley & Sons.

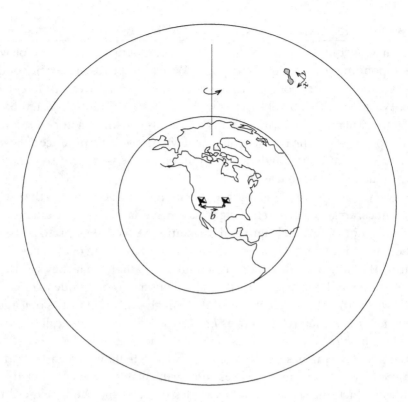

FIGURE 6.13   Baseline, $\vec{b}$, with a random orientation at some latitude on the Earth's surface obtains visibility data of a source at some location in the sky. The x and y axes lie in the sky plane at the source; y points north and x points west.

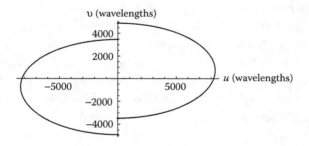

FIGURE 6.14 uv tracks of a single baseline over 12 h consist of two half-ellipses. The tracks are calculated for a source declination of 30°, an observing wavelength of 6 cm, and a 502-m baseline, tilted 5.7° relative to the east–west line and located at the equator.

With an array, each antenna pair obtains data along a different uv track. An array of four antennas, for example, has six baselines (baselines 1–2, 1–3, 1–4, 2–3, 2–4, and 3–4). In general, for N_A antennas in the array the number of baselines is

$$N_b = N_A\left(N_A - 1\right)\big/2 = \left(N_A^{\,2} - N_A\right)\big/2 \qquad (6.24)$$

$N_A(N_A - 1)$ is the total number of ways to combine N_A antennas, taken two at a time, but this overcounts the number of baselines by a factor of 2; for example, baseline 1–2 is same as baseline 2–1. Hence, we divide by 2 to remove the redundant counting. The number of baselines is roughly proportional to the *square* of the number of antennas. With a large number of baselines, each one yielding visibilities over a distinct ellipse in the *uv*-plane, the total *uv*-coverage of a typical 12-h aperture synthesis observation can be quite extensive. With 6 antennas, for example, 15 *uv* tracks are created. Figure 6.15 shows 15 times greater *uv*-coverage compared to the single baseline of Figure 6.14. The number of visibilities measured, and their coverage of the *uv*-plane, determine the number and distribution of the Fourier components that form the image. Hence, the quality of the resulting image increases dramatically as more antennas are used.

The Jansky VLA (shown in Figure 5.2), for example, has 27 antennas. It has 351 baselines, and hence samples 351 points in the *uv*-plane simultaneously. Even in a short "snap-shot" observation of just a few minutes, the *uv*-coverage is sufficient to produce a reasonable quality image. The ALMA telescope in Chile (shown in Figure 5.1) is an even more extreme example. With 54 antennas, the instantaneous *uv*-coverage is 1431 baselines.

Example 6.3:

The Sub-Millimeter Array (SMA) in Hawaii is composed of eight 6-m dishes, while the Very Large Array in New Mexico has twenty-seven 25-m dishes. Which array would benefit more from the addition of another dish?

Answer:

The SMA has $(8 \times 7)/2 = 28$ baselines. With one more antenna it would have eight additional baselines (one with each of the eight original antennas); or, as a check, $(9 \times 8) \div 2 = 36$ baselines. This is an increase of 29%. The VLA has $(27 \times 26) \div 2 = 351$ baselines. One additional antenna would give it 27 more baselines, for a total of 378. This is an increase of 7.7%. When the array is larger to start with, each additional antenna will provide more new baselines, but the percentage increase in the number

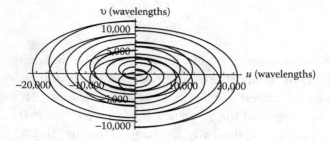

FIGURE 6.15   Plot of the *uv*-coverage in a 12-h observation of a source at a declination of 30° using an array with six antennas. The number of tracks = 6(5)/2 = 15; this doubles to 30 when we include the complex conjugate tracks.

of baselines will be smaller. Thus, other considerations being equal, the SMA would benefit more than the VLA by adding another antenna.

The placement of the antennas within the array also needs careful planning to optimize the *uv*-coverage. Although some redundancy in baselines can be useful, in general the array design should minimize the number of duplicate baseline vectors. This, in turn, will minimize the number of redundant Fourier components, which do not give new information about the source structure. Array design is essentially an optimization problem (often with various constraints applied) in which one tries to balance competing factors. In Question 5, we provide a link to a Mathematica program that readers can use to design an array.

6.10 INTERFEROMETERS AS SPATIAL FILTERS

If we decompose the sound of a choir into its constituent Fourier components, we will need a range of temporal frequencies to accurately reproduce the singing voices. The tenors and baritones will especially need low temporal frequencies while the altos and sopranos will especially need high temporal frequencies. But always, some range of frequencies will be needed. If the range of frequencies is truncated for some reason, we may not be able to faithfully reproduce the music being sung.

If we decompose a sky brightness distribution into its Fourier components, as in the case of a choir where we will need a range of temporal frequencies, we need a range of *spatial frequencies* to produce the image. If large structures are present in the field, we will need low *spatial frequencies* to reproduce them, while if small structures are present, we will need high *spatial frequencies* to reproduce them. Short baselines sample low spatial frequencies while long baselines sample high spatial frequencies, as we illustrate below. As in the audio situation described above, if a crucial range of frequencies is left out, we will not be able to faithfully image the brightness distribution.

We can relate the angular size of structures in the image plane to the corresponding *uv* spacing of the visibilities by using the reciprocal nature of the Fourier transform that we discussed in Section 6.8. In particular, the angular size in the image domain, measured in radians, is proportional to the *inverse* of the *uv* spacing in the visibility domain. If we represent angular scale on the sky by θ_{xy}, then

$$\theta_{xy} = \frac{\lambda}{\vec{b} \cdot (\vec{x} + \vec{y})} \tag{6.25}$$

The dot product in the denominator is just the projected baseline length.

This relation allows us to quickly estimate two important parameters for any interferometric observation: the *largest angular size* and the *resolution*. Because of the inverse relation between the image and the *uv*-planes, the largest angular structures are probed by the smallest *uv* spacings. Equivalently, the largest structure that an array can reliably reformulate is determined by the shortest baselines, so $\theta_{xy}(\text{largest}) \sim \lambda/b_{min}$, where b_{min} is the shortest baseline. Conversely, the smallest angular structures (which determine the source sizes we can resolve) are determined by the largest *uv* spacings. So the angular resolution of the observations is given approximately by θ_{xy}(smallest) ~ λ/b_{max}. The actual resolution in the map (given by the synthesized beam) depends on the distribution of the data in the uv-plane and how the uv data are weighted in the mapping process (discussed in Section 6.13). Nevertheless, λ/b_{max} gives a reasonable approximation to the angular resolution of the observations.

Note the similarity of θ_{xy}(smallest) ~ λ/b_{max} to the expression for the resolution of a single telescope, θ_{res} ~ λ/D. Recall that the resolution of a single-dish telescope is determined by the interference of the incoming waves incident at opposite edges of the dish. It makes intuitive sense, then, that when synthesizing a telescope via interferometry we get the same relation between the angular size scale probed and furthest distance between incoming waves. Note that for a single-dish telescope one not only has baselines as large as the diameter of the primary reflector, but also all of the shorter baselines—including zero. It is for this reason that single-dish telescopes are sensitive to emission even from very large sources.

As we showed in the previous section, sampling the uv-plane at many locations will permit us to measure more Fourier components, and hence improve the image quality. The crucial point to understand is that any array must have a maximum and minimum baseline length, so only a certain range of spatial frequencies can be sampled. Lower and higher spatial frequencies, sampled by baselines shorter than b_{min} and longer than b_{max}, respectively, *will not be present in the data*. In some cases, the absence of these spatial frequencies may have a profound effect on the image we obtain, as we illustrate in the following discussion.

To provide for a large range of baselines—and hence spatial frequencies—interferometer arrays are often designed so that the antennas can occasionally be moved about. The Jansky VLA, for example, has four different configurations in which it can operate. The most compact D-configuration has baselines up to 1 km in length. By design, each subsequent configuration (denoted as C, B, and A) is about 3.3 times larger than previous one, so these configurations have maximum baseline lengths of about 3.3, 11, and 36 km.

To see the effect of observing in the different configurations, examine Figure 6.16. We show the visibility amplitudes as a function of uv distance from a VLA B-configuration observation (left) and a D-configuration observation (right) both at a wavelength of 3.6 cm and of the same field. The longest baselines in the D-configuration are about 25 kλ while those of the B-configuration are about 10 times longer—about 250 kλ. Similarly, the *hole* in the center of the uv-plane is about 1 kλ for the D-array, and about 5 kλ for the B-array. The hole in the B-array is less than 10 times larger than the D-array hole because of the projection of the baselines onto the sky plane. Examining the visibility amplitudes, we see that they rise sharply for shorter uv distances. On the right, we see that baselines of about 5 kλ (the shortest baselines sampled) have amplitudes of about 0.3 Jy. On the left, we see that at 5 kλ baselines the D-array also detects about 0.3 Jy. But the D-array has data for even shorter baselines, where the amplitudes rise sharply, reaching values of over 1 Jy. Such behavior in the visibility amplitudes means that there is a large, bright object in the field—as seen in Figure 6.17. Because these short-baseline data

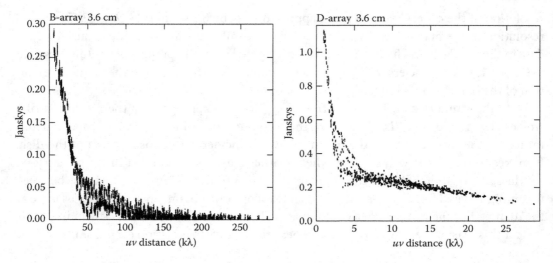

FIGURE 6.16 Plots of the visibility amplitude versus uv distance for the same field of view but for datasets with quite different ranges of projected baselines. Note that the VLA B array data (left) contains about 10 times larger spacings than the VLA D array data (right). In fact, many of the D array points fall within the central *hole* of the B array, so that the D array is sampling low spatial frequencies that are absent from the B array data. In the overlap region (at uv distances of about 5–10 kλ) both arrays measure a flux of about 0.3 Jy. Shorter uv distances have substantially higher fluxes, arising from a bright, extended source that is invisible to the B array. Likewise, the B array data contain information of structure on angular scales smaller than can be resolved by the D array data (cf. Figure 6.17).

were missing from the B-array observations, those data cannot image the large structure in the field. In this case, the original observations were in the B-array, so the observers did not realize that the small (8″) source they had detected was embedded within a much larger (2′) envelope of ionized gas. In some cases, this can lead to a greatly different astronomical analysis.

The salient point of Figures 6.16 and 6.17 is that by excluding a certain range of baseline lengths, the image will not show the corresponding structures, *even if they are physically present in the field.* To borrow a phrase from computer programmers, this is a *feature*, not a *bug*. For example, if you know that there is bright, extended emission in the field of view, and *if it is of no interest to you* (because, for example, you are interested in some compact foreground object), then by observing with long baselines, the extended emission will be *filtered out*, and will never appear in your image. Removing bright, unwanted emission in this way can result in a much higher quality image of the compact emission, devoid of the contaminating, extended emission.

Thus, when observing with an interferometer array, it is important to have some idea what the angular size of your source will be. When planning an observation, you must be sure that the shortest baseline of the array will be adequate to image the largest angular size that you want to image. Similarly, you must ensure that the longest baseline will be adequate to resolve the smallest structure that you need to image. The first criterion is

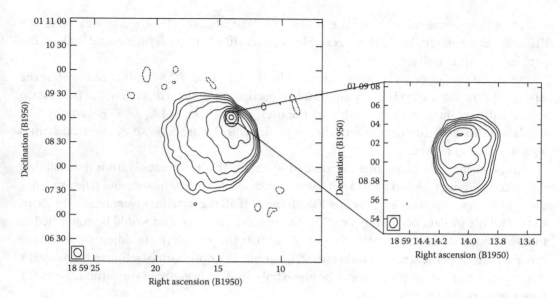

FIGURE 6.17   Shown here are two images of the galactic photoionized nebula G35.20–1.74 obtained with the Jansky VLA. On the left is the image obtained using the shorter baselines of the D-array. Although the compact core is unresolved, the image shows a rather large (2 arcmin) structure surrounding the compact core. On the right is the image obtained with the longer baselines of the B-array. Although the angular resolution of this image is about 10 times higher than the D-array image, the large, 2′ envelope of emission is completely absent. In both images, the synthesized beam size (corresponding to the angular resolution of the image) is shown in the lower left corner.

perhaps the more crucial one: if you do not have adequate short baselines, you would not be able to detect the object at all. The second criterion is a little less serious: if you do not have adequate long baselines, you can at least detect the object, even if you do not resolve it. If you are unsure of the size of the object, it is usually better to be conservative, and observe first with relatively short baselines. It is easier to justify more telescope time, with longer baselines, to resolve a detected object than it is to justify re-observing an *empty* field, using shorter baselines, in the hope that a larger object might be present.

Because any particular array will have a longest and a shortest baseline, we can express the *range* of angular sizes the array can detect by the ratio b_{max}/b_{min}. In the case of the VLA, for example, within a single configuration the ratio of the longest to shortest baseline is about 40, so the largest angular size that can be imaged is about 40 times larger than the resolution. In some cases, the astronomical object may contain structures over a wider range of angular sizes than can be obtained in a single configuration. In this case—and provided that the source's emission is not time-variable—we can observe the field in one array configuration, then wait several weeks or months until the array is moved to a different configuration and re-observe. The visibilities from the two observations will sample different spatial frequencies (and hence different angular scales in the image plane) but there is no problem in combining them after-the-fact; the different Fourier components need not be measured simultaneously.

Let us do a couple of numerical examples to illustrate how the angular size versus uv distance relationship works. These examples will also illustrate a useful method for locating problems in your visibility data.

Imagine a pair of point sources separated by $\theta_0 = 30$ arcsec (1.45×10^{-4} radians) in the east–west direction. The visibility amplitude is oscillatory in the u direction, but is constant in the v direction. The u-wavelength of the oscillations is $\lambda_u = 1/\theta_0 = 6900$ radians^{-1}. A display of this visibility function in the uv-plane, as in Figure 6.18, shows vertical stripes with a period of 6900 radians^{-1}.

Now, consider what baselines are needed to obtain this information from the visibility data. Study Figure 6.18 and consider what uv tracks would reveal the sinusoidal dependence on u, with a spacing wavelength of 6900 radians^{-1}? If all the baselines were relatively short, so that all the uv data occur at $u \ll 6900$, then all the visibility data would be contained in the white band down the middle of Figure 6.18 (near the peak of the middle stripe) and the general pattern of the stripes would not be apparent. The optimum baseline to discover the stripe pattern in the uv-plane would be one whose uv distances are of the same scale as the wavelength of the stripes, that is,

$$d_{uv} \sim 1/\theta_0$$

or, since the uv spacing is proportional to the baseline length measured in wavelengths, we require baselines of length

$$b \sim \lambda/\theta_0$$

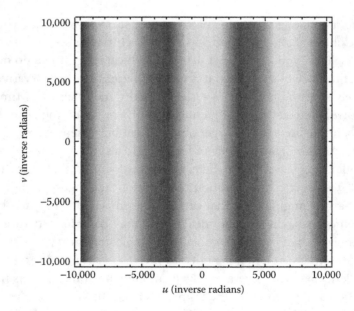

FIGURE 6.18 Two-dimensional visibility plot, showing the amplitude of $V(u,v)$ in gray scale, with u on the horizontal axis and the v on the vertical axis. The plot is for a pair of point sources separated by 30 arcsec in the east–west direction.

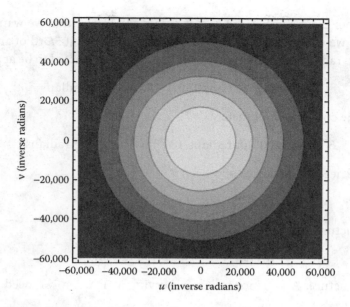

FIGURE 6.19 Approximate visibility plot, $V(u,v)$, for a single resolved, circular source with a 5 arcsec diameter.

Now imagine a single source of circular shape that is 5 arcsec in diameter. Approximating the source as a uniform disk, we can use the concept discussed in Section 5.7 to estimate the visibility amplitude plot. Figure 6.19 displays such a plot. Note that the visibility amplitude goes to zero at approximately 50,000 radians^{-1} (this is 1.22 times the inverse of the angular size of the source, in radians). Consider what uv tracks, overlaid on Figure 6.19, would reveal the structure of this source. The best array will be one that contains a range of baseline distances that reveal the exact shape of the amplitude profile, up to the approximate distance of the first null at $b/\lambda \sim 50,000$. If these observations were made at a wavelength of 2 cm, then we would want an array with baselines as long as 1000 m. What if we had only baselines whose uv distances were significantly larger than 1000 m? At these large uv distances, the visibility amplitude is extremely small and so the source becomes undetectable. We say that such a source is *resolved out* at the long baselines.

Example 6.4:

We learn of a new radio source reported to have an angular size of approximately 3 arcmin with a brighter central core component of about 5 arcsec. To obtain a reasonable map of this source at 1.35-cm wavelength, what baseline lengths should we use?

Answer:

We should obtain visibilities over a range of uv distances that encompass the inverses of the angular sizes of the source and the core (in radians). If we wish to see structure within the core, then we will need even longer baselines. A common guideline is to have about three synthesized beams across any source that we want to resolve, which means we would like baselines corresponding to about 5/3rd of an arcsecond. Converting to radians, the angular size of the source is expected to be approximately

$$3'(1°/60')(\pi \text{ radians}/180°) \sim 8.73 \times 10^{-4} \text{ radians}$$

and that of the core is approximately

$$5''/3 \ (1°/3600'')(\pi \text{ radians}/180°) \sim 8.07 \times 10^{-6} \text{ radians}.$$

Therefore, we need *uv* distances of order

$$1/(8.73 \times 10^{-4}) \text{ radians}^{-1} \sim 1150$$

for the 3' structure and

$$1/(8.07 \times 10^{-6}) \sim 12,400 \text{ radians}^{-1}$$

for the 5'' structure. And, since we are observing at 1.35 cm, we need baselines of lengths from

$$1150 \times 0.0135 \text{ m} \sim 15 \text{ m}$$

to

$$41,300 \times 0.0135 \text{ m} \sim 1670 \text{ m}$$

At a *minimum,* the array should contain baselines from approximately 15 to 1700 m. In practice, even shorter and longer baselines should be included to make a good-quality image. The reason for this is that just a few visibilities in a particular range of baselines are usually not enough to make a good quality image. In terms of our audio analogy at the beginning of the section, including a single tenor or soprano in a large choir is not enough; the high and low notes must be well represented in order to faithfully reproduce the music.

6.11 SENSITIVITY AND DETECTION LIMITS

The success of an observation depends critically on careful planning by the observer, including a calculation of the sensitivity requirements needed to achieve the scientific goals. Sometimes the goal of an observation is simply to detect a source while other times one needs to obtain a high-quality image to reveal detailed information about the source structure. In the former case, a 5σ detection may be adequate, while in the latter case much higher sensitivity may be needed.

In Chapter 4, we discussed the noise involved when observing with a single antenna. Here, we consider noise in two other contexts: that involved when observing with a pair of antennas, forming a single baseline and producing a single visibility, and that involved when combining the visibilities from many different baselines to produce an image. The former type of noise is most important for the calibration of the data (which we describe in Section 6.12), while the latter type is relevant for the quality of the final image.

6.11.1 Noise in a Visibility

Interferometer noise calculations use nearly the same principles that we discussed in Section 4.7, with the primary difference that interferometers do not need to be switched (since the amplifier noise of the two telescopes is uncorrelated). This removes a factor of 2 from our earlier calculation. We begin by rewriting Equation 4.33 without this factor of 2, obtaining

$$\sigma(F_v) = \frac{2k}{A_{\text{eff}}} \frac{T_{\text{sys}}}{\sqrt{\Delta t \Delta v}} \quad (6.26)$$

where:

A_{eff} is the effective aperture of each antenna

T_{sys} is the system temperature

Δt and Δv are the integration time for the visibility and the bandwidth of the observation, respectively

This is the noise of a total power radiometer, attached to a single antenna of effective area $A_{\text{eff}} = \eta_A A_{\text{geom}}$ (Equation 4.17). It can be shown that for a two-element interferometer, the noise level in the cross-correlated signal is

$$\sigma(F_v) = \frac{\sqrt{2}k}{A_{\text{eff}}} \frac{T_{\text{sys}}}{\sqrt{\Delta t \Delta v}} \quad (6.27)$$

From Equations 6.26 and 6.27, we see that a *single* antenna of effective area A_{eff} has a noise level that is $\sqrt{2}$ times larger than the noise level of a *pair* of antennas with a combined effective area of $2A_{\text{eff}}$. Said another way, the sensitivity of a single cross-correlation interferometer pair is $\sqrt{2}$ times *worse* than a single-dish total power system with the same collecting area. The reason for the factor of $\sqrt{2}$ lower performance is that the interferometer does not make use of the *auto*-correlation of the signals; that is, some information is lost.

Because the effective area varies with frequency, it is common to use $\eta_A A_{\text{geom}}$ instead of A_{eff} and the frequency dependence is absorbed in the η_A. Several additional factors, arising mostly from the electronics of the correlator and the digital processing, can affect the sensitivity at levels from a few percent to a few tens of percent. The biggest effect is from the number of bits used in the digital correlator. All of these effects can be lumped into a single efficiency term that we call the *correlator efficiency*, η_c. Typical values are in the 0.8–0.9 range; precise values should be available in the telescope documentation. Introducing these quantities into Equation 6.27, we obtain

$$\sigma(F_v) = \frac{\sqrt{2}k}{\eta_c \eta_A A_{\text{geom}}} \frac{T_{\text{sys}}}{\sqrt{\Delta t \Delta v}} \quad (6.28)$$

This is the noise expected for visibilities produced by a single interferometer pair, in an integration time Δt, over a bandwidth Δv. If two orthogonal polarizations are observed, then the noise in a total intensity map will be smaller by a factor of $\sqrt{2}$. The parameters controlled by the observer are Δt and Δv.

6.11.2 Image Sensitivity

To determine the noise level in an image, we must account for all the observed visibilities that are used to make it. As we describe later (Section 6.13), the image may still require some processing before reaching its final form, and this processing can profoundly affect the image noise. Nevertheless, there exists a lowest possible noise level that can be achieved in the map. This level is often called the *thermal noise*, and it depends on how many visibilities are used to make the image and their *rms* noise level.

If t_{int} is the integration time spent on each visibility, then replacing Δt in Equation 6.28 with t_{int} gives the *rms* noise for a *single* visibility. To account for the total number of visibilities we divide the right-hand side of Equation 6.28 by $\sqrt{N_{vis}}$. The calculation is straightforward. Each baseline contributes one measurement (i.e., one visibility) for each integration time. Thus, the total number of visibilities recorded by one baseline is $\Delta t/t_{int}$ where Δt is the total time spent observing the source. To find the expected number of visibilities, we need only multiply by the number of baselines present in the array. For N antennas, there will be $N(N-1)/2$ baselines, each contributing a visibility every t_{int} seconds. For a total observing time Δt, we will obtain $N(N-1)/2 \times (\Delta t/t_{int})$ visibilities. The expected *rms* noise level *per synthesized beam area* in the image, then, is

$$\sigma(F_v) = \frac{\sqrt{2}k}{\eta_c \eta_A A_{geom}} \frac{T_{sys}}{\sqrt{(1/2)N(N-1)(\Delta t/t_{int})} \; t_{int}\Delta v} = \frac{2k}{\eta_c \eta_A A_{geom}} \frac{T_{sys}}{\sqrt{N(N-1)\Delta t \Delta v}} \quad (6.29)$$

Equation 6.29 assumes that a single polarization component is being measured; the *rms* noise in the image will be $\sqrt{2}$ lower if both polarizations are measured and combined.

In Chapter 4, we showed how the effective area of the antenna combines with the system temperature and the Boltzmann constant to give the telescope sensitivity in terms of the *system equivalent flux density* (SEFD). Using this form of the telescope sensitivity, Equation 6.29 becomes

$$\sigma(F_v) = \frac{SEFD}{\eta_c \sqrt{N(N-1)\Delta t \Delta v}} \quad (6.30)$$

Many different factors can affect a map to lower its quality and result in a higher noise level. Nevertheless, Equation 6.30 is a very useful planning tool because it tells us the lowest noise we can hope to achieve in a given amount of observing time, in the absence of other factors. Although we will not know until after the observation has occurred the precise number of visibilities we will have, we can still use this calculation when *planning* an observation by using the number of visibilities we *expect* to obtain.

Example 6.5:

We want to detect an unresolved source, observing at 22.2 GHz. A previous 22.2-GHz survey with a sensitivity limit of 0.300 Jy did not detect this source. We will use an array of three antennas, all with system temperatures of 50 K, effective apertures of 30.4 m^2, observing a single polarization with a bandwidth of 50 MHz. Assume a correlator efficiency $\eta_c = 0.8$.

1. If we fail to detect the source, we would at least like to halve the upper limit on its flux density. Using 5 times the *rms* noise as a conservative upper limit for a null detection, what is the minimum amount of time that we should observe the source?
2. What is the minimum flux density that will be detected in the cross-correlations for an integration time of 2 min?
3. If we desire a 20σ upper limit, instead of 5σ as in (1), for how much time should we observe?
4. What is the SEFD of the telescope?

Answers:

1. Using Equation 6.29, keeping Δt as a variable and inserting $\sigma(F_\nu) = 0.03$ Jy per synthesized beam (to match the minimum SNR of 5 in detecting a flux density half the current limit) along with all the given parameters, we have

$$0.03 \text{ Jy} = \frac{2\left(1.38 \times 10^{-23} \text{J K}^{-1}\right)}{0.8 \times 30.4 \text{ m}^2} \frac{50 \text{ K}}{\sqrt{3(3-1)\Delta t \ 50.0 \times 10^6 \text{ Hz}}}$$

and so

$$\Delta t = \frac{1}{6\left(5 \times 10^7 \text{ Hz}\right)} \left[\frac{50 \text{ K}}{0.03 \text{ Jy}} \frac{2\left(1.38 \times 10^{-23} \text{J K}^{-1}\right)}{0.8 \times 30.4 \text{ m}^2} \frac{1 \text{ Jy}}{10^{-26} \text{ J s}^{-1} \text{ m}^{-2} \text{ Hz}^{-1}} \right]^2$$

$$= 119 \text{ s}$$

or about 2 min. This is the minimum total observation time, just to meet the detection limit.

2. Using Equation 6.28, we have

$$\left(F_\nu\right)_{\min} = \frac{\sqrt{2}\left(1.38 \times 10^{-23} \text{ J s}\right)}{0.8 \times 30.4 \text{ m}^2} \frac{50 \text{ K}}{\sqrt{120 \text{ s} \left(50.0 \times 10^6 \text{ Hz}\right)}}$$

and so the minimum flux density detected in the cross correlations is

$$\left(F_\nu\right)_{\min} = 5.18 \times 10^{-28} \text{ W m}^{-2} \text{ Hz}^{-1} = 0.0518 \text{ Jy}$$

This is less than one-fifth of the current known flux-density limit of the source.

3. Moving from a 5σ to a 20σ limit means that we should have an *rms* noise that is 4 times lower. The noise is proportional to $1/\sqrt{\Delta t}$, so to achieve 4 times less noise we must observe 16 times longer, or 2 min × 16 = 32 min.

4. Referring to Chapter 4, or comparing Equations 6.29 and 6.30, we see that

$$\text{SEFD} = \frac{2kT_{\text{sys}}}{A_{\text{eff}}} = \frac{2 \times 1.38 \times 10^{-23} \text{J K}^{-1} \times 50 \text{ K}}{30.4 \text{ m}^2} \cdot \frac{1 \text{ Jy}}{10^{-26} \text{J/m}^2} = 4500 \text{ Jy}$$

6.11.3 Brightness Sensitivity

It is worthwhile to consider the effect that angular resolution can have on the sensitivity of an observation. In particular, for a *resolved* source, the required integration time to achieve a particular SNR depends on the synthesized beam area—and can increase dramatically as the beam size decreases. Consider the following situation.

A circular source, 5 arcsec in diameter, of uniform surface brightness, has a flux density of 10 mJy. Following Equation 6.30, and assuming an SEFD of 500 Jy, an array of 27 antennas, observing a single polarization, a correlator efficiency of 0.9, a bandwidth of 1 MHz, and an integration time of 1 h, we will obtain a noise in the image of

$$\sigma\left(F_{\nu}\right) = \frac{500 \text{ Jy}}{0.9\sqrt{1 \times 27 \times 26 \times 10^6 \text{Hz} \times 3600 \text{ s}}} = 0.35 \text{ mJy/beam}$$

If our array produces a circular synthesized beam of diameter 5 arcsec (the same angular size as the source), then the brightness of the source will be 10 mJy/beam and we would expect an SNR of about 10 mJy/beam/0.35 mJy/beam = 29. Suppose that we re-observe the source with longer baselines, to obtain better angular resolution. If the synthesized beam is now 1 arcsec in diameter, any single beam will capture only a fraction of the 10 mJy flux. In particular, one beam will receive a fraction $(\Omega_{\text{syn}}/\Omega_{\text{src}})$ of the flux. These solid angles are proportional to the angular size squared, so in our example a 1 arcsec beam (observing a uniformly bright source) would see an intensity of 10 mJy × $(1/5)^2$ = 0.4 mJy/beam.

A 1-h observation, then, will not even detect the source! Each beam area, presenting 0.4 mJy of flux, will have a noise of about 0.35 mJy, so the SNR will be only 1.2, which is insufficient to claim detection. To obtain the same SNR of 29 that we had in the previous observation, a noise of 0.4/29 or 0.014 mJy/beam would be needed. Repeating our sensitivity calculation above, but solving for the time required to give σ = 0.014 mJy/beam, we obtain an on-source observing time of 820 h. We see, then, that as we begin to resolve a source, the time requirements to observe it can climb enormously. This phenomenon, of not having sufficient sensitivity to detect well resolved, low surface-brightness emission, is often referred to as *resolving out* the source. Note that it is fundamentally different from spatially filtering a source, as we discussed in Section 6.10. In the former case, the proper range of *uv* data is present, but is of insufficient sensitivity. In the latter case, data from the proper *uv* range are not present at all.

6.12 CALIBRATION

In Chapter 4, we explained how the voltages detected by single-dish telescopes are ultimately converted into flux density expressed in units of janskys. Likewise, the correlations, which are the output of the interferometer, must also be converted into flux densities. This is one sort of *calibration*, which is essentially a scaling factor, or units conversion. For interferometers, an additional form of *calibration* is needed. This new form of calibration is independent of units, and rather serves as a correction factor to adjust flawed data.

The *initial calibrated* visibilities are corrupted from their true values for numerous reasons, arising from ionospheric or atmospheric effects and from instrumental variations. Common atmospheric effects, for example, include signal attenuation from clouds passing above one antenna but not another, or radio *seeing* in which tropospheric water vapor causes phase shifts in the data recorded by one antenna but not another. Similarly, one antenna might have a defective electronics module that causes large phase drifts or delay jumps that affect data on all baselines to this antenna, but not on other baselines. The timescales for these effects can be from seconds to hours, depending on the nature of the effect involved. In this type of calibration we correct the measured visibilities to obtain the *true* visibilities.

In this calibration, we must find multiplicative factors for the amplitudes and additive factors for the phases, which get applied to the raw visibilities. We can find these correction factors by observing a source whose visibilities have a known form. If the amplitude or phase of the measured visibilities differs from the expected value, then the correction factor is easily determined. These correction factors contain both amplitudes and phases and so are called *complex gains*.

The ideal source to use for this calibration is a point source located at the phase center of the field: all baselines should then measure the same amplitude (equal to the flux density of the source) and zero phase. In addition to being point like, a calibrator should be strong, so that only a short observation is needed to detect it, and it should be near to the astronomical source on the sky, so that the atmospheric path followed by the electromagnetic waves is the same for both the calibrator and the source.

It would be helpful if the calibrator were stable in flux density, so that it always presented the same visibility amplitudes. Unfortunately, most point sources are variable, and the few that are stable are unlikely to be nearby to the program source on the sky. As a result, most interferometric observations use *two* calibrators: a *primary* or *flux calibrator*, and a *secondary* or *phase calibrator*. The flux calibrator is a source that is relatively stable in flux density (on a timescale of years) but possibly far away on the sky, while the phase calibrator may be variable on much shorter timescales but is nearby on the sky. An observing procedure commonly employed is to first observe the flux calibrator and use this to calibrate the amplitudes of the phase calibrator. The phase calibrator, then, serves to calibrate the visibility amplitudes and phases for the remainder of the observation. We generally intersperse short scans of the phase calibrator throughout the observing period. Figure 6.20 displays the calibration correction measurements in an observation in which the phase calibrator source was observed every 4 min. The idea is that just before and just after observing the program source, we can determine the gains that will correct the

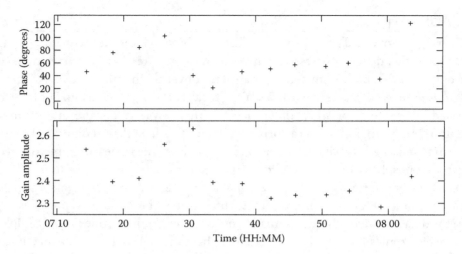

FIGURE 6.20 Calibration solution plots. The calculated correction factors for the phase (top) and amplitude gain (bottom) are obtained from observations of the phase calibrator source. Each point shows the average of 6 integration times, of 2-min duration. Corrections applied to the integrations on the program source, which occur between the points shown, are calculated as interpolations between these points.

data for atmospheric and instrumental effects. At times between the calibration scans, when we are observing the program source, we apply interpolated values of the gains to calibrate the visibility data.

Quasars are commonly used as calibrator sources and the initial flux calibrators are usually well-studied sources whose flux densities are regularly monitored. Planets can also be used as the initial flux calibrator, particularly at millimeter wavelengths.

How often one observes the phase calibrator depends on the observing frequency (more frequent scans at higher frequencies), the baseline length (more frequent scans when observing with longer baselines) and the weather (more frequent scans in poor weather conditions). Put another way, if the instrument or the atmosphere changes too drastically between calibrator scans, the assumption that we can track these changes by linear interpolation will no longer be valid.

Although calibrator scans are essential to observing with an interferometer, the time spent observing calibrators is time *not* available to observe the program source. Thus, we do not want to observe the calibrators for more time than is necessary. We can use the results of Section 6.11 to calculate how much time is needed. The basic requirement is that each baseline should detect the calibration source. A common guideline is that the calibrator scan should be long enough to ensure an SNR of 5 on *each baseline*.

Example 6.6:

A 0.8-Jy source is to be used as a phase calibrator. We wish to observe with correlation integration times of 3 s, but Equation 6.28 predicts that a single 3-s integration of this source will have a noise of 0.5 Jy. How many integrations, and for how long, should we observe the calibrator?

Answer:

To obtain an SNR > 5 the scan must have a noise < 0.8 Jy/5 = 0.16 Jy. This noise level is a factor of 0.5 Jy/0.16 Jy = 3.13 lower than the prediction using a 3-s integration. Therefore, we need an integration that is 3.13^2 = 9.8 times longer. We should succeed using this source as our phase calibrator by observing it for 10 integrations, or 30 s scans. This numerical calculation is equivalent to the solution used in Example 6.5, part 3.

Note that *each* calibrator scan should have *SNR* > 5. For the program source we can combine many scans to improve the SNR, but for calibration each scan alone must meet the SNR requirement.

When observing in spectral-line mode (with many channels at slightly different frequencies), a *bandpass calibration* is also needed. In addition to observations in which spectral information is desired, spectral-line mode is also often used for continuum observations. By channelizing the signals, RFI can be excised by deleting individual channels and bandwidth smearing can be avoided (with large-bandwidth observations) by processing subsets of the channels independently. The bandpass calibration is accomplished by observing a bright continuum source of known spectral shape and then calculating frequency-dependent complex gain factors that will make the observed spectrum agree with the true spectrum. To avoid adding noise to the target spectrum, the bandpass calibration spectrum should have a higher SNR, so a bandpass calibrator much brighter than the target source is desirable. The planning of bandpass calibration scans follows the same guiding principle as the flux and phase calibration—or any other kind of calibration, for that matter. The calibrators must be observed on timescales shorter than the variations that affect the data.

6.13 IMAGE FORMATION

If the observed visibilities completely covered the *uv*-plane, the Fourier transform could be done quickly and easily by a computer program and an image could be created almost at once. In practice, the imaging process is much more involved than this. The main complicating factor is the incomplete *uv*-coverage, which negatively affects the image in several ways.

The first step of the process is converting from visibilities as a function of time for each baseline, $V_{ij}(t)$ (where *ij* represents the baseline between antennas *i* and *j*) to a function of *uv*-space, $V(u,v)$. This is done easily, and often invisibly to the user, by the computer program. The precise locations of all the antennas and the position of the map center are the only information needed, and so a computer calculation of the *uv* position for each data point is straightforward.

Now, we use the computer to Fourier transform the data from the uv-plane to the image plane to form an image of the sky. But a computer does not calculate the Fourier transforms analytically; rather the transforms are calculated by numerical techniques. A better representation of Equation 6.23, as it is implemented in the computer, is

$$I(x, y) = \sum \sum V(u_i, v_j) e^{-i2\pi(u_i x + v_j y)} \Delta u \Delta v \tag{6.31}$$

We call this a *discrete Fourier transform*. One implication of using discrete Fourier transforms is that the visibilities must be regularly spaced in the uv-plane, corresponding to the quantities u_i and v_j. To achieve this, we first define a grid of regularly spaced cells, with dimensions $\Delta u \times \Delta v$. Each cell will meet one of three conditions: if there are no data in the cell, it will be assigned a zero value; if there is one visibility, the cell will assume that value; if there are multiple visibilities, they are averaged and the cell is assigned the averaged value.

The grid size and dimensions are intimately related to the size and dimensions of the image we wish to produce. Because our final goal is an image, we will first discuss how its dimensions and pixel size are chosen, and then show how these parameters relate to the uv gridding.

6.13.1 Image Dimensions and Gridding Parameters

Like all digital images, the map we produce will be an $N_x \times N_y$ array of pixels; the array is often square, with $N_x = N_y$, but this is not a necessary condition. Each pixel will have an angular width and height, which we will call Δx and Δy; again, typically, but not necessarily, square. Thus, the angular dimensions of the map on the sky will be $N_x \Delta x \times N_y \Delta y$. Usually x is chosen to be the axis in right ascension while y is declination, but other coordinate systems can be used, after a suitable regridding of the data.

There are no universal rules for selecting the parameters N_x, N_y, Δx, and Δy. There are only guidelines. In fact, the values selected often depend on what the map is to be used for. Primarily, Δx and Δy determine how finely the sky-plane is sampled while $N_x \Delta x$ and $N_y \Delta y$ determine how large a sky area will be imaged. There is an interaction between the two sets of parameters, of course, through the products $N_x \Delta x$ and $N_y \Delta y$: the same sky area can be imaged with fewer pixels if those pixels are larger.

To make a high-quality image, usually the best procedure is to first establish the pixel size (using criteria that we give below) and then choose values for N_x and N_y to cover the desired sky area with the chosen pixel dimensions. So we begin by examining how pixel size relates to the sampling of the uv data.

The largest uv distances in the visibility data determine the smallest angular sizes that can be imaged. To satisfy the Nyquist sampling theorem, we must sample at least two points within each spatial period, so we should choose pixel sizes so that $\Delta x < \dfrac{1}{2u_{max}}$ and $\Delta y < \dfrac{1}{2v_{max}}$. This is approximately equivalent to requiring 2 pixels across the FWHM of the main lobe of the synthesized beam. In practice, some degree of over-sampling is helpful to the mapping process. Usually pixel sizes that place 3 to 5 pixels across the FWHM of the beam are used. Even smaller pixels (5–10 across the beam) are sometimes used to improve the performance of various image processing procedures (see Section 6.13.4).

Now, since small *uv* distances correspond to large angles on the sky, the sizes of the *uv* cells are determined by the maximum angular distances asked for in the map, that is,

$$\Delta u = \frac{1}{N_x \Delta x} \tag{6.32}$$

and

$$\Delta v = \frac{1}{N_y \Delta y} \tag{6.33}$$

The smallest *uv* distance determines the largest scale structure within the field of view that can be properly reconstructed.

Example 6.7:

In a particular observation we have $u_{max} = v_{max} = 50$ kλ. The antennas of the array have a primary beam FWHM of 9 arcmin. If we desire to just meet the Nyquist sampling rate, with no oversampling, and to image the full primary beam of the antennas, what should the pixel and map sizes be?

Answer:

To have two pixels each in the highest spatial frequency, we need $\Delta x = \Delta y = 1/(2 \times 50,000) = 1/100,000$ radians, or 2 arcsec. To image a 9 arcmin primary beam, we will want $(9 \times 60'') \times 1/(2''$ per pixel$) = 270$ pixels on a side. If we choose to over-sample, we would use somewhat smaller pixels, and then require a larger $N_x \times N_y$ to map the same area.

Although large maps made with small pixels give us the best of both worlds, such maps come at a high price: they require large amounts of storage space on the computer and they are time-consuming both to make and to transfer. Often a *quick look* map can be made with larger pixels to cover the sky area with fewer pixels, thus reducing the computer requirements.

When observing an unknown field, it is often advisable to image the full primary beam of the antennas. If the source has been observed before, so that we know its precise position and that there are no other nearby objects, then a much smaller map may be sufficient. If two (or more) sources are present, but widely separated from one another, it may be preferable to make two small maps, one centered on each source, rather than one huge map, that is mostly devoid of emission.

6.13.2 Dirty Map and Dirty Beam

In the discrete Fourier transform sum (Equation 6.31), the *uv*-cells that contain no visibility data are skipped and so the Fourier transform effectively interpolates over these pixels leading to an imaging error.

The effect of these holes in *uv* coverage, though, can be determined and then removed. Recall our discussion in Chapter 4 when we explained how the observed map in a single-dish observation is the convolution of the sky intensity with the telescope's beam pattern. We mentioned that sometimes the beam is deconvolved from the observed map to reveal a truer image. Essentially, the same phenomenon occurs in aperture synthesis. Apart from the primary beam of each antenna, the interferometer as a whole has a beam pattern on the sky, and what we observe is the convolution of the sky intensity with this beam pattern. This beam pattern is determined by the combination of all the different baselines that go into making the image, and is called the *synthesized beam*.

In Figure 6.21, we show two examples of the *uv*-coverage obtained with the VLA. The example in the bottom row of Figure 6.21 shows somewhat sparse *uv*-coverage, while the example in the top row of Figure 6.21 shows fairly complete *uv*-coverage.

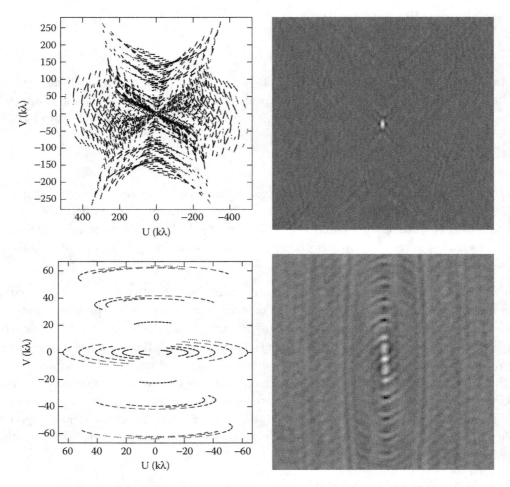

FIGURE 6.21 Plots of the *uv*-coverage (left column) and corresponding dirty beams (right column) from two observations with different *uv* distributions. The top row is an example of good *uv*-coverage, while the bottom demonstrates sparse coverage. The corresponding dirty beams show that sparse *uv*-coverage yields larger dirty beam with more significant sidelobes.

The synthesized beam pattern is determined by the Fourier transform of the points in the *uv*-plane for which visibilities have been measured. One can think of this as a *sampling function,* which has value 1 at *uv* points where data are recorded and value zero elsewhere. It is precisely the image pattern produced when observing a point source at the field center.

By Fourier transforming the points in the *uv*-plane, we obtain a sensitivity pattern (see Figure 6.22). This sensitivity pattern is known as the *dirty beam*. To understand its name, consider that the beam we would *like* to have would be a symmetric two-dimensional elliptical Gaussian pattern with no additional structure. The beam that we *actually get* may have complex additional structures. These structures are called *sidelobes*, in analogy to the sidelobes of a single-dish antenna that we encountered in Chapter 4. Flux density from an astronomical source—even if it is very localized in position—will be scattered throughout the field, according to this sidelobe pattern.

What we actually measure with an interferometer is the *convolution* of the true sky intensity with the sensitivity pattern of the array, that is, with the dirty beam. This observed convolution is called the *dirty map* or *dirty image*. If the sidelobe levels are sufficiently weak (say, at the noise level of the image; see Section 6.11), then the dirty image may be of sufficient quality to meet the science goals of the observation. In this case, the dirty image may be used directly. More often, the sidelobe levels are so strong that they badly distort

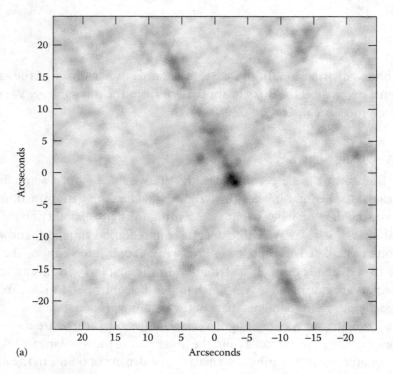

(a)

FIGURE 6.22 (a) A dirty image from an observation with the Janksy VLA. The structure of the dirty beam clearly affects the image. *(Continued)*

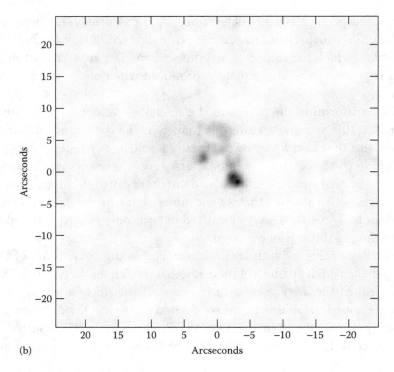

(b)

FIGURE 6.22 (Continued)   (b) The CLEAN image of the same observation, after the dirty beam has been deconvolved.

the sky brightness distribution. In this case, to obtain a usable image of the astronomical source we must *deconvolve* the dirty beam pattern from the dirty image. We discuss this deconvolution procedure in Section 6.13.4.

6.13.3 *uv* Weighting Schemes

The Fourier transform can be performed with data *weighted* so that some data contribute more (or less) to the final image. The weighting factors are determined in two different ways. First, the data may be weighted according to their reliability, usually based on instrumental factors, such as system temperature, integration time, or bandwidth. These weights are often calculated and applied online, without user intervention. The second way to weigh the data depends on their distribution in the *uv*-plane. In this case, the way in which the weights are calculated is very much under the control of the user, and the choice of weighting scheme is a fundamental part of the imaging process.

Most interferometer arrays are designed in such a way that there are more short baselines than long ones. As a result, cells from the inner part of the *uv*-plane will contain more visibilities than other cells. The inhomogeneity in the density of points in the *uv*-plane will affect the form of the synthesized beam, and the weighting schemes that we discuss are different approaches for dealing with the nonuniform *uv*-coverage.

One approach is to let the number of data points per cell determine the weighting. Thus, any cell having multiple data points will contribute more to the final image than cells with a single visibility. This is a very common approach, and is called *natural weighting*. This weighting scheme produces maps with the highest SNR, because all of the data receive the same weight. The scheme has two drawbacks, however. First, because there are usually more data points at small *uv* distances, natural weighting will be biased toward the short baseline data, which has the effect of reducing the resolution. Second, the *uv*-coverage tends to have more and larger holes at larger *uv* distances (see Figure 6.16 in Section 6.10), which is exacerbated by the bias toward shorter baselines, so the dirty beam has a more complicated structure. Despite these weaknesses, natural weighting is a useful weighting choice, and is especially desirable for mapping low flux-density sources, where maximum SNR is required.

Another popular weighting scheme is a clever response to the weaknesses of natural weighting. Natural weighting produces poorer resolution and a more complex dirty beam by allowing the parts of the *uv*-plane with more data to dominate the image. To avoid this bias, we give the same weight to all cells, regardless of the number of visibilities they contain. This is called *uniform weighting*. The result is a simpler dirty beam structure and enhanced resolution. The drawback of this approach is that it produces lower SNRs. A *uv* cell with a single visibility is weighted equally to a *uv* cell with multiple visibilities, which has a higher SNR because of averaging. Moreover, the cells with few visibilities tend to be the longer baseline data, which often have smaller SNRs, and so the lower SNR visibilities are given higher weights. Uniform weighting is especially useful when mapping strong sources, where SNR is not a limiting factor, or when slightly higher angular resolution is needed.

Another *uv* weighting adjustment is to down-weight data points from longer *uv* spacings. This is called *tapering* and is useful to reduce the resolution of the observation (to increase the surface brightness sensitivity or to match the resolution of another observation) or to reduce imaging problems caused by sparse *uv*-coverage from long baselines.

6.13.4 CLEANing the Map: Deconvolving the Dirty Beam

If there were no holes in the *uv*-plane, the dirty beam would have much simpler structure, and would appear as the beam of a filled aperture of very large diameter. Deconvolving this beam from the dirty image would be straightforward. Because of the holes, the sidelobes can be quite strong and have much structure. The most arduous step in the mapping process is to deconvolve this dirty beam. There are different approaches available; here we will discuss only the most popular method, an algorithm called *CLEAN*, that was developed by Jan Högbom in 1974.

The basic idea of CLEAN is, by an iterative process, to slowly subtract the dirty beam pattern associated with each small piece of flux density (treating the map like an array of point sources) until nothing is left in the map but noise. The pieces of subtracted flux density are then added back into the map, but after convolving them with a new beam pattern that reflects the resolution of the observation and is free from sidelobes. There are many steps involved, most of which happen automatically, after the user decides on some input parameters. After a number of iterations, the user checks on the progress and either halts the CLEAN or lets it continue, possibly with some modifications to the previous parameters.

There are various implementations of the CLEAN algorithm. In general terms, most of them perform the following steps:

1. The brightest pixel in the map is determined and some fraction of the flux density in the pixel is inferred to be real emission. The fraction used is set by the user and is called the *gain*. The CLEAN algorithm stores the fractional flux density and location of this point source in a list of *clean components*.
2. The clean component is convolved with the dirty beam and this convolution centered on the location of the clean component is subtracted from the map, yielding a *residual map*. The program then returns to step 1, and searches for the brightest pixel in the residual map.
3. Steps 1 and 2 are repeated until the user decides that the residual map has no real structure left, that is, it looks like a map of random noise.
4. The algorithm calculates a *clean beam*, which is a 2D elliptical Gaussian with no sidelobes. It is calculated by fitting a 2D Gaussian to the central lobe of the dirty beam.
5. All of the clean components are then convolved with the clean beam and added back into the residual map to produce the *clean map*.

There are many parameters that the user can adjust, but the main four are the *gain*, which is typically between 5% and 20%, the *clean boxes*, which are the areas in the dirty map in which real structure should be looked for, *when to pause* CLEAN to check on the progress, and *when to stop* the CLEAN.

In Figure 6.22a and b, we show an example of a dirty map and its clean map.

One of the most common doubts that new CLEAN users have is deciding when to stop the CLEAN. Sometimes this is obvious from visual inspection of the map, but this is not always the case. Moreover, it is not unusual to do a noninteractive CLEAN, in which the user forgoes the option of periodic inspection, but rather letting the program run until some predetermined criterion is met. A common (but arbitrary) approach is to pick some large number of clean components, and stop when that number of peak pixels have been found and subtracted. Because the pixels in an interferometric map can be negative as well as positive, another approach is to monitor the clean components, and when the program starts to select about as many negative components as positive components (so that they tend to cancel one another out, and very little additional flux is CLEANed), then it is time to stop. If the image is noise limited, then a good criterion is to stop the CLEAN when the maximum pixel in the residual map is less than some small multiple (e.g., 3) of the expected noise. Because imaging errors often show themselves as some percentage of the brightest feature of the map, the faintest features in the map that can be reliably imaged may be limited by the contrast with the brightest feature. An image's *dynamic range* is defined as the ratio of the brightest to the faintest reliably mapped features. If the image is dynamic range limited, then the criterion to stop the deconvolution is when the maximum residual is less than some small fraction of the maximum pixel in the original dirty map.

6.13.5 Self-Calibration and Closure Phase and Amplitude

Even after deconvolution, the quality of a map can often be improved substantially. There are many factors that degrade the image besides incomplete *uv*-coverage, such as erroneous data, residual calibration errors, and baseline-based errors. A common problem, for example, is that the troposphere was less stable during the observations than we had planned for, so that the calibration gain factors changed substantially (and probably nonlinearly) while we were observing the target source. A simple linear interpolation would not be adequate to track these gain changes, so the gains we applied were not quite correct. Often, the worst of these effects can be removed through a clever process called *self-calibration*.

Let us first revisit the idea of phase calibration from Section 6.12. The greatest source of error in the visibility phases is variations in the travel time of the radio signals. Whether due to a changing atmosphere, thermal expansion of a cable, differing delays in the digital electronics, or some other instrumental effect, the signal arrival times at the correlator may be variable and require correction on short timescales. As a result, the relative delay on a baseline may not be exactly the same as what was calculated for the fringe function. Consider what effect a small delay error will have on the visibilities. Equations 6.17 and 6.22 are helpful here. If the delay is not exactly equal to that of the theoretical fringe pattern, that is, $\vec{b} \cdot \hat{n}_0 / c$, then the difference will end up in the exponent of the complex visibility, that is, in the visibility phase, ϕ. Let us rewrite Equation 6.17 with an explicit delay error, $\Delta\tau$, to see this more clearly:

$$\Gamma = \iint E^2(\vec{x}, \vec{y}) e^{i2\pi\left[\left(\vec{b}\cdot\hat{n}_0 - c\Delta\tau\right)/\lambda\right]} e^{i2\pi\left[\vec{b}\cdot(\vec{x}+\vec{y})/\lambda\right]} d\vec{x}d\vec{y}$$

$$= e^{i2\pi\left(\vec{b}\cdot\hat{n}_0/\lambda\right)} \iint E^2(\vec{x}, \vec{y}) e^{\left[-i(2\pi c\Delta\tau/\lambda)\right]} e^{i2\pi\left[\vec{b}\cdot(\vec{x}+\vec{y})/\lambda\right]} d\vec{x}d\vec{y}$$

The inferred visibilities, which we get by dividing by the fringe function, will then be

$$V(u, v) = \Gamma e^{-i(2\pi/\lambda)\cdot\left(\vec{b}\cdot\hat{n}_0\right)} = \iint I(x, y) e^{i\left[2\pi(ux+vy)-(c\Delta\tau/\lambda)\right]} dxdy$$

So indeed, the delay errors appear in the exponent and hence are errors in the visibility phases.

Now, these errors are *antenna based*, meaning that a constant shift in phase $\Delta\phi_i$ occurs in all visibilities involving antenna *i* (at any given time). On the *ij*th baseline, then, a delay error for antenna *i* causes a shift in the visibility phase while an erroneous delay for antenna *j* causes a reverse shift in phase. As a result, we can consider each measured visibility phase as being altered by the *difference* in the phase errors of the two antennas, that is

$$\phi_{ij}^{\text{measured}} = \phi_{ij}^{\text{true}} + \Delta\phi_i - \Delta\phi_j \qquad (6.34)$$

where:

$\Delta\phi_i$ and $\Delta\phi_j$ are the phase errors of antennas *i* and *j*, respectively

Fortunately, there exists a neat trick for removing these errors. We consider a triplet of antennas and define the *closure phase* at each integration time as the sum of the visibility phases at that time over all three baselines around the triplet. That is, if Φ_{ijk} is the closure phase around the triplet of antennas i, j, and k, then

$$\Phi_{ijk} = \phi_{ij} + \phi_{jk} + \phi_{ki} \qquad (6.35)$$

Substituting the measured phases of Equation 6.34 into Equation 6.35, we find that closure phase of the *ijk* triplet is

$$\Phi_{ijk}^{\text{measured}} = \phi_{ij}^{\text{true}} + \Delta\phi_i - \Delta\phi_j + \phi_{jk}^{\text{true}} + \Delta\phi_j - \Delta\phi_k + \phi_{kl}^{\text{true}} + \Delta\phi_k - \Delta\phi_l$$

When we add the terms together, we see that the antenna-based errors cancel, that is,

$$\Phi_{ijk}^{\text{measured}} = \phi_{ij}^{\text{true}} + \phi_{jk}^{\text{true}} + \phi_{kl}^{\text{true}}$$

and so the measured closure phases are free of these errors. The trick to removing these errors is to somehow make use of the closure phases. The important point to recognize is that the *closure* phases are more accurate than the individual visibility phases.

The method for improving a map through use of the closure phases is called *hybrid clean*, which was developed by Readhead and Wilkinson in 1981. This is also an iterative process but treats the mapping and self-calibration together as one iteration. To start, an initial clean map is made which will serve as a *model* for the following steps. The model is Fourier transformed back to the *uv*-plane to obtain *model visibilities*. Then the measured visibilities are compared to the model visibilities. We know that the measured visibilities contain phase errors that are not seen in the phase closures. Additionally, the model phases are probably better than the measured errors because the CLEAN process pushes the data to be more consistent with a reasonable map. Next, in a step called *self-calibration*, the measured visibility phases are allowed to vary, to best agree with the model, but preserving the closure phases. For example, in the data at some moment in time, ϕ_{12} might change by $+10°$ and ϕ_{23} by $-6°$ as long as ϕ_{31} also changes by $-4°$, so that Φ_{123} stays constant. If this produces a better fit between the data and model, then this should improve the data. For each time interval (of duration chosen by the user) of the observation, all the phases are allowed to vary with the constraint that all the closure phases are preserved, to produce the minimum *rms* deviation between data and model. The result of the self-calibration is an improved data set. It has been *self-calibrated*. This data set is then used to make a new clean map. The new clean map serves as a new model, which is used in another round of self-calibration to produce an even better data set. This entire process is repeated over and over until the fit between the model and the data no longer shows any improvement.

In Figure 6.23, we show the final clean map, made from several rounds of hybrid cleaning with self-calibration, of the observation shown in Figure 6.22. Note the significant reduction in the noise in the map.

One reason that self-calibration works is that mathematically speaking, the problem is highly over-determined. The ratio of closure phases (constraints) to phases (knowns) is

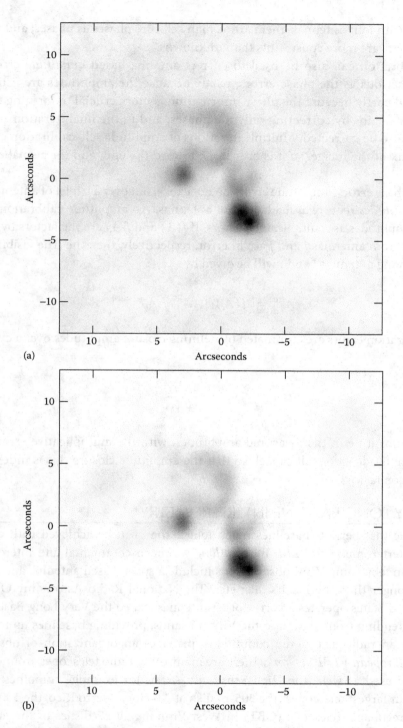

FIGURE 6.23   (a) Original CLEAN image with no self-calibration (same image as Figure 6.22b, but at slightly larger scale) and (b) the same field, after several rounds of self-calibration.

$(N - 2)/2$. With four antennas, there are as many closure phases as phases, and with more antennas there are more constraints than unknowns.

Self-calibration can also be used to correct antenna-based *amplitude* errors. These are not as serious as the phase errors, partly because the amplitudes are usually more reliable, and partly because the phase information is more crucial to forming the image. Usually one begins by correcting only the phases, and in the final iterations the amplitudes may also be corrected. Multiple iterations of amplitude self-calibration can change the total flux of the image, so it is only done toward the end, and for a smaller number of iterations.

Unlike phase errors, amplitude errors do not cancel out over a triplet of antennas; rather, *closure amplitudes* require a quadruplet of antennas. An amplitude calibration alters the visibility amplitudes as multiplicative factors. If $f(A_i)$ and $f(A_j)$ are the factors by which the amplitudes from antennas i and j are in error, respectively, then the true visibility amplitudes involving antennas i and j will be given by

$$A_{ij}^{\text{true}} = \left[f(A_i)f(A_j) \right]^{-\frac{1}{2}} A_{ij}^{\text{measured}}$$

These calibration errors are eliminated by defining closure amplitudes over a cycle of four antennas as

$$A_{ijkl} = \frac{A_{ij} A_{kl}}{A_{ik} A_{jl}} \tag{6.36}$$

When we substitute in the measured amplitudes, with the multiplicative errors, we find that the amplitude errors all cancel, so that the amplitude closure A_{ijkl} is independent of the correction factors.

6.14 VERY LONG BASELINE INTERFEROMETRY*

A technique that permits baselines much longer than can be achieved with connected-element interferometry is *VLBI*. By *very long* we mean continental and intercontinental distances. In fact, some VLBI observations include a space-based antenna, and so contain baselines longer than the Earth's diameter. The National Radio Astronomy Observatory in the United States operates an array of 10 antennas, called the Very Long Baseline Array (VLBA) extending from Hawaii to the Virgin Islands, providing baselines up to 8000 km. The European radio astronomy community manages an organization of observatories, called the *European VLBI Network*. There are many other radio telescopes around the globe that can be used as well. The High-Sensitivity Array, for example, combines the VLBA with the four largest telescopes (the 305-m dish at Arecibo, Puerto Rico, the 100-m Robert C. Byrd Greenbank Telescope (GBT) in West Virginia, all 27 dishes of the Jansky VLA working in unison, the 100-m dish in Effelsberg, Germany, and the 50-m Large Millimeter Telescope [LMT] in Mexico). VLBI observations can involve almost any combination of

VLBI-compatible telescopes, subject only to the condition that the source must be simultaneously observable by at least two telescopes at a time.

These very long baselines produce extremely high-resolution maps, which is what makes VLBI a valuable technique. On the other hand, antennas separated by such great distances cannot be connected by cables, so it is not immediately obvious how their signals can be cross-correlated. We will look at how the correlation is done in just a bit.

6.14.1 VLBI Resolution and Sensitivity

First, let us consider the angular resolution that can be achieved with VLBI. For observations at a wavelength of 1 cm, and baselines on the scale of the Earth's diameter, that is, $\sim 10,000$ km, the angular size of the structure probed is $\lambda/b \sim 10^{-2}$ m$/10^7$ m $\sim 10^{-9}$ radians, or 6×10^{-8} degrees, which is 0.0002 arcsec, or less than a *milliarcsecond*. The resolution of the Hubble Space Telescope, for comparison, is about 50 milliarcsec. The resolution of VLBI is equivalent to being able to read the date on a penny at a distance of *150 miles*.

There is a price to be paid for this high angular resolution, however. Recall from Section 6.11 that interferometer sensitivity is specified as *brightness* (i.e., an intensity). So it is the flux density *per synthesized beam* that determines if we can detect a source. Because the synthesized beam of VLBI observations is so small, many astronomical sources, particularly if they are spatially extended, will not present sufficient flux density per beam to be detected. Thus, VLBI techniques are generally useful only for objects that have significant flux on very small angular scales. Typically, only sources with brightness temperatures of 10^5 K or more can be detected with VLBI observations.

6.14.2 Hardware Considerations for VLBI

Because the individual elements of a VLBI array are not physically connected, the signals captured by each antenna must be stored for cross-correlation at a later time. To do this, each antenna is equipped with an atomic clock and large-volume data storage. The signal detected by each antenna + receiver, along with the time stamps from the atomic clock, is digitally recorded. The data storage medium is physically transported to a processing center where the data are read and synchronized according to the time stamps; these synchronized data are then correlated.

Actually, it is not quite that simple. Although the accuracy of the atomic clocks is very high—about 1 part in 10^{13}—the *synchronization* of the clocks is much worse, and is insufficient to successfully perform the cross-correlations. Without properly synchronized clocks, incorrect delays will be used. As we explained in Section 6.7, the nonzero bandwidth causes the cross-correlation amplitude to decrease, almost to zero, if the wrong delays are used.

The trick that makes VLBI work is to find the real fringes in a process called *fringe searching* or *fringe fitting*. Basically, the atomic clocks serve to synchronize the data rather coarsely, and then the fringe searching process provides a precise synchronization of the signals.

6.14.3 Fringe Searching or Fringe Fitting

The trick is to take advantage of the effect of observing with a range of frequencies, which we presented in Section 6.7 as a complicating factor. Recall that for most delays the cross-correlation was decreased to the point of being undetectable; the solution was to equalize the delays by inserting additional *path length*—either by cables or by digital means. With VLBI, the antennas are not physically connected, so this solution is not possible. Nevertheless, VLBI can still be done, ironically, by turning the *problem* of nonzero bandwidth into a solution. To see the implications of this, study the plot in Figure 6.6 (which shows the effect of the finite bandwidth). This plot shows that, for each baseline, the peak in the cross-correlation occurs when the delay is exactly equal to zero, while for delays far from zero the cross-correlation is below the level of the system noise. Therefore, with VLBI, we can determine the correct relative times of each signal by varying the relative times (which varies the delays), searching for the expected fringes, and then setting the relative times to be when the fringe amplitudes are maximum. This is called *fringe searching*. Essentially, the VLBI correlator calculates the cross-correlation *as a function of delay* and then locates the maximum cross-correlation for each baseline. To be sure that the correlator is finding actual fringes, and not just noise spikes, the signal is fitted to the fringe function (Equation 6.10). Hence, this procedure is also called *fringe fitting*.

QUESTIONS AND PROBLEMS

1. List and explain the three multiplicative factors involved in the resultant cross-correlation of two antennas in interferometry. What is each factor due to?

2. What is the *synthesized beam* and what is the *primary beam*? What shapes do you expect for each, and what do these shapes depend on?

3. What are *uv tracks*? What are they due to? How do they come into play in aperture synthesis data processing, that is, in converting the correlation function to a map of the source's intensity distribution?

4. What characteristic of the observations does the design of the *array* affect? What is the main issue of consideration when one designs the placements of antennas? Suppose that you have four antennas distributed along a line. Suggest a possible placement position for each antenna that will maximize the number of nonredundant baselines.

5. Go to the online materials www.crcpress.com/product/isbn/9781420076769 and download the Mathematica program uvplot.nb (seventh link down the page). This program calculates and displays the *uv* tracks for an array of antennas located at the equator and with no elevation change between antennas. Use this program to design an array that maximizes *uv* coverage and yields a resolution of 1 arcsec when observing at 6 cm. The equations for *u* and *v* in this program are

$$u = \frac{b}{\lambda} \cos\gamma \cos(-\omega_E t) \qquad (6.37)$$

$$v = \frac{b}{\lambda}[\cos\gamma\,\sin\delta\,\sin(-\omega_E t) + \sin\gamma\,\cos\delta] \qquad (6.38)$$

where:

b is the baseline length

γ is the north–south tilt angle of the baseline

ω_E is the angular rotation frequency of the Earth

δ is the declination of the source

t is the hour angle, in seconds, of the source relative to the midpoint of the baseline

You must provide numbers in the first eight lines (dec: declination of source in degrees, lambda: wavelength, nant: number of antennas, nohrs: duration of observation in hours, $x[1], y[1], x[2], y[2]$: east–west and north–south coordinates of the first two antennas), and add a pair of lines for the coordinates of each antenna needed to make the total number: nant. (Note: There are two lines that are too long and continue onto the next line. These lines should not have carriage returns.)

6. Show that by combining Equations 6.37 and 6.38 to eliminate $\omega_e t$ one obtains Equation 6.39 for the uv track which is, indeed, an ellipse.

$$u^2 + \left(\frac{v}{\sin\delta} - \frac{b\sin\gamma\cos\delta}{\lambda\sin\delta}\right)^2 = \left(\frac{b\cos\gamma}{\lambda}\right)^2 \qquad (6.39)$$

7. Consider a visibility function at $\lambda = 1$ cm that is a delta function at the origin with flux density = 500 Jy when the total field of view of the map is 30′. Describe the source responsible in terms of shape and brightness temperature.

8. Outline the data reduction process in aperture synthesis from cross-correlation to a quality clean map.

9. How are the *holes* in the synthesized aperture dealt with?

10. Explain how, in aperture synthesis, the dirty beam of a particular observation is determined.

11. Explain what closure phases are and how they are used to improve aperture synthesis maps.

12. What determines the desired upper limit to the integration time in an aperture synthesis observation? What determines the upper limit to the observing bandwidth? For both, discuss the cause and the general rule-of-thumb calculations.

13. Consider an observation using an array with maximum baseline of 5.00 km and minimum baseline of 10.0 m, at $\lambda = 18.0$ cm, with a desired field of view of 2.00′. What are the limits to the integration time and bandwidth to avoid image degradation?

14. Under what circumstance (i.e., at what point in the sky) does a source have its maximum fringe rate? When does it have its minimum fringe rate?

15. The antennas of your array are 10.0 m in diameter and you are observing at a wavelength of 18.0 cm. The baselines of the array are sufficient to give a synthesized beam of 1.00 arcsec. Choose appropriate pixel and map sizes to image the primary beam of the antennas with slightly better than Nyquist sampling.

16. Show that the number of gridded uv cells is necessarily the same as the number of pixels in the map.

17. Demonstrate that Equation 6.36 is correct; that is, show that the closure amplitude is independent of the calibration gain factors, as we demonstrated for the closure phase of Equation 6.35.

Appendix I

Constants and Conversions

Speed of light $= c = 3.00 \times 10^8$ m s^{-1} $= 3.00 \times 10^{10}$ cm s^{-1} $= 3.00 \times 10^5$ km s^{-1}

Boltzmann constant $= k = 1.38 \times 10^{-23}$ J K^{-1} $= 1.38 \times 10^{-16}$ erg K^{-1}

Planck constant $= h = 6.626 \times 10^{-34}$ J s $= 6.626 \times 10^{-27}$ erg s

Stefan–Boltzmann constant $= \sigma = 5.67 \times 10^{-8}$ J m^{-2}s^{-1}K^{-4} $= 5.67 \times 10^{-5}$ erg cm^{-2}s^{-1}K^{-4}

Universal Gravitation constant $= G = 6.67 \times 10^{-11}$ m^3 kg^{-1}s^{-2} $= 6.67 \times 10^{-8}$ cm^3 g^{-1}s^{-2}

Mass of proton $= m_p = 1.67 \times 10^{-27}$ kg $= 1.67 \times 10^{-24}$ g

Mass of electron $= m_e = 9.11 \times 10^{-31}$ kg $= 9.11 \times 10^{-28}$ g

Charge of electron $= e = 1.60 \times 10^{-19}$ Coul $= 4.80 \times 10^{-10}$ esu

Solar mass $(M_\odot) = 1.99 \times 10^{30}$ kg $= 1.99 \times 10^{33}$ g

Solar luminosity $(L_\odot) = 3.9 \times 10^{26}$ J s^{-1} $= 3.9 \times 10^{26}$ W $= 3.9 \times 10^{33}$ erg s^{-1}

π radians $= 180°$

Electron volt (eV) $= 1.602 \times 10^{-19}$ J $= 1.602 \times 10^{-12}$ ergs

Parsec (pc) $= 3.086 \times 10^{16}$ m $= 3.086 \times 10^{18}$ cm

Light-year (ly) $= 9.461 \times 10^{15}$ m $= 9.461 \times 10^{17}$ cm

Astronomical unit (AU) $= 1.496 \times 10^{11}$ m $= 1.496 \times 10^{13}$ cm

$$1 \text{ year} = 3.16 \times 10^7 \text{ s}$$

$$\text{Jansky (Jy)} = 10^{-26} \text{ J s}^{-1} \text{ Hz}^{-1} \text{ m}^{-2} = 10^{-26} \text{ W Hz}^{-1} \text{ m}^{-2} = 10^{-23} \text{ ergs s}^{-1} \text{ cm}^{-2} \text{ Hz}^{-1}$$

$$\text{Permittivity of free space} = \varepsilon_0 = 8.85 \times 10^{-12} \text{ C}^2 \text{ N}^{-1} \text{ m}^{-2}$$

$$\text{Thomson scattering cross section} = \sigma_T = 6.65 \times 10^{-29} \text{ m}^2 = 6.65 \times 10^{-25} \text{ cm}^2$$

Appendix II

Derivation of Beam Pattern

IN THIS APPENDIX, WE derive the beam pattern for a telescope with a uniformly illuminated one-dimensional aperture. We start by assuming that the telescope's receiver, located at the focus of the reflector, is equally sensitive to incoming radiation from all directions. Thinking of the telescope as a transmitter, this means that the primary reflector is uniformly illuminated. Although such a primary illumination pattern is nearly impossible to produce (see Section 3.1.3), we make this simplification to reduce complexities in our derivation.

Considering just the primary reflector, we can make use of the fact that it is a parabola. Therefore, for any set of parallel light rays, the distances from the front plane of the reflector to the point where the rays meet are equal, as demonstrated in Figure II.1. Therefore, considering the diffraction of the waves, which results because of different path lengths, we can ignore the distance from the front plane to the focus in our derivation. This also enables us to visualize the front plane of the reflector as an *aperture* (or opening) through which the waves travel, so we can model the diffraction, as it would occur on a screen beyond the aperture.

For simplification purposes, we will derive the diffraction in one dimension, in which case we can think of the telescope as a slit, much like the slit used in the introduction of diffraction in freshman physics texts (i.e., a long and narrow rectangle). Imagine, now, a chain of radio waves passing through a long thin slit of width D.

A common way of understanding diffraction is to imagine the slit of width D as a continuum of many little slits, each with an infinitesimal width d, as shown in Figure II.2. The waves that pass through each tiny slit, then, interfere with the waves passing through the other slits.

We have, then, a continuum of waves, each carrying an oscillating electric field vector $\vec{E}$ entering a tiny slit. Let us consider the radiation coming from a specific direction in the sky at an angle θ relative to the straight-on position. The total electric field, $\vec{E}$, that emanates from the whole slit and detected at the focus (where all the waves meet) is the sum,

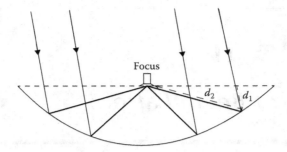

FIGURE II.1 For any ray, the total distance from the front plane to the focus is the same ($=d_1 + d_2$).

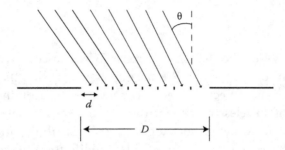

FIGURE II.2 A wave passing through a slit can be viewed as a series of waves passing through a series of infinitesimally thin slits.

$$\vec{E} = \sum_{k=1}^{n} \vec{E}_k$$

where:

  k is the index counter of the tiny slits and there are $n = D/d$ tiny slits across the whole slit.

At the end, we will solve for $\vec{E}$ from the total slit by turning the sum into an integral by letting $nd = D$ as $d \to 0$ and $n \to \infty$. Now, as shown in Figure II.3, the wave chain entering the second slit travels a slightly longer distance than the wave chain entering the first slit. This extra distance causes the phase of the second wave chain to be shifted relative to the phase of the first wave chain. This is the basic principle of interference.

Figure II.4 shows that the extra distance traveled relates to the direction of incoming waves, θ, by

$$\text{extra distance} = d\sin\theta$$

causing a phase difference between the two wave chains when they meet beyond the slits. Since the phase of a wave cycles through a full 2π radians over a distance of one full wavelength, the relative phase difference between the waves passing through the first and second slits is given by

$$\Delta\phi_{12} = \frac{2\pi d \sin\theta}{\lambda} \tag{II.1}$$

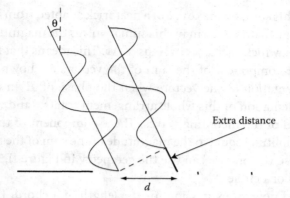

FIGURE II.3 Two wave chains traveling in parallel paths and approaching the slits at an angle θ relative to the vertical travel different distances. One path includes an extra distance of travel.

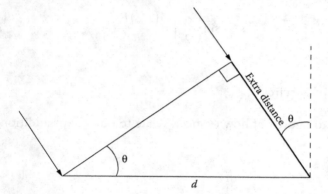

FIGURE II.4 The extra distance traveled by the second wave chain equals $d\sin\theta$.

Likewise the distance between the kth slit and the first slit is $(k - 1)d$, so the phase shift between the waves in the kth slit and the first slit is

$$\Delta\phi_{1k} = \frac{(k-1)2\pi d\sin\theta}{\lambda}$$

Now, we want the expression for $\vec{E}_{\text{total}}$, which is the vector sum of all the $\vec{E}_k$s. To set a zero point for the phases, we can choose some time, which we call $t = 0$, when the phase, ϕ, of the wave chain in the first slit $= 0$. We have, then,

$$\vec{E}_{\text{total}} = \vec{E}_1 + \vec{E}_2 + \cdots + \vec{E}_k + \cdots$$

$$= \vec{E}_0 + \vec{E}_0 \cos\left[\frac{2\pi d\sin\theta}{\lambda}\right] + \cdots + \vec{E}_0 \cos\left[(k-1)\frac{2\pi d\sin\theta}{\lambda}\right] + \cdots$$

$$= \vec{E}_0 \sum_{k=1}^{n} \vec{E}_0 \cos\left[(k-1)\frac{2\pi d\sin\theta}{\lambda}\right]$$

Now, to accomplish this sum, we make use of a neat trick adapted from Feynman, Leighton, and Sands (1963).[*] Note that each term in this sum involves the magnitude of a vector times the cosine of an angle, which increases in steps of $\Delta\phi$. This means that this sum is the same as determining the x-component of the sum of the vectors as shown in Figure II.5. This figure shows the orientation of the vectors when the phase of $\vec{E}$ in the first slit is 0. As time changes, the orientation of this whole arrangement rotates, and will cycle all the way around in one period of the incoming waves. The x-component of the vector sum, then, will oscillate with amplitude equal to the magnitude of the sum of these vectors. Therefore, to get the sum of these vectors, we can use the geometry in Figure II.5. The vector sum is, essentially, the chord of a circle.

Therefore, we need now an expression for the length of a chord. Let the length of the chord be E_c, and let the angle that it subtends at the center of the circle be α, as shown in Figure II.6. Then, by trigonometry,

$$\sin\left(\frac{\alpha}{2}\right) = \frac{(1/2)E_c}{R} \qquad (\text{II.2})$$

where:

R is the radius of the circle

So, we now need to figure out how α and R relate to our vector addition for the $\vec{E}_k$s from the slit.

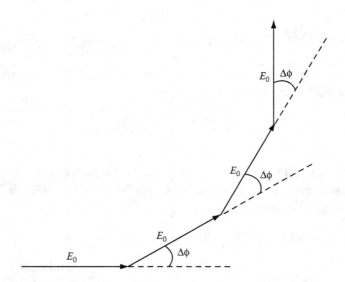

FIGURE II.5 The total electric field of a series of electric field vectors with phase differences of $\Delta\phi$ is identical to summing vectors with directions shifted by $\Delta\phi$.

[*] Feynman, Leighton, and Sands (1963). *The Feynman Lectures on Physics*, Vol. I. Menlo Park, CA: Addison-Wesley.

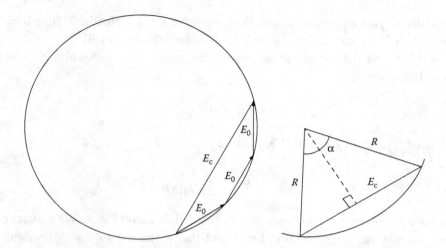

FIGURE II.6   The sum of the electric field vectors is a chord of length E_c, which subtends an angle α in a circle of radius R, where $\sin\alpha = 1/2(E_c/R)$.

We will find an expression for α first. Since the chords are perpendicular to the radial lines which bisect them, the angle between successive radii is equal to the angle between successive chords, which, as was shown in Figure II.5 above, equals $\Delta\phi$, the shift in phase between successive $\vec{E}_k$s. Therefore, as shown in Figure II.7, the angle between each pair of successive chords is also equal to the angle $\Delta\phi$.

Now, since there are n vectors to add and the angle between each vector is given in Equation II.1 as $2\pi \, (d\sin\theta/\lambda)$, the total angle subtended by all the $\vec{E}_k$ vectors is

$$\alpha = n\Delta\phi_{12} = n\frac{2\pi d\sin\theta}{\lambda} \tag{II.3}$$

We can use Equation II.3 to substitute in for α in Equation II.2. We have, then,

$$\sin\left(\frac{n\pi d\sin\theta}{\lambda}\right) = \frac{(1/2)E_c}{R} \tag{II.4}$$

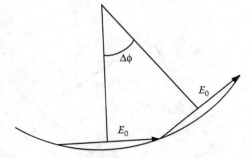

FIGURE II.7   The angle subtended at the center of the circle by consecutive vectors, in the sum of electric field vectors, equals $\Delta\phi$, the phase shift between $\vec{E}_k$s passing through contiguous slits.

We will return to this equation shortly. First, we solve for R. Consider Figure II.8. In general, the angle α in radians is defined by α (radians) $= s/R$, or $s = \alpha R$.

We can determine the arc length because in the limit where $d \to 0$ and $n \to \infty$, the arc length $s = nE_0$. Therefore,

$$nE_0 = \alpha R \tag{II.5}$$

By combining Equations II.3 and II.5, we have

$$R = \frac{n}{\alpha} E_0 = \frac{\lambda n E_0}{n 2\pi d \sin\theta} \tag{II.6}$$

We now have two expressions (Equations II.4 and II.6) involving R. We can manipulate Equation II.4 to yield an expression for R and then equate it to the right-hand side of Equation II.6 (eliminating the variable R). This gives us

$$E_c = nE_0 \frac{\sin\left(n\pi d \sin\theta/\lambda\right)}{\left(n\pi d \sin\theta/\lambda\right)} \tag{II.7}$$

In the limit, where $d \to 0$ and $n \to \infty$, we have $nd = D$. Also, nE_0 is the total possible amplitude when there is no interference, that is, nE_0 is the maximum signal, which occurs for waves entering with $\theta = 0$ (i.e., straight-on to the antenna). Rewriting Equation II.7 with these two changes, we have

$$\vec{E}_{\text{Total}} = \vec{E}_{\max} \frac{\sin\left(\pi D \sin\theta/\lambda\right)}{\left(\pi D \sin\theta/\lambda\right)} \tag{II.8}$$

This is known as a *sinc* function (where $\mathrm{sinc}\,\theta = \sin\theta/\theta$). In the limit, where $\theta \to 0$, $\mathrm{sinc}\,\theta \to 1$.

The telescope, though, detects the power of the radiation (as we discussed in Chapter 3), which is proportional to the square of the electric field, $P \propto E^2$. Therefore, the telescope's sensitivity is proportional to $\mathrm{sinc}^2(\pi D \sin\theta/\lambda)$, which has a maximum of 1 at $\theta = 0$ (the direction that the telescope is pointed). We find then the normalized beam pattern for a uniformly illuminated one-dimensional aperture is given by

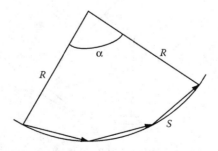

FIGURE II.8 The total angle subtended, α, in radians, is the total arc length, s, divided by R.

$$P_{\text{bm}}(\theta) = \frac{\sin^2(\pi D \sin\theta/\lambda)}{(\pi D \sin\theta/\lambda)^2} \tag{II.9}$$

This same beam pattern is derived in Section 3.1.5 using a Fourier relation.

Graphically, this looks something like that shown in Figure II.9.

Note that, by Equation II.9, $P_{\text{bm}} = 0$ when $\sin(\pi D \sin\theta/\lambda) = 0$, or when $(\pi D \sin\theta/\lambda) = m\pi$, where m is an integer. This occurs when $\sin\theta = m\lambda/D$. The first null, therefore, occurs where $\sin\theta = \lambda/D$. For small angles, $\sin\theta \approx \theta$, so the first null occurs at

$$\theta_{\text{first null}} = \frac{\lambda}{D}$$

The angular distance between first nulls (on either side of the midpoint), then, is

$$\theta_{\text{between first nulls}} = 2\left(\frac{\lambda}{D}\right)$$

The *width* of the main beam is generally defined as the distance between the points half-way up the beam (full width at half maximum, or *FWHM*). In Equation II.9 and Figure II.9, the *FWHM* is

$$\theta_{\text{FWHM}} = 0.886\left(\frac{\lambda}{D}\right)$$

Now, for a uniformly illuminated circular, two-dimensional, aperture, the beam pattern turns out to involve a Bessel function

$$P_{\text{bm}}(\theta) = \frac{\left[J_1(\pi D \sin\theta/\lambda)\right]^2}{(\pi D \sin\theta/\lambda)^2} \tag{II.10}$$

where:

J_1 is the first order Bessel function of the first kind

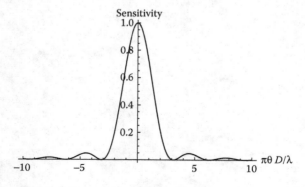

FIGURE II.9   The sensitivity function given in Equation II.9.

This function looks very similar to the ordinary sinc function of Equation II.9 at small angles, except that it is slightly wider. The *FWHM* of the beam of a circular telescope with 100% sensitivity to all parts of the dish is

$$\theta_{\text{FWHM}} = 1.02 \left(\frac{\lambda}{D} \right) \qquad \text{(II.11)}$$

Appendix III

Cross-Correlations

A CROSS-CORRELATION FUNCTION IS A measure of how similar two functions are to each other *as a function of a shift* in the independent variable in one of the functions. Cross-correlations commonly occur when the independent variable is time, so for the sake of discussion, we will consider two functions of time.

The mathematical definition of a cross-correlation, between two functions of time, $f(t)$ and $g(t)$, is given by

$$(f \times g)(\tau) = \int_{-\infty}^{\infty} f(t)g(t-\tau)dt \qquad \text{(III.1)}$$

where:

τ is a variable shift in time between the two functions

By this definition, the cross-correlation yields a function of the time shift τ, called *delay*. Note that Equation III.1 is similar to the autocorrelation in Section 3.5.2 and the convolution in Section 4.5.1.

Let us consider a concrete example. Imagine a short signal in time, called a *pulse*, which has a Gaussian profile, such that the signal as a function of time is given by

$$f(t) = A\exp\left[-\left(\frac{t}{w}\right)^2\right] \qquad \text{(III.2)}$$

where:

A is the amplitude, or peak, of the pulse

w is a measure of the width of the pulse in time (to be precise, w is the half-width at a height $1/e$ times the peak)

Figure III.1 shows a plot of this function with $w = 1$ and $A = 1$.

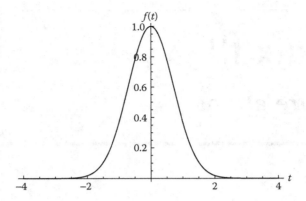

FIGURE III.1   The function, $f(t)$, given in Equation III.2 with $w = 1$ and $A = 1$.

Now, let us consider detecting this signal with two different receivers in which one is slightly further away. We represent this mathematically by accounting for an extra time, Δt, for the signal to reach the second receiver. We let f_1 represent the function of the signal that is delayed. We then need to *add* this extra time to f_2 in order to correctly represent the fact that this signal peaks at an *earlier* value of t. Therefore, the signal arriving at the second receiver is given by

$$f_2(t) = A\exp\left[-\left(\frac{t+\Delta t}{w}\right)^2\right]$$

Now we calculate the cross-correlation, which we will denote with the symbol Γ, by plugging in the expressions for our two functions in Equation III.1. We get

$$\Gamma(\tau) = \int_{-\infty}^{\infty} A\exp\left[-\left(\frac{t}{w}\right)^2\right] A\exp\left[-\left(\frac{t+\Delta t-\tau}{w}\right)^2\right] dt$$

$$= A^2 \exp\left(-\frac{(\Delta t-\tau)^2}{w^2}\right) \int_{-\infty}^{\infty} \exp\left[\frac{2t(\Delta t-\tau)-2t^2}{w^2}\right] dt$$

Substituting in the solution of this definite integral, we find that the cross-correlation is

$$\Gamma(\tau) = 1.25wA^2 \exp\left[-\frac{(\Delta t-\tau)^2}{2w^2}\right] \tag{III.3}$$

Equation III.3 is a general expression for the cross-correlation of the Gaussian-shaped pulse defined by Equation III.2; this expression is helpful for the discussion in Chapter 6.

A plot of this function is shown in Figure III.2. The first thing to notice is that the cross-correlation peaks when $\tau = \Delta t$, that is, when the earlier signal is delayed so that when the signals are combined, they arrive at the same time, as we should expect. But, when τ is far from Δt, then the cross-correlation is very small. For example, when $\tau = \Delta t + 4w$, there is essentially no overlap between the arrival of the two signals, so the cross-correlation is insignificant.

Note also that as a function of delay, the cross-correlation is also Gaussian shaped, but with a width that is $\sqrt{2}$ times larger than the width of the input signal, $f(t)$. This is discussed in Chapter 4 when we convolve a Gaussian shaped intensity distribution with a Gaussian beam, leading to Equation 4.28.

In Figure III.3, we show our two Gaussian functions with a delay set at $2w$. With this delay, the overlap area is relatively small. Remember that the cross-correlation is the integral of the product of the two functions and therefore is related to the area of the overlap as illustrated in Figure III.3.

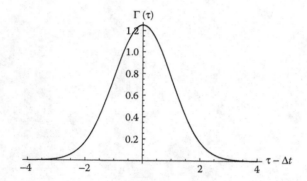

FIGURE III.2   The cross-correlation for a Gaussian-shaped pulse, as given in Equation III.2, detected by two receivers as a function of the time delay, τ, to the second receiver. In this example, we assumed that $A = 1.0$ and $w = 1.0$.

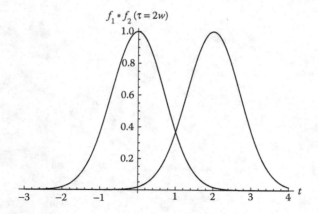

FIGURE III.3   The cross-correlation of functions f_1 and f_2 is related to the overlap area when the two functions are displayed on the same axes. In this figure, $\tau = 2w$ and $w = 1$.

Appendix IV

Complex-Exponential Form of Wave Functions

A N INTRODUCTORY DISCUSSION OF waves usually involves a cosine or a sine function, or a combination of both, that is,

$$y = A\cos(kx - \omega t + \phi)$$

(IV.1)

or

$$y = A\sin(kx - \omega t + \phi)$$

(IV.2)

or

$$y = A\cos(kx - \omega t) + B\sin(kx - \omega t + \phi)$$

(IV.3)

These are all considered *wave functions* because they meet all the criteria needed for a mathematical description of waves. What are these criteria? The physical conditions that define a wave are provided by what physicists call the *wave equation*. This is a partial differential equation that describes the relation between how the wave changes with time (i.e., its oscillation) and how it changes with distance (i.e., its shape). The wave equation in two dimensions is given by

$$\frac{\partial^2 y}{\partial x^2} = \frac{1}{c^2}\frac{\partial^2 y}{\partial t^2}$$

where:
  c is the speed of the wave

This equation describes a function that has a repeating pattern in time as well as in distance and the variations of these patterns are related. The speed, c, is what connects these patterns; as the wave moves in the $+x$ direction, the y-motion at any fixed x (as a function of time) obeys simple harmonic motion, and this results because of the passing of a wave with a simple cosine-shape in the *x*-direction.

The requirement for the mathematical form of the wave function, $y(x,t)$, is that it must satisfy the wave equation, meaning that when you substitute the $y(x,t)$ function for y, above, and calculate the partial derivatives, the two sides of the equation are then found to be equal. Additionally, because the equation involves second derivatives, the connection between the equation and the function involves two steps of integration, which introduces two constants of integration (which are generally determined by the *initial conditions*, such as the initial position and initial velocity). The solution, then, must also contain two free parameters, which depend only on the initial conditions. The free parameters in the cosine and sine functions are the amplitude, A, and the phase ϕ. With Equation IV.3, the free parameters are A and B. We do not need ϕ in this case, and in fact, we are not allowed to have a third free parameter, or the equation is not solved. The k ($=2\pi/\lambda$) and ω ($=2\pi\nu$) are not free parameters, because they are constrained by the relation $\omega/k = c$, which is implicitly given in the wave equation.

The cosine and sine functions (Equations IV.1 and IV.2), independently, are both good solutions because they do indeed fit in the wave equation. In addition, the combination of cosine and sine also works. There is, though, another mathematical function that also works, which is the complex exponential form, that is,

$$y = Ae^{\pm i(kx - \omega t + \phi)}$$

(The exponent can be positive or negative.) Let us plug this function into the wave equation to see that it works. (If you are not familiar with partial derivatives, just trust us here and skim through the rest of this paragraph.) Using the positive exponent function, for demonstration purposes, we have

$$\frac{\partial^2 y}{\partial x^2} = \frac{\partial}{\partial x}\left(ikAe^{i(kx-\omega t+\phi)}\right) = -k^2 Ae^{i(kx-\omega t+\phi)}$$

and

$$\frac{\partial^2 y}{\partial t^2} = \frac{\partial}{\partial t}\left(-i\omega Ae^{i(kx-\omega t+\phi)}\right) = -\omega^2 Ae^{i(kx-\omega t+\phi)}$$

Then,

$$\frac{\partial^2 y}{\partial x^2} = \frac{k^2}{\omega^2}\frac{\partial^2 y}{\partial t^2}$$

Since $k/\omega = 1/c$, then

$$\frac{\partial^2 y}{\partial x^2} = \frac{1}{c^2}\frac{\partial^2 y}{\partial t^2}$$

and the wave equation is satisfied. This function also has two free parameters, A and ϕ. This function, therefore, *must* be a legitimate mathematical expression of waves.

In fact, this function is very closely related to the cosine and sine functions. A famous equation, known as *Euler's formula* (which is usually derived in a complex numbers math class) is

$$Ae^{\pm i\phi} = A\cos\phi \pm iA\sin\phi$$

Therefore, one can also write the complex exponential form in a slightly more familiar form as

$$y = Ae^{\pm i(kx-\omega t+\phi)} = A\cos(kx-\omega t+\phi) \pm iA\sin(kx-\omega t+\phi)$$

The complex exponential form, therefore, is actually the same thing as the wave function that includes both the cosine and sine functions, where B, the parameter in front of the sine, is set equal to iA, and a phase constant is added. (We need the ϕ, now, since B is set by A and therefore is not a free parameter.)

As shown in Euler's formula, the complex exponential function can always be broken up into a sum of its real and imaginary parts, where $A\cos\phi$ is the real part and $\pm A\sin\phi$ is the imaginary part. In addition, if you make a graph in which the x-axis represents real numbers, and the y-axis represents imaginary numbers, as shown in Figure IV.1, then Euler's formula shows that the complex exponential function can be viewed as describing a vector of magnitude A that makes an angle ϕ with respect to the real-numbers axis. If you calculate the components of this vector, what do you get? Answer: Exactly what you have in Euler's formula—the component along the horizontal axis is $A\cos\phi$ and the component along the vertical axis is $iA\sin\phi$. With this vector image in mind, the A in front of the exponential is called the *amplitude* and the argument of the exponent, not including the i, is called the *phase angle*.

The two-dimensional system defined by the graph in Figure IV.1 is called the *complex plane*. You can always consider the mathematics that you have come to know using just

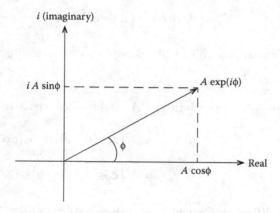

FIGURE IV.1   Graphical representation of a complex number in the *complex plane* in which the real part is plotted on the x-axis and the imaginary part is plotted on the y-axis.

real numbers as the same math, which occurs in the complex plane, but is done solely on the *real axis*.

Specific to our case here, consider how much easier the complex exponential form is to work with than cosines and sines. The product of two exponentials is simply the exponential of the sum of the exponents, that is,

$$e^A e^B = e^{A+B}$$

while the product of two cosines is more complex,

$$\cos(A)\cos(B) = 1/2[\cos(A+B) + \cos(A-B)]$$

Note the complexity of working with cosines in the derivation of Equation 5.5, in Section 5.3.1, for example. Also, the derivatives are much easier with exponentials. The derivative of an exponential is still the same exponential, but multiplied by the derivative of the exponent (following the chain rule), whereas the derivative of a cosine turns it into a sine. Therefore, hopefully, you can see that if you get used to the complex exponential, the math does, indeed, get simpler.

There is another subtle but important issue that we need to introduce here. In a cross-correlation, we multiply two functions of time. If these functions are represented as complex exponentials and we multiply them as if they were real numbers, we can get strange results. Due to the presence of i in complex numbers, there are some arithmetic pitfalls to learn about and avoid. For example, consider what happens when you square a complex number, which we will call X. Let us write it in the vector-component form in the complex plane, that is,

$$X = A\cos\phi + iA\sin\phi$$

and then calculate X^2. Since X is a vector, you might expect X^2 to equal the sum of the square of the components, that is,

$$X^2 = (A\cos\phi)^2 + (A\sin\phi)^2 = A^2$$

But, if we used the näive approach of just taking two copies of the expression for X and multiplying, we get

$$(A\cos\phi + iA\sin\phi)(A\cos\phi + iA\sin\phi) = A^2\cos^2\phi + 2iA^2\cos\phi\sin\phi + i^2A^2\sin^2\phi$$

$$= A^2\left(\cos^2\phi + 2i\cos\phi\sin\phi - \sin^2\phi\right)$$

$$= A^2\left(\cos^2\phi - \sin^2\phi\right) + i2A^2\cos\phi\sin\phi$$

Well, this certainly is different. Consider if ϕ is between $\pi/4$ and $\pi/2$. Then, even though both the real and imaginary parts of X are positive, the real part of our näive solution for X^2 is negative! The reason for the difference is the i^2. If we actually want to get an answer of A^2, we can compensate for this -1 by defining another arithmetic process in which the imaginary part of one of the numbers changes sign. If we did this with X, then, we get

$$\left(A\cos\phi+iA\sin\phi\right)\left(A\cos\phi-iA\sin\phi\right) = A^2\cos^2\phi-i^2A^2\sin^2\phi$$

$$= A^2\cos^2\phi+A^2\sin^2\phi = A^2$$

And that looks like what we want. This, now, is the same as finding the square of a vector. The complex number in which the sign of the imaginary part is switched is called the *complex conjugate* of the original number, and is often symbolized with an asterisk superscript, for example, X^*.

When using the complex exponential form, then, when multiplying two numbers, or two functions, one often really wants to multiply one number by the complex conjugate of the other. The correct expression to obtain the amplitude of a complex number, for example, is not to simply square it but to multiply it by the complex conjugate, that is,

$$XX^* = A^2$$

This also applies, in our case, to the cross-correlation in aperture synthesis (see Chapter 6).

Example IV.1:

1. Write the number $5\exp(-i\pi/3)$ in terms of the real and imaginary parts.
2. Write the number $2 - i3$ in the complex exponential form.
3. What is the complex conjugate of $-32 - i16$?
4. Multiply the number $-20\exp(i\pi/4)$ by its complex conjugate.
5. Calculate the square of the amplitude of the number $-3 - i4$.

Answers:

1. Using Euler's formula, we have

$$5\exp(-i\pi/3) = 5\cos(\pi/3) - i5\sin(\pi/3) = 2.5 - i\,4.33$$

   Note that we get the same answer if we absorb the negative sign in the exponent into the phase angle, that is, $5\exp(-i\pi/3) = 5\exp[i(-\pi/3)]$.
   Then we have $5\exp[i(-\pi/3)] = 5\cos(-\pi/3) + i5\sin(-\pi/3) = 2.5 - i\,4.33$.

2. In order to convert to the complex exponential form, we need to calculate the amplitude and the phase angle, as depicted in Figure IV.1. The amplitude is simply the quadrature sum of the components, that is,

$$A = \sqrt{2^2 + 3^2} = 3.61$$

The phase angle is given by

$$\phi = \tan^{-1}\left(\frac{-3}{2}\right)$$

There is an ambiguity if we just plug in the numbers and use the calculator. We need to determine which quadrant this angle is in. The real part is positive and the imaginary part is negative so this corresponds to the quadrant below the positive real axis (i.e., the fourth quadrant). Therefore, we can express θ either as a negative angle or as a positive angle between $3\pi/2$ and 2π. The former choice is more compact, so we have

$$\theta = -\tan^{-1}(1.5) = -0.983 \text{ radians} = -0.313\pi \text{ radians}$$

The final answer to the question, then, is

$$2 - i3 = 3.71e^{-i0.313\pi}$$

Note: Following the comment in (1), we could just as easily have chosen to leave the negative sign for the exponent, and calculated the phase angle by using a positive value for the imaginary part.

3. The complex conjugate is given by changing the sign of the exponent, which is the same as changing the sign of the imaginary part. Therefore, the complex conjugate of $-32 - i16$ is $-32 + i16$.

4. Multiplying this number by its complex conjugate yields

$$[-20\exp(i\pi/4)]\,[-20\exp(-i\pi/4)] = 400$$

5. To get the square of the amplitude, we can multiply the number by its complex conjugate, which is

$$(-3 - i4)(-3 + i4) = 9 + 16 = 25$$

We could, also, obtain the square of the amplitude by using the vector concept of the number, that is,

$$(-3)^2 + (-4)^2 = 25$$

Appendix V

Primer on Fourier Transforms, with Focus on Use in Aperture Synthesis

I N ADDITION TO THE following discussion, you may find the interactive computer application for experimenting with Fourier transforms, called *Tool for Interactive Fourier Transforms* (TIFT) to be helpful. This java package is available for free download and easy installation at www.crcpress.com/product/isbn/9781420076769. Download TIFT.zip, extract all files, and then find and click on TIFT.jar. Click on Full Complex TIFT and click Instructions.

The Fourier transform is an important and a commonly used process that occurs in many situations in nature. Although you may not realize it, you are already quite experienced with Fourier transforms, even if you have not seen them discussed mathematically. A classic example of a Fourier transform is converting a time-varying signal measured as a function of time into a spectrum, that is, an amplitude as a function of frequency, and vice versa. When a sound wave entering your ear causes your eardrum to oscillate very rapidly, your brain quickly interprets that as a high pitch, or a high-frequency sound. You know naturally the frequencies present in the sound waves, even though your brain receives the signals in the time domain. The Fourier transform is a mathematical tool that permits decomposing the time-dependent signal into its various frequency components, a task that we perform every day when we listen to sounds.

V.1 MATHEMATICAL DEFINITION

Consider a function of time, $f(t)$. Its Fourier transform is performed by multiplying the function by a complex exponential containing 2π, t, and the inverse of t, which is frequency, v, that is, $e^{-i2\pi vt}$, and then integrating over all t.

$$F(v) = \int_{-\infty}^{\infty} f(t)e^{-i2\pi vt} dt \qquad (V.1)$$

The function $F(\nu)$ represents the spectrum of the signal $f(t)$. Similarly, $f(t)$ can be obtained from $F(\nu)$ by performing an inverse Fourier transform, which is the same process except that the positive exponent is used, that is,

$$f(t) = \int_{-\infty}^{\infty} F(\nu)e^{i2\pi\nu t}\,d\nu \qquad (V.2)$$

V.2 TRANSFORM OF A SIMPLE COSINE FUNCTION

Consider a cosine function of time of a single frequency, that is,

$$f(t) = A\cos(2\pi\nu_0 t) \qquad (V.3)$$

where:

A is the amplitude

ν_0 is the frequency, and we have chosen a zero phase, $\phi = 0$

The Fourier transform of this function, given by Equation V.2 converts the function of time, $f(t)$, to $F(\nu)$, a function of inverse time, that is, frequency. Using the expression for $f(t)$ from Equation V.3 in Equation V.1 we have

$$F(\nu) = \int_{-\infty}^{\infty} A\cos(2\pi\nu_0 t)e^{-i2\pi\nu t}\,dt$$

$$= \int_{-\infty}^{\infty} A\cos(2\pi\nu_0 t)\big[\cos(2\pi\nu\, t) - i\sin(2\pi\nu\, t)\big]\,dt \qquad (V.4)$$

$$F(\nu) = \int_{-\infty}^{\infty} A\cos(2\pi\nu_0 t)\cos(2\pi\nu\, t)\,dt - i\int_{-\infty}^{\infty} A\cos(2\pi\nu_0 t)\sin(2\pi\nu\, t)\,dt$$

Focus, first, on the last integral. The product of a cosine function and sine function will be negative half the time and positive half the time. Therefore, when integrated over all time, the total will be zero. We can, then, cross out the second integral.

Now, consider the first integral. Its value depends on the relation between ν and ν_0. If ν and ν_0 have different values, then the integrand is the product of two cosine functions of different frequencies. In this case, there will, again, be an equal amount of time that the product is negative as when it is positive, and so the integration over all time will again be zero. But, when $\nu = \nu_0$ then we have the same cosine function multiplied by itself, which will always be positive. This will also work if $\nu = -\nu_0$ (since $\cos(-\theta)=\cos(\theta)$). We see, then, that $F(\nu)$ is zero for all values of ν except at $\nu = \pm\nu_0$. The Fourier transform of our single-frequency cosine function versus time, then, is a nonzero function only at $\nu = +\nu_0$ and $-\nu_0$. The negative frequency, again, is equivalent to the positive frequency, so these two nonzero points of $F(v)$ are, essentially, the same function, and therefore we find that $F(v)$ contains one frequency. Of course, that is what we set up.

To solve for the magnitude of $F(v)$ at $v = \pm v_0$, we can substitute in $v = v_0$ and continue with the integration. We have, then,

$$F(v_0) = A \int_{-\infty}^{\infty} \cos^2 (2\pi v_0 t) dt$$

Since the integrand is non-negative for all times, this integral diverges. This really just means that our $F(v)$ is a *delta function*—zero everywhere except at a single point, where it is infinite. In practice, we do not get infinite values in our spectral measurements. The reason for our calculated infinite value at $v = v_0$ involves the fact that this purely theoretical example is not really possible because there can be no such thing as a single frequency wave train that has existed from the beginning of time, $t = -\infty$, and continues until $t = +\infty$. The main point to grasp here is that the spectrum of this simple cosine function, which represents a single wave chain, is a pair of narrow spikes at $v = \pm v_0$. A narrow spike is represented mathematically by $\delta(v - v_0)$. The integral in Equation V.6 yields delta functions at $\pm v_0$, and this is multiplied by A so that our entire function, $F(v)$, is given by

$$F(v) = A\delta(v \pm v_0)$$

This means that the spectrum shows a signal only at the frequency of v_0. This result is useful to remember, so we will restate it for emphasis.

The Fourier transform of a simple, single-frequency wave v_0 is a pair of delta functions located at $v = \pm v_0$.

Let us now address the issue of obtaining both positive and negative frequencies in the solution. It is tempting, but would be incorrect, to simply dismiss the negative frequency part, since it is essentially the same function as the positive frequency part. If we were to ignore this part of the solution, though, then the inverse Fourier transform would not work. Partly for demonstration, but also because it leads to a fundamentally important concept useful in aperture synthesis, let us follow through on this exercise. Consider a delta function at only $v = +v_0$ and take the inverse Fourier transform. We represent the delta function at v_0 by $F(v) = A\delta(v - v_0)$. The inverse Fourier transform is then

$$f(t) = \int_{-\infty}^{\infty} A\delta(v - v_0) e^{i2\pi v t} dt$$

By the definition of the delta function, this integrates to

$$f(t) = Ae^{i2\pi v_0 t} \qquad\qquad\qquad (V.5)$$

which is the same thing as

$$f(t) = A\left[\cos(2\pi v_0 t) + i\sin(2\pi v_0 t)\right]$$

and so we see that the inverse Fourier transform of a single delta function has both a real part and an imaginary part, which is not what we started with (in Equation V.3). But, if we include the negative frequency delta function, then in the integration, we get two identical imaginary terms but of opposite sign that cancel, so we do recover Equation V.3.

V.3 FOURIER TRANSFORM AND INVERSE TRANSFORM

To demonstrate another important aspect of Fourier transforms, let us calculate the inverse Fourier transform of our simple cosine function (Equation V.3). To distinguish the Fourier transform from the inverse Fourier transform, we will denote our initial function by $F(t)$ and its inverse Fourier transform by $f(v)$. We have then

$$f(v) = \int_{-\infty}^{\infty} A\cos(2\pi v_0 t) e^{i2\pi vt} dt$$

which leads to almost the same solution as Equation V.4 except that the second term is positive. However, the integral in the second term still equals zero, so we are again left with only the first term, and we end up with the same solution as before. We find, then, that the Fourier transform and inverse Fourier transform of $\cos(2\pi v_0 t)$ are identically equal.

Let us consider, now, a simple sine function and calculate both its Fourier transform and inverse Fourier transform. So, we now calculate the Fourier transform of $f(t) = A\sin(2\pi v_0 t)$ and the inverse Fourier transform of $F(t) = A\sin(2\pi v_0 t)$. We have then

$$F(v) = \int_{-\infty}^{\infty} A\sin(2\pi v_0 t) e^{-i2\pi vt} dt \text{ and } f(v) = \int_{-\infty}^{\infty} A\sin(2\pi v_0 t) e^{i2\pi vt} dt$$

Following the same math that led to Equation V.4, we get in this case

$$F(v) = \int_{-\infty}^{\infty} A\sin(2\pi v_0 t)\cos(2\pi v\, t) dt - i \int_{-\infty}^{\infty} A\sin(2\pi v_0 t)\sin(2\pi v\, t) dt$$

and

$$f(v) = \int_{-\infty}^{\infty} A\sin(2\pi v_0 t)\cos(2\pi v\, t) dt + i \int_{-\infty}^{\infty} A\sin(2\pi v_0 t)\sin(2\pi v\, t) dt$$

and now we find, in both transforms, that the first term integrates to zero, leaving only the second term, which, again, is nonzero only at $v = \pm v_0$. We have, then,

$$F(v) = -iA\delta(v \pm v_0) \text{ and } f(v) = +iA\delta(v \pm v_0)$$

Therefore, the Fourier transform and inverse Fourier transform of $A\sin(2\pi\nu_0 t)$ will be of the same form, but opposite in sign. In general, the inverse Fourier transform and the Fourier transform of any function yield identical real parts, but imaginary parts of opposite sign.

V.4 TRANSFORM OF DELTA FUNCTION

Focus now on our solution of the inverse Fourier transform of $A\delta(\nu - \nu_0)$. Equation V.5 gives it simply, in complex exponential form,

$$f(t) = Ae^{i2\pi\nu_0 t}$$

In this form, we see that it has *constant amplitude* equal to A. Note that we get the same amplitude if we use the negative exponent, that is, if we do the forward Fourier transform. This is shown visually in Figure V.1, in which a delta function occurring at 0.1 Hz is shown in Figure V.1a, and its Fourier Transform is shown in Figure V.1b. The real part of the

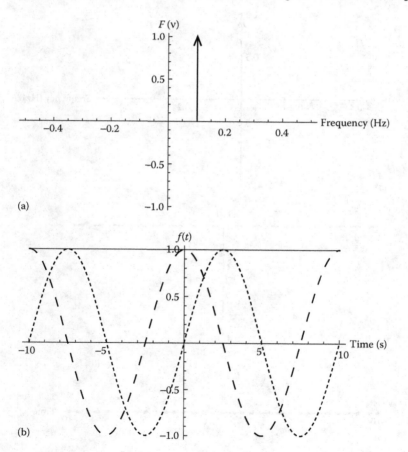

FIGURE V.1   (a) Delta function at frequency 0.1 Hz and (b) its Fourier transform. The real part of the transform is shown as a long-dashed line, the imaginary part as a short-dashed line, and the amplitude is displayed as a solid line.

Fourier transform is a cosine wave with a period $T = 1/0.1$ Hz $= 10$ s, and the imaginary part is a sine wave of the same frequency. The amplitude given by the square root of the sum of the squares of the real and imaginary parts is constant.

Also of interest is the phase, ϕ, of the complex exponential, that is, the expression in the exponent. In this case, we have

$$\phi = 2\pi v_0 t \tag{V.6}$$

What if the delta function occurred at $v = 0$, as shown in Figure V.2a? Contemplate, first, what this $F(v)$ represents physically. This is a signal that has a frequency of zero, which is just another way of describing a non-oscillating function, or a signal that has a constant power in time. Now we apply the inverse Fourier transform and see what $f(t)$ looks like. We have $F(v) = A\delta(v - 0)$, and the inverse Fourier transform is

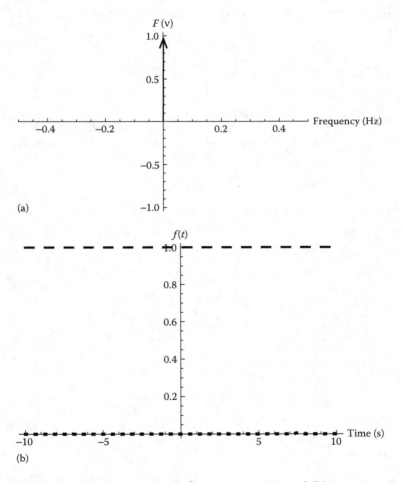

(a)

(b)

FIGURE V.2 (a) Function containing a single frequency at 0 Hz and (b) its inverse Fourier transform. The real part of the transform, shown as a long-dashed line, is constant at 1, while the imaginary part, the short-dashed line, is 0.

$$f(t) = \int\limits_{-\infty}^{\infty} A\delta(0)e^{i2\pi vt} dv$$

which is, simply,

$$f(t) = Ae^{i0} \tag{V.7}$$

shown in Figure V.2b. This is indeed a constant where the real part is equal to A and the imaginary part is zero.

In addition, the phase of the Fourier transform of a delta function at the origin, we see, is zero.

In sum, we find that the (inverse) Fourier transform of a delta function has a constant amplitude, equal to the amplitude of delta function, and a phase that is proportional to the location of the delta function.

V.5 TRANSFORM OF SUM OF FUNCTIONS

Now, let us consider the sum of two delta functions, as in Figure V.3a. To simplify the math, we place one of the delta functions at $v = 0$. The inverse Fourier transform is given by

$$f(t) = \int\limits_{-\infty}^{\infty} \left(\delta(v)e^{i2\pi vt} + \delta(v - v_0)e^{i2\pi vt} \right) dt$$

which can be broken up as

$$f(t) = \int\limits_{-\infty}^{\infty} \delta(v)e^{i2\pi vt} dt + \int\limits_{-\infty}^{\infty} \delta(v - v_0)e^{i2\pi vt} dt$$

We see that the total inverse Fourier transform is the sum of the two individual transforms shown in Figures V.1 and V.2. Therefore, the real part of the inverse Fourier transform, in this case, is the sum of the cosine function with period $1/v_0$ and the constant term. This sum oscillates about a positive value and is never negative, as shown in Figure V.3b. The imaginary part, though, oscillates about zero with period $1/v_0$. The most interesting result, though, is that now the *amplitude* oscillates, with minima separated by $1/v_0$.

Let us follow through with the math of this particular transform. For the sake of generality, let us include factors in front equal to F_1 and F_2. The inverse Fourier transform, then, is given by

$$f(t) = \int\limits_{-\infty}^{\infty} \left[F_1\delta(v) + F_2\delta(v-v_0) \right] e^{i2\pi vt} dt$$

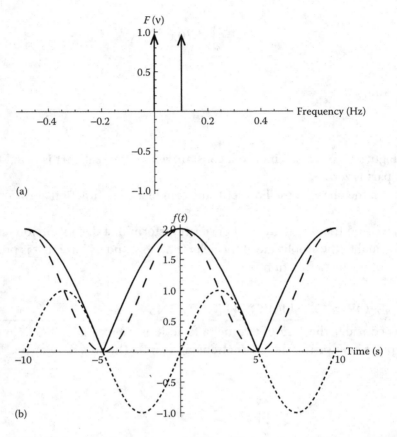

(a)

(b)

FIGURE V.3 (a) Two delta functions: one at 0.1 Hz and the other at 0 Hz and (b) the inverse Fourier transform (real as long dashes, the imaginary as short dashes, and the amplitude as solid line). The amplitude of the total inverse Fourier transform, now, oscillates with a period of 10 s.

or

$$f(t) = \int_{-\infty}^{\infty} F_1 \delta(v) e^{i2\pi vt} dt + \int_{-\infty}^{\infty} F_2 \delta(v - v_0) e^{i2\pi vt} dt$$

which becomes

$$f(t) = F_1 e^{i0} + F_2 e^{i2\pi v_0 t}$$

or

$$f(t) = F_1 + F_2 \cos(2\pi v_0 t) + i F_2 \sin(2\pi v_0 t)$$

The point of interest here is the amplitude, A.

$$A = \sqrt{\left[F_1 + F_2 \cos(2\pi v_0 t)\right]^2 + \left[F_2 \sin(2\pi v_0 t)\right]^2}$$

$$A = \sqrt{F_1^2 + F_2^2 + 2F_1 F_2 \cos(2\pi v_0 t)}$$

(V.8)

This amplitude oscillates in time with a period $= 1/v_0$, and looks like that shown in Figure V.4. (We trust that you will accept that this is true even when neither delta function is placed at the origin. The algebra is simpler by setting one frequency to zero.)

Example V.1:

A time-varying signal like that shown in Figure V.5 is detected.
   This signal is represented mathematically by

$$P(t) = P_{max} \cos\left[2\pi(5.0\text{Hz})t\right]\cos\left[2\pi(105\text{Hz})t\right]$$

where:
   P_{max} is the maximum power in the time-varying signal

This signal is sent to a spectrometer. What should the detected spectrum look like?

Answer:

Before following through with the straightforward application of the integral, we will point out a quicker way, which some students might recognize. The plot in Figure V.5 looks like that of a beat signal, which is created by combining two simple waves of slightly different frequencies. When a wave of frequency v_1 is combined with a wave of frequency v_2 and of equal amplitude, the sum is

$$P(t) = A\cos(2\pi v_1 t) + A\cos(2\pi v_2 t)$$

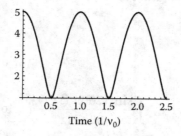

FIGURE V.4   The amplitude of the inverse Fourier transform of two delta functions, $3\delta(v) + 2\delta(v-v0)$. The amplitude oscillates with a period $= 1/v_0$.

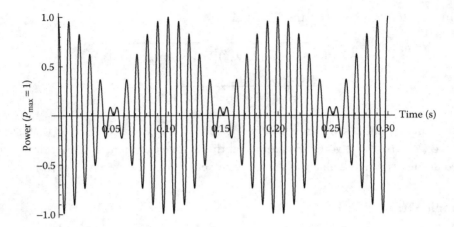

FIGURE V.5   Detected power versus time of the signal in Example V.1 with P_{max} set equal to 1.

which can be manipulated with some trigonometric identities to become

$$P(t) = 2A\cos\left[2\pi\left(\frac{v_1 - v_2}{2}\right)t\right]\cos\left[2\pi\left(\frac{v_1 + v_2}{2}\right)t\right]$$

This fits the signal above if $v_1 = 110$ Hz and $v_2 = 100$ Hz. Therefore, we expect that the spectrometer should indicate a signal at just two frequencies: 100 and 110 Hz.

Now, we will perform the Fourier transform to show that this is indeed the case. We start by substituting in the given $P(t)$ for $f(t)$ in Equation V.1, giving us

$$F(v) = \int_{-\infty}^{\infty} P_{max}\cos\left[2\pi(5\text{Hz})t\right]\cos\left[2\pi(105\text{Hz})t\right]e^{-i2\pi vt}\,dt$$

This integral can be solved quickly using tricks taught in a complex analysis course, but can also be accomplished using integration by parts (by choosing $u(t) = e^{-i2\pi vt}$ and $dv\,(t) = \cos[2\pi (5.0 \text{ Hz})t]\cos[2\pi (105 \text{ Hz})t]dt$).

Skipping to the answer, the solution is

$$F(v) = \begin{cases} 0 & \text{for } v \neq 100, 110 \text{ Hz} \\[2ex] 200\text{Hz}\pi\, P_{max} \displaystyle\int_{-\infty}^{\infty} \frac{\sin^2(200\pi \text{ Hz } t)}{400\pi \text{ Hz}}\,dt, & \text{for } v = \pm100 \text{ Hz} \\[2ex] 220\text{Hz}\pi\, P_{max} \displaystyle\int_{-\infty}^{\infty} \frac{\sin^2(220\pi \text{ Hz } t)}{440\pi \text{ Hz}}\,dt, & \text{for } v = \pm110 \text{ Hz} \end{cases}$$

We find that $F(v)$, indeed, has non-zero values only at ± 100 and ± 110 Hz. This is, in short, the sum of two delta functions, at the two frequencies used to create the beat.

You may have been tempted to pick 5.0 and 105 Hz as the fundamental frequencies in this signal, since the signal is described, in the question, as a simple product of these two frequencies. It is instructive to ask why this is wrong. The answer is that these two cosine functions are multiplied, not added. The cosine (or sine) functions that make up a basis for any function must be added (just as the unit vectors that make up a basis for three-dimensional vectors, are added, after being multiplied by their respective lengths). Just remember that when you are assembling a structure (such as a house or model) you keep *adding* the fundamental pieces in doing so.

We have, now, another useful rule of Fourier transforms, as demonstrated in Example V.1, which we can apply to aperture synthesis data as follows.

The Fourier transform of a sum functions is the sum of the Fourier transforms of the individual functions.

V.6 FOURIER TRANSFORM OF GAUSSIAN-PROFILE FUNCTION

Finally, let us consider a Gaussian-shaped function of t and ask what its Fourier transform is. We express $f(t)$ as

$$f(t) = \exp\left[-4\ln 2\left(\frac{t}{w}\right)^2\right]$$

where:
  w is the *FWHM*

Putting $f(t)$ into the Fourier transform, we have

$$F(v) = \int_{-\infty}^{\infty} e^{-4\ln 2\,(t/w)^2} \cdot e^{-i2\pi vt}\,dt = \int_{-\infty}^{\infty} e^{-4\ln 2(t/w)^2 - i2\pi vt}\,dt$$

The solution of this integral is

$$F(v) = \sqrt{\pi}\,\frac{w}{2\sqrt{\ln 2}}\exp\left[-\frac{(\pi w v)^2}{4\ln 2}\right]$$

or

$$F(v) = \frac{W}{\sqrt{\pi}}\exp\left[-4\ln 2\,\frac{v^2}{W^2}\right]$$

where:

$$W = \frac{4\ln 2}{\pi} \frac{1}{w}$$

is the *FWHM* of F(v). We see, then, that *the Fourier transform of a Gaussian is another Gaussian, and, as you might have expected, the widths of the two Gaussians are inversely related.*

V.7 TWO ADDITIONAL RULES ABOUT FOURIER TRANSFORMS

1. Since the argument of an exponential function must be unitless, the independent variables of functions which are Fourier transforms of each other must have units the inverse of each other. For example, frequency = 1/time. If $F(u)$ is the Fourier transform of $f(x)$ then the units of u must be 1/(the units of x).

2. In the Fourier transform, since all cosine functions are even about the origin and sines are odd, that is, $\cos(-x) = \cos(x)$, while $\sin(-x) = -\sin(x)$; the imaginary part of the transform is needed to pick up asymmetries in the initial function. That is, the limitation of a cosine-only transform is that it cannot produce an $f(x)$ in which $f(-x)$ does *not* equal $f(x)$.

Appendix VI

Wiener-Khinchin Theorem

THE WIENER-KHINCHIN THEOREM STATES that the Fourier transform of the autocorrelation function is the power spectrum.

The Fourier transform of a correlation of two functions can be shown to be equivalent to the product of the Fourier transforms of the two functions. Mathematically, this is given by

$$\mathscr{F}(a) \times \mathscr{F}(b) = \mathscr{F}(a * b)$$

where: $a * b$ represents the correlation of functions a and b, and $\mathscr{F}(a)$ represents the Fourier transform of function a. If the correlation is to occur over the time domain, then the correlation $h(\tau)$, of the functions $a(t)$ and $b(t)$, is defined by

$$h(\tau) = \int a(t)b(t-\tau)dt$$

where the independent variable, τ, is called the *delay*. An *autocorrelation* is the correlation of a function with itself.

Applying this to $E(t)$, the electric field entering the spectrometer (see Section 3.5.2), the autocorrelation function (ACF) is

$$\mathrm{ACF}\big[E(\tau)\big] = \int E(t)E(t-\tau)dt$$

and by taking the Fourier transform, we have

$$\mathscr{F}(\mathrm{ACF}[E(t)]) = \mathscr{F}[E(t)] \times \mathscr{F}[E(t)]$$

By the Wiener-Khinchin Theorem this equals the power spectrum $P(v)$. In a digital spectrometer the power spectrum is obtained by calculating $\mathscr{F}[(E(t)] \times \mathscr{F}[E(t)]$.

Appendix VII

Interferometer Simulation Activities

For those schools that have a Haystack Very Small Radio Telescope (VSRT),[*] we provide instructions and data analysis software for a set of table-top interferometry labs, at our webpage (web page and instructions given below). These labs provide first-hand experience regarding the use of interferometers and help students to develop an intuitive sense of the basics. For students with minimal mathematical backgrounds, we recommend performing these labs before starting Section 5.2.

For schools that do not have a Haystack VSRT, the VSRTI_Plotter package[†] can also be used to perform the following activities, which simulate the results that are obtained with the telescope. As with the labs, we feel that students unfamiliar with interferometry and aperture synthesis will find that performing the activities *before* embarking on the long, complex discussion of how aperture synthesis works will provide an intuitive appreciation of the relationship between aperture synthesis data and the radio-emitting objects, which will help to make the explanation of aperture synthesis more concrete and easier to follow.

To get the data analysis and simulation applications along with the lab instructions you need to download (free) the following zip files, which contain java packages that are easy to install and run on any operating system. The two zip files, *VSRTI_Plotter. zip* and *TIFT.zip*, are located at www.crcpress.com/product/isbn/9781420076769. After downloading, unzip the files. In Windows, right click on the .zip file and click Extract all. On a Macintosh, just double click the zip file, and it will naturally unzip. Enter the extracted folders and find and create shortcuts on your desktop for the *VSRTI_Plotter. jar* and *TIFT.jar* files. When you double click on these shortcuts, the applications will start up. With VSRTI_Plotter, you should see a small window that looks like Figure VII.1 appearing on the upper left of your screen. With TIFT, you should see a small window, as shown in Figure VII.2, appear.

---

[*] The development of the VSRT was funded by the NSF through a CCLI grant to MIT Haystack Observatory.
[†] The development of the VSRTI_Plotter and TIFT java packages was funded by the National Science Foundation, IIS CPATH Award #0722203

FIGURE VII.1   The initial VSRTI_Plotter window.

FIGURE VII.2   The initial TIFT window.

VII.1 VISIBILITY FUNCTION

The activities involve interpreting plots of interferometric data, so we first explain the parameters that are plotted. The independent variable (plotted on the horizontal axis), known as the *spacing*, is defined as b/λ, where λ is the wavelength of the radiation and b is the *baseline* of the interferometer. The dependent variable (plotted on the vertical axis) is known as the *visibility*. In general, the aperture synthesis data that lead to the formation of an image of the astronomical radio source is the *visibility function*, $V(b/\lambda)$.

In an actual aperture synthesis observation, the visibility function has complex values (real and imaginary parts) and is a function of two dimensions (east–west and north–south components of the baselines). To keep these activities relatively simple and straightforward, we consider only the *amplitudes* of the visibility function and in only one dimension, so that it can easily be displayed in a simple graph.

VII.2 VSRTI_PLOTTER ACTIVITIES

VII.2.1 Activity 1: Interferometric Observations of a Single Source

1. Open VSRTI_Plotter, maximize the window, and click on Plot Visibilities.

2. To simulate the observation of just one source, set *T2* to 0. (*T1* and *T2* are brightness measures of two sources.)

3. Let us start by seeing what an observation of a very small source looks like, so set Φ1 to 0.001 (this is the angular size of the source in radians), and click on Update. (Leave λ set to 2.5.)

4. Click on Display Model Visibilities. Describe the shape of the curve displayed in the graph. This is the description of the visibility amplitudes for a single, unresolved source.

5. Change Φ1 to 0.01 and click on Update. How has the visibility amplitude function changed? This is the visibility amplitude function for a single, marginally resolved source.

6. Change Φ1 to 0.1 and click on Update. Describe the shape of the plot now.

7. Take note of the value of b/λ where this visibility function first reaches zero. Write down this value along with the current value of Φ1. (Remember that Φ1 is in radians.)

8. Right click on the plot window and save this plot as a .jpg file to view later.

9. Change Φ1 to 0.2 and click on Update. How did the plot change? Record the current values of Φ1 and the b/λ where *this* visibility function first reaches zero.

10. Change Φ1 to 0.05 and click on Update; record the new values of Φ1 and the b/λ where the visibility function first reaches zero.

11. Consider your three sets of values of Φ1 versus b/λ. Do you notice a relationship?

VII.2.2 Activity 2: Interferometric Observations of a Pair of Unresolved Sources

1. Open VSRTI_Plotter, maximize the window, and click on Plot Visibilities, or, if continuing from the previous section, click on Reset.

2. Set *T1* and *T2* both to 10, and Φ1 and Φ2 both to 0.001, and leave θ = 0, for now. (θ is the angular separation of the sources, in radians.) Click on Update and Display Model Visibilities.

3. You should see a curve very similar to what you saw with a single unresolved source. Since we currently have θ = 0, the two sources are in the same position and so act like a single source.

4. Change θ = 0.1 and click on Update. Describe the visibility amplitude curve. Note and record the distance in b/λ between two places where the curve reaches zero. Write the current value of θ as well. (Remember that θ is in radians.)

5. Right click in the plot window and save as a .jpg file to view later.

6. Change θ = 0.2 and click on Update. What is the b/λ-distance between zeros now? Record this value along with the current value of θ.

7. Change θ = 0.25 and click on Update, and note and record the new b/λ distance between zeros along with the value of θ.

8. Is there a simple relationship between θ, the angular distance (in radians) between the sources and the distance between zeros in the periodic pattern of the visibility amplitude function?

VII.2.3 Activity 3: Interferometric Observations of a Pair of Resolved Sources

We will now simulate the observation of a pair of sources, which are large enough to be resolved.

1. First, make a prediction: Considering the plots you encountered in the previous two activities, make a prediction about what you think the visibility plot will look like when you have two sources separated by 0.2 radians and which each have angular size of 0.02 radians. Sketch your predicted plot.

2. Open VSRTI_Plotter and click on Plot Visibilities.

3. Set *T1* and *T2* both to 10, Φ1 and Φ2 both to 0.02, and θ to 0.2. Click on Update. (Note that θ should be larger than Φ1 and Φ2 since the separation between the sources must be greater than the sizes of the sources.)

4. How does the plot compare to your prediction? Describe the plot and explain the relation between the features and the radio source.

VII.2.4 Activity 4: Inferring the Structure of a Mystery Source from Its Visibilities

Figure VII.3 shows the visibility amplitude function from an observation of an unknown source. Use your conclusions from the previous exercises about the relations between the features in the visibilities and the structure of the source to develop a theory about the structure of this source. How does the visibility function appear with a resolved source? How does the visibility function appear for a single pair of sources separated by a given angular

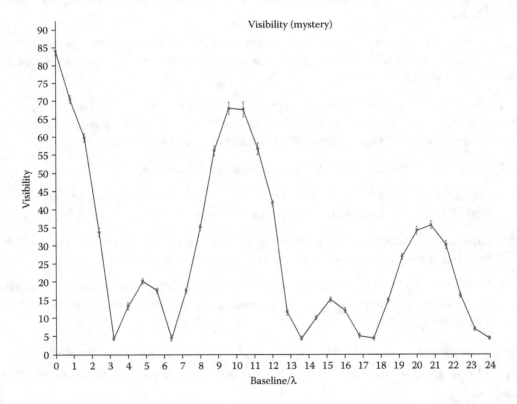

FIGURE VII.3   VSRTI_Plotter Visibility plot from an observation of a mystery source (structure to be determined from analysis of the plot).

distance? By extrapolating, can you apply these principles to the general case and figure out the structure of the more complicated source of the observation shown in Figure VII.3?

On a piece of paper, explain your model. (Note: You will not be able to reproduce this plot using the VSRTI_Plotter model visibilities. This source is slightly more complicated than what is available in the VSRTI_Plotter models.)

Here are some principles from the previous activities, which we give here as hints for you:

1. When observing a single resolved source, the longer baselines resolve out more of the structure of an extended source. This means that the zero-baseline visibility should contain the flux from the entire source and the visibility amplitude will decrease with b/λ at a rate that is inversely proportional to the angular size of the source.

2. Each pair of sources produces a periodic visibility function with period given by $P(b/\lambda) = 1/\theta$. In addition, each periodicity must have a maximum at $b/\lambda = 0$. To identify periodicities, start at $b/\lambda = 0$ and look for the next peak after that.

These activities (hopefully) have demonstrated that details of the structure of a source are contained in the visibility data. The angular size of a source is contained in the rate that the visibility amplitudes decrease with b/λ; information of the existence of a pair of sources is provided by an oscillation in the visibility amplitude, and the separation angle of the pair is inversely related to the periodicity of the oscillations. By extrapolation, you might be able to grasp that even much more complex structures can be inferred by clever methods of teasing out the structure details from the visibility data. The math to do that, it turns out, involves a *Fourier transform*. We do not expect most readers to be familiar with Fourier transforms, and so we provide some more activities; these focused on demonstrating how Fourier transforms reproduce the relation between the visibility data and the structure of the source. No mathematical experience with Fourier transforms is expected at this point. In Section 6.5, we bring Fourier transforms into the discussion of the mathematics behind aperture synthesis.

VII.3 ACTIVITY 5: EXPLORING THE FOURIER-TRANSFORM RELATION BETWEEN THE INTENSITY DISTRIBUTION FUNCTION AND THE VISIBILITY FUNCTION

Here, we discuss exercises using the TIFT.jar application. This tool creates the Fourier Transforms of functions, which can be input either as an expression, or by moving data points on a plot. These are designed to help explain the relation of the visibility functions to the source structure.

VII.3.1 Brief Introduction to Fourier Transforms

Fourier transforms are extremely useful and far more common than most people realize. Since there are extensive presentations of Fourier transforms in numerous textbooks and we provide a longer discussion in Appendix V, we will not bother to give a mathematical explanation here. We will, instead, just convey the basic idea.

A proper Fourier transform involves complex numbers, so the software we will use in these activities must also contain complex numbers. However, since the goal of these exercises is to enhance your understanding of the VSRTI_Plotter exercises, you will only need to focus on the *amplitudes* of these numbers.

There are two aspects of Fourier transforms that are important to know, and remember.

1. The independent variables of the two functions—such as time and frequency must have inverse units of each other. For example, note that the unit of frequency is s^{-1}.

2. If the Fourier transform of $f(t)$ is $F(v)$, then $f(t)$ is the inverse Fourier transform of $F(v)$. And, the mathematics of inverse Fourier transform is nearly identical to the Fourier transform, so one can say that $f(t)$ and $F(v)$ are Fourier transforms of each other.

VII.3.2 Getting Familiar with the TIFT Window Buttons and Controls

1. Start by clicking on the TIFT.jar shortcut. You will then see a small pop-up window with two buttons. Click on "Full Complex FT".

2. A new display window will pop up with four graphs. The top two graphs show the magnitude and phase for any complex function of time, $f(t)$, and the lower graphs show the magnitude and phase for a corresponding function of frequency, $F(v)$. The lower graphs represent the *Fourier Transform* of the function represented in the top two graphs. Similarly, the function represented in the top two graphs is the *inverse Fourier Transform* of the function shown in the bottom graphs.

3. One can choose to change the display to plot the real and imaginary parts of both functions by clicking on "Show Rectangular", and return to magnitude and phase by clicking on "Show Polar". For the exercises we discuss here, though, you should keep the display set to show *magnitudes* and ignore the phase plots.

4. To the right are two sets of boxes for inputting parameters. The inputs of the top two boxes affect the Fourier transform calculation (one sets the step size along the x-axis of the input function and the other sets the number of points). The bottom boxes only affect how much of the display windows are shown and can be used for zooming in on either or both graphs. Be sure to click on "Update" after changing a value in any input box.

5. Scroll the mouse over the topmost graph and click when the cursor becomes a + sign. Notice that this moves the data point at that x-value to the position of the cursor. Alternatively, you can grab a data point and pull it up or down.

6. Try tracing a curve (rather than changing every individual point) by holding the left button down, while you move the mouse (slowly) along the desired curve.

7. You will find that as you change the function in the upper graph(s), changes occur instantly in the lower graphs. This provides an interactive way for you to see the relation between the Fourier transform and the initial function.

8. The x and y values of any data point can be read by moving the cursor onto the data point.

9. The "Reset" button undoes any changes you made to any data points, resetting the data values to that given by the equation (if an equation was input), but does not change the values of the input parameters. The "Full Reset" erases the input function and sets all data values to zero.

10. If you get a plot that you would like to save, you can right click on the graph and save the image as either a .jpg or an .eps file.

Now, we discuss some activities using TIFT that help to show that the Fourier transform of $I(\theta)$ is $V(b/\lambda)$, in the same way that $F(v)$ is the Fourier transform of $f(t)$. For the purposes of simplifying the instructions, we will only focus on the amplitudes and ignore the phases.

VII.3.3 Observing a Single Source of Varying Size

1. Open TIFT and click on "Full Complex FT."

2. At the right, set "Min. frequency" to 0 and "Max. time" to 1.0.

3. In the topmost graph, move the point at $t = 0$ upward, to a magnitude ~ 1. Note what happens to the magnitude of $F(v)$ (the third graph).

4. Move the second point in the top graph up to the same magnitude as the first point, and note, again, how the magnitude of $F(v)$ changes.

5. Keep moving subsequent points up to the same magnitude, watching the magnitude of $F(v)$ as you do so.

6. Consider the top graph as representing the radio source in your VSRTI_Plotter exercises in Appendix VII.2.1, of interferometric observations of a single resolved source. Recall the shape of the visibility amplitude function with a single source and how visibility amplitudes depended on the angular size of the source. Does the magnitude of $F(v)$ look like the visibility amplitude function you found in Appendix VII.2.1?

VII.3.4 Observing a Double Source of Varying Angular Separation

1. Open TIFT and click on "Full Complex FT," then set "Min. frequency" to 0 and "Max. time" to 1.0. Or, if continuing from the previous activity in Appendix VII.3.3, first click on "Reset" to move all the points back to zero.

2. In the topmost graph, move the point at $t = 0$ upward (to a magnitude ~ 1) and then move the point at $t = 0.1$ to about the same height. What is the shape of the magnitude of $F(v)$? What is the period of the oscillations in $F(v)$?

3. Move the data point at $t = 0.1$ back to zero and move the point at $t = 0.2$ up to the same level as the $t = 0$ point. How does this change the magnitude of $F(v)$? What is the new period of the oscillations in $F(v)$?

4. Move the data point at $t = 0.2$ back to zero and move the point at $t = 0.25$ up to the same level as the $t = 0$ point. What is the period of the oscillations in $F(v)$ now?

5. What is the relation between the period of oscillation in $F(v)$ and the positions of the spikes in $f(t)$?

6. Considering the $f(t)$ plot as representing the distribution of unresolved point sources, as in the exercise in Appendix VII.3.3, does the magnitude of $F(v)$ look like the visibility amplitude function you found in that exercise?

VII.3.5 Observing a Pair of Resolved Sources

1. Open TIFT and click on "Full Complex FT," then set "Min. frequency" to 0 and "Max. time" to 1.0; or, click on "Reset" if continuing from the previous activity in Appendix VII.3.4 and make sure that the Min. frequency is still set to 0 and that Max. time is set to 1.0.

2. Set the first three points, those for $t = 0, 0.01$, and 0.02 to 1. In addition, do the same for the points at $t = 0.20, 0.21, 0.22$. This will now represent two resolved sources (of width 0.02 radians) separated by 0.2 radians. The plot of the magnitude of $F(v)$, then, should look like the visibility amplitude function you found in Appendix VII.2.3. Is this true?

VII.3.6. Fourier Transforms and the Visibility Function

1. Discuss the relation between the visibility function, $V(b/\lambda)$, and the distribution of intensity in the radio source, $I(\theta)$, and how this compares with the Fourier relation between $f(t)$ and $F(v)$.

2. Consider the independent variable in the visibility function. With regard to the units, does it have the same relation to the independent variable in the intensity distribution function, $I(\theta)$, as that between the independent variables of $f(t)$ and $F(v)$? (Hint: Remember that the resolution of a single-dish telescope is proportional to D/λ, where D, a distance, is the diameter of the reflector.)

3. Considering that any function's Fourier transform can be inverted to yield the original function, discuss how the radio source intensity distribution can be recovered from the visibility function, that is, what must radio astronomers do to obtain an image, of even complex sources, from the interferometric data?

We hope that you found these activities instructive in helping to demonstrate why aperture synthesis works. We provide mathematical explanations for these results in Chapter 5. A summary of the main conclusions demonstrated in these activities, and discussed in Chapter 5, is provided in Section 5.9.

Index

3C catalog/3CR catalog, 3

A

Absorption line, 9
Airy disk/Airy pattern, 83–84
Aliasing, 113
ALMA, *See* Atacama Large Millimeter/
    submillimeter Array
Altitude, 13, 18
Amplifier, 95, 96, 100, 102–105
    DC amplifier, 102
    gain, 96, 103–105
    IF amplifier, 95, 100
    noise, 96, 102–103
    noise temperature, 103–105, 128
    RF amplifier, 95, 100, 102
Analog filter-bank spectrometer, 110
Angular resolution, *See* Resolution
Angular size, 17–18
Antenna, 24, 76, 80, 85, 96
    antenna beam, *See* Beam pattern
    antenna solid angle, 137–138, 142, 169
    array, 120–121, 181, 183, 196, 205, 222, 249–252
    dipole antenna, 117–120
    feed, 21, 23, 24, 76, 77, 85–88, 96, 139, 155, 174–175
    ground plane, 120
    very low frequency antenna, 116–121
    Yagi–Uda antenna, 117–118, 120
Antenna pattern, *See* Beam pattern
Antenna temperature, 103–104, 128, 131, 136,
    144–145, 159–160, 162
    measurement of, 131
    of resolved/extended source, 146–148
    uncertainty in, 131–134
    of uniform source that fills the sky, 149–150
    of unresolved source/point source, 145
    vs. brightness temperature, 150–151
Antenna theorem, 138, 141–142
Aperture synthesis, 181
    angular resolution, 183, 205, 252–253, 277
    array of antennas, 181, 183, 196, 205, 222, 249–252

bandpass calibration, 265
bandpass function, 237, 239, 245–246
bandwidth smearing/effect of non-zero
    bandwidth, 209–211, 237–240
baseline, 184, 187, 229, 232–234, 244–245,
    249–252
calibration, 263–265
closure amplitude, 276
closure phase, 274
coherence of radiation
    coherence time, 210, 240
    spatial coherence, 186, 208
    temporal coherence, 209–211, 237–240
complex gain, 263–264
cross-correlation, 185, 223–226, 241–243, 245
delay, 186–187, 211, 214, 232
delay center/delay tracking center, 212
delay tracking, 211–212, 240
detection limit, 260–262
dirty beam, 267–269
dirty image/dirty map, 267–269
discrete Fourier transform, 266
earth rotation synthesis, 249–252
field of view, 222
Fourier transform, 229–231, 243, 252, 265–266
fringe function, 192, 196, 233, 234–235, 237–238,
    243, 245
    fringe frequency, 234–235
    fringe phase, 195
    fringes, 192, 221
    fringe spacing, 192–193
image formation, 265–276
    CLEAN, 271–272
    clean beam, 272
    clean box, 272
    clean component, 272
    clean map, 272
    closure amplitude, 276
    closure phase, 274

Printed in the United States
by Baker & Taylor Publisher Services